➤➤➤➤➤

# The ACS Style Guide

THIRD EDITION

# The ACS Style Guide

## Effective Communication of Scientific Information

Anne M. Coghill
Lorrin R. Garson
*Editors*

AMERICAN CHEMICAL SOCIETY
Washington, DC
2006

OXFORD
UNIVERSITY PRESS

Oxford University Press, Inc., publishes works that further
Oxford University's objective of excellence
in research, scholarship, and education.

Oxford New York
Auckland Cape Town Dar es Salaam Hong Kong Karachi
Kuala Lumpur Madrid Melbourne Mexico City Nairobi
New Delhi Shanghai Taipei Toronto

With offices in
Argentina Austria Brazil Chile Czech Republic France Greece
Guatemala Hungary Italy Japan Poland Portugal Singapore
South Korea Switzerland Thailand Turkey Ukraine Vietnam



Developed and distributed in partnership by the
American Chemical Society and Oxford University Press

Published by Oxford University Press, Inc.
198 Madison Avenue, New York, NY 10016
www.oup.com



*Library of Congress Cataloging-in-Publication Data*

The ACS style guide : effective communication of scientific information.—3rd ed. /
Anne M. Coghill [and] Lorrin R. Garson, editors.
p. cm.
Includes bibliographical references and index.
ISBN-13: 978-0-8412-3999-9
1. Chemical literature—Authorship—Handbooks, manuals, etc. 2. Scientific literature—Authorship—Handbooks, manuals, etc. 3. English language—Style—Handbooks, manuals, etc. 4. Authorship—Style manuals.
I. Coghill, Anne M. II. Garson, Lorrin R. III. American Chemical Society
QD8.5.A25 2006
808'.06654—dc22 2006040668

5 7 9 8 6
Printed in the United States of America
on acid-free paper

➤➤➤➤➤

# Contents

## *Part 2. Style Guidelines*

➤➤➤➤➤

# Foreword

I fell in love with chemistry when I was 13. I fell in love with writing at the age of four when I learned to read. Indeed, my love of writing, and of writing well, was inspired by my love of reading. Perhaps that is true for all writers.

Fortunately for me, I have been able to combine my love of chemistry with my love of reading and writing in a long career as a science communicator and journalist. Most recently, I served for eight and a half years as editor-in-chief of *Chemical & Engineering News*, the flagship newsmagazine of the American Chemical Society. This gave me ample opportunity to read all of the stories in *C&EN* every week, not once but twice and sometimes three times; write weekly editorials and occasionally longer stories; and indulge my love of chemistry vicariously, as I read the scientific papers we highlighted in *C&EN*.

But writing is not as easy as reading. Writing and communicating take a great deal of skill and effort. One of my favorite quotations on the subject of writing comes from the novelist John Irving, who observed in *The World According to Garp* that a writer never reads for fun. It's true for me. When I read a sentence that is well crafted or even better, a scientific paper that is full of well-crafted sentences, I am always trying to figure out how the author managed to express a complicated idea with such ease and grace.

The goal of *The ACS Style Guide* is to help authors and editors achieve that ease and grace in all of their communications. To my mind, there's no reason why scientific papers should not be as easy to read as a good novel. That's a tall order, I realize, but if you read through this style guide, you will have all the tools you need to help you achieve that goal. It's a wonderful reference book that I keep on my bookshelf and refer to often. I hope you will as well.

MADELEINE JACOBS
*Executive Director and Chief Executive Officer*
*American Chemical Society*

# Preface

Since publication of the second edition of *The ACS Style Guide* in 1997, much has changed in the world of scientific communication—and yet, many things remain the same.

During the past eight years, electronic dissemination of scientific, technical, and medical (STM) information has come to fruition. In chemistry, both the American Chemical Society and the Royal Society of Chemistry have made their scientific journals available on the World Wide Web and have digitized their respective publications back to the 19th century. Commercial publishers, who publish most of the world's chemical information, have likewise made their publications available on the Web. Publications in other scientific disciplines, engineering, and medicine have also taken this digital pathway. Whereas traditional journals continue to be printed and used, electronic delivery has greatly expanded the availability and reading of STM information far beyond what could have ever been envisioned with paper journals. Most manuscripts are now written with de facto standard word-processing software and adhere to formats developed for electronic creation and processing. Most manuscripts are submitted electronically, principally via the Internet on the Web. Communications among editors, reviewers, and authors are now largely electronic, as is communication between editors and production facilities and printers.

Regardless of the mode of information creation and delivery, the necessity for accurate information communicated in a clear, unambiguous manner, coupled with the ethical behavior of all participants, remains the same. As Janet Dodd wrote in the preface to the second edition, "In the midst of all this change, the comforting thought is that one goal of authors and editors has not changed: to communicate information in the most understandable and expedient fashion in publications of the highest quality. To accomplish that goal, we need guidelines. This book is intended to guide and answer questions for authors and editors, to save them time, and to ensure clarity and consistency."

## Third Edition

The third edition aims to continue such guidance while broadening the scope of the book to accommodate changes in technology and the homogenization of international scientific publishing. New topics in the third edition include chapters on

- ethics in scientific communication;
- submitting manuscripts via the Web;
- preparing and submitting publisher-ready figures, tables, and chemical structures, including information about various software programs to create artwork;
- formatting manuscript references to electronic resources and information on reference-management software; and
- markup languages, in anticipation of the classification and capture of scientific information in yet-to-be-defined structures.

The chapters on peer review, copyright, the editorial process, and writing style and word choice have been extensively rewritten. Although language certainly evolves with time, there have not been substantial changes in English during the past seven years. The chapters on grammar, punctuation, spelling, and conventions in chemistry remain largely the same as in the second edition. The use of typefaces, superscripts and subscripts, Greek letters, special symbols, numbers, mathematics, units of measure, and names and numbers for chemical compounds are generally unchanged, although some of the existing rules have been clarified. Some new rules and examples have been added to reflect new fields in chemistry, such as combinatorial chemistry and chemical biology. In all chapters, errors have been corrected (and almost certainly new errors inadvertently introduced!), and some changes have been made to reflect changes in practice, particularly as related to electronic issues.

Several features have been added to the third edition to improve the readers' ease of use:

- The contents are reorganized into two sections. The first section, "Scientific Communication", contains chapters giving readers information on broad topics such as ethics in scientific communication, writing style and word usage, and submission of manuscripts using a Web-based system. The second section, "Style Guidelines", contains chapters that give specific rules and examples. For instance, in these chapters readers will find information on such topics as grammar, punctuation, and spelling; formatting numbers and specialized chemical conventions; when to use special typefaces; how to format references; and how to create figures, tables, and chemical structures.

- Throughout the book, the arrowlike icon (➤) precedes rules. These rules may refer to grammar, word usage, or punctuation rules. Also, the icon may precede rules for creating publisher-ready artwork, rules about styling chemical terms, or rules about formatting names and chemical compounds. Examples are given under the rule to further illustrate it.
- Attention is drawn to particularly important topics by the use of reminders and boxes. Reminders are bounded by horizontal rules and are identified with a small pencil icon (✐); they contain a brief note on a single topic. Boxes are numbered sequentially within each chapter and contain more extensive information on a specific topic. Reminders and boxes that contain ACS-specific information are identified by a small ACS phoenix icon (◆). We believe that identification of these key issues in this manner will be helpful to readers.

Because of the desire on the part of the publisher to increase the use of the third edition of *The ACS Style Guide,* it is being made available on the World Wide Web. It is expected that periodic updates will be made to the electronic edition, which would not be feasible for the printed version. Additionally, if readers would like to request clarification of rules, they may do so by contacting the publisher at styleguide@acs.org or by addressing correspondence to The ACS Style Guide, Books Department, American Chemical Society, 1155 Sixteenth Street, NW, Washington, DC 20036.

Although *The ACS Style Guide* is written with an emphasis on chemistry and, to some extent, a focus on ACS journals, we believe that it has wide applicability to the sciences, engineering, medicine, and other disciplines. Chemistry is a mature science that cuts across virtually all basic and applied sciences.

Science in its broadest sense has always been an international activity. However, there is an increasing trend toward internationalization of scientific communication. For example, for the past several years, the majority of authors publishing in ACS journals reside outside North America. English has become the lingua franca of science in the same way that French once was the international language of diplomacy and commerce. The venerated *Beilsteins Handbuch der Organischen Chemie* has been published in English for a number of years. The prestigious journal *Angewandte Chemie: International Edition in English* conveys internationalization and the English language merely by its title. The premier publications *Science* and *Nature,* both published in English, have broad international authorship and readership. We believe that *The ACS Style Guide* will be a useful tool for the international scientific community using this common language.

## Acknowledgments

The editors would like to thank all the chapter authors and reviewers who contributed to this project. In particular, we would like to thank our colleagues in Columbus who provided assistance with all the style guidelines in the book, namely, Toddmichael Janiszewski, Diane Needham, "Ram" Ramaswami Ravi, Teresa Schleifer, and Joe Yurvati. A special thank you goes to Betsy Kulamer and Paula M. Bérard for their skilled editorial efforts. We certainly could not have completed this project without their capable assistance. We want to thank Sue Nedrow, who prepared an in-depth index that we think will be very useful to the readers. We also wish to express our appreciation to Bob Hauserman at the ACS for his suggestions and help.

Finally, we would like to express our indebtedness to Janet S. Dodd, who edited the first and second editions of *The ACS Style Guide*. Janet was more than the editor; she wrote much of the first two editions. Her contributions persist in the third edition.

ANNE M. COGHILL
LORRIN R. GARSON
*April 2006*

# Contributors


Frank H. Allen
*Cambridge Crystallographic Data Centre*

Paula M. Bérard
*Chattanooga, Tennessee*

Sarah C. Blendermann
*ACS Publications Division, Office of Journal Support Services*
*American Chemical Society*

Barbara A. Booth
*Department of Civil and Environmental Engineering*
*University of Iowa*

Karen S. Buehler
*ACS Publications Division, Copyright Office*
*American Chemical Society*

C. Arleen Courtney
*ACS Publications Division, Copyright Office*
*American Chemical Society*

Janet S. Dodd
Chemical & Engineering News
*American Chemical Society*

Gordon G. Hammes
*Department of Biochemistry*
*Duke University*

Stephen R. Heller
*Division of Chemical Nomenclature and Structure Representation*
*International Union of Pure and Applied Chemistry*

Betsy Kulamer
*Kulamer Publishing Services*

Derek Maclean
*KAI Pharmaceuticals*

Alan D. McNaught
*Division of Chemical Nomenclature and Structure Representation*
*International Union of Pure and Applied Chemistry*

Peter Murray-Rust
*Unilever Centre for Molecular Informatics*
*Department of Chemistry*
*University of Cambridge*

Henry S. Rzepa
*Department of Chemistry*
*Imperial College London*

Leah Solla
*Physical Sciences Library*
*Cornell University*

Eric S. Slater
*ACS Publications Division, Copyright Office*
*American Chemical Society*

Antony Williams
*Advanced Chemistry Development, Inc.*

PART 1

➤➤➤➤➤

# Scientific Communication

➤➤➤➤➤ CHAPTER 1

# Ethics in Scientific Publication

Gordon G. Hammes

The principles that govern the ethics of scientific publication are no different than for any other endeavor: complete and accurate reporting and appropriate attribution to the contributions of others. However, as always, "the devil is in the details." The ethical responsibilities of authors and reviewers are sufficiently important and complex that the editors of the American Chemical Society journals have developed a detailed document outlining these responsibilities. (This document, "Ethical Guidelines to Publication of Chemical Research" is presented in Appendix 1-1.) The purpose of this chapter is not to duplicate this document, but rather to discuss some of the important underlying principles and situations that often arise.

Scientific research, perhaps more than most professions, crucially depends on the integrity of the investigators. Most research consists of a series of complex experiments or theoretical calculations that cannot (or will not) be duplicated easily elsewhere. Moreover, it is usually extremely difficult to determine in detail if the results are correct and can be trusted. Published results generally are accepted at face value. Very often related work eventually may be done by others that tests the results, so that checks and balances exist within the system. This is usually a long process, however, and the advance of science may be significantly delayed if published results are not correct. The bottom line is that we depend on the integrity of the investigators reporting the results. We assume that the description of the work is accurate and honest unless proven otherwise. This places a considerable burden on the authors to ensure that the system works.

Research is by its nature exploratory, and honest mistakes may occur. Errors due to human fallibility are unfortunate, but not unethical. Research inevitably pushes the boundaries of existing methodology and theory, so that errors in judgment and interpretation are bound to occur. This is a normal part of the scientific establishment. An often-quoted adage is that the only way never to make a mistake in print is never to publish. Errors due to carelessness or haste are poor science; they represent irresponsible, but not unethical, behavior.

Errors due to fabrication and falsification clearly are unethical and cannot be tolerated under any circumstances. Breakdowns in the system that are not honest mistakes have occurred; some examples are published by the Office of Research Integrity of the U.S. Department of Health and Human Services at http://ori.dhhs.gov. Fortunately, these breakdowns seem to be relatively few.

It is the responsibility of each author to ensure the quality and integrity of the research that is reported. The ethical principles governing the conduct of science should be well understood by all participants. This chapter considers only some aspects of this subject. An excellent introductory publication is available online from the National Academy of Sciences; see "On Being a Scientist: Responsible Conduct in Research" at http://www.nap.edu/readingroom/books/obas/.

## When To Publish: Significance and Timeliness

When is it time to publish? Research is open-ended, so the answer to this question is not always obvious and requires authors to balance significance and timeliness to arrive at a high-quality manuscript.

---

✐ **Reminder:** Research should be published in a timely manner when enough work has been done to yield significant results.

---

Researchers must decide when enough work has been done to make a significant contribution to a field. "Significant" is in the eye of the beholder, and sometimes reviewers and authors will differ markedly with regard to this judgment. The give and take between authors and reviewers is part of the normal process of science and undoubtedly improves the quality of published work. Clearly neither science nor scientific publishing are enhanced by a continual stream of short, incomplete descriptions of a research project. A publication should describe a project that is complete unto itself and represents a true advance in the field. (An exception to this rule occurs when a very unusual result is obtained that is of great interest and significance—in this case, publication as a preliminary note may be justified.)

Scientists also have an obligation to publish their research results in a timely manner. Unpublished research results constitute research not done in the eyes of other scientists. Unnecessary delays can result in duplication of efforts and may hinder the advancement of science. Under no circumstances should a manuscript be submitted and then held up in the revision or page proof stage for reasons not directly related to the research—for example, because of patent considerations.

Given the "publish or perish" mentality that sometimes exists, researchers may be tempted to maximize their number of publications by publishing many short, somewhat repetitive research reports. This practice serves no useful purpose for science or the investigator. In truth, the reputation of an investigator is ultimately determined by the quality of research done over an extended time. Beginning independent investigators are often told that a research reputation can be thought of as a product of quantity times quality of published work. If only one publication appears every 10 years, they may be advised, it had better be a good one. On the other hand, a large number of low-quality publications is not of benefit to the individual or the profession.

Investigators may be tempted to publish the same material, or material only slightly different, multiple times. This practice is unethical. The manuscript should clearly describe prior work that has been done by the authors. It is the obligation of the corresponding author to inform the journal editor of any related manuscripts that have been submitted and/or published elsewhere, including preliminary communications and symposium volumes. There are no exceptions. Moreover, although the review process can be lengthy, under no circumstances should a manuscript be submitted simultaneously to multiple journals.

## What To Publish: Full Disclosure

Unfortunately, because of space limitations, the trend in publishing research results is to provide less and less detail. Although brevity is admirable, it is important that the results be described fully and accurately. Moreover, all of the results should be reported, not just those supporting the underlying hypotheses of the research. If necessary, most journals allow the possibility of submitting supporting documentation as supplementary information. Although this material does not appear in the printed version, it is readily available online. The rule of thumb is that sufficient information should be provided so that other investigators could repeat the experiments if they so desired. The necessity for providing sufficient detail has to be balanced with the need to conserve publication space. As might be expected, considerable variation exists in practice as to what this entails. The manuscript review process plays a tempering role, balancing these two factors.

Representative data and/or calculations are an important part of any scientific presentation. Obviously, not all of the data, derivations, and calculations can be presented. It is acceptable for the "typical data and/or calculations" that are presented to be among the best, but all the data should be included in the analyses. The reproducibility of the results is an implicit assumption for published work. However, first-rate research often involves difficult measurements at the edge of existing methodology, and the difference between signal and noise may be hard to distinguish. It is acceptable to report results for which this is the case, as long as the appropriate qualifications are clearly stated. A critical assessment of the research should be made by the investigator, including an error analysis. No one should be more critical of the research that is reported than the authors.

## Who Are Authors?

Generally speaking, all authors of a publication should have made significant and substantial intellectual contributions to the work being reported. Unfortunately, this principle is often breached, as evidenced by manuscripts with tens, even hundreds, of authors. Some laboratories put the names of everyone in the laboratory on the published work, and some individuals put their names on every publication coming out of a laboratory, even if their participation was only nominal.

If a colleague prepared buffers or did routine computer programming, these contributions should be acknowledged, but they are not sufficient contributions for authorship. General discussion with colleagues or within research groups is rarely sufficient for inclusion in authorship. Despite some arbitrariness in defining what constitutes a significant intellectual contribution, the guiding ethical principle is clear and should be adhered to. Usually the question of authorship can be decided by discussion among the participants in the research. Occasionally, a third party may be required to adjudicate this issue. In any event, this matter should be fully resolved before submission of a manuscript.

A question that often arises concerns the order of the authors' names. This is not really an ethical issue, and practice varies from place to place. Most often the first author is assumed to have made the major contribution to the work, and the senior and/or corresponding author is listed last. However, many variations to this theme exist, such as putting the authors in alphabetical order. In some cases, the specific contributions of each author are described. Ideally, the order of authorship should be decided amicably among the authors, but perceptions sometimes differ between the individuals involved. Authors should not become obsessed with this matter. Ultimately, a researcher's scientific reputation rests on the totality of publications and the significance of contributions to the field.

It is often said that all authors are responsible for the entire content of a manuscript. This is a meritorious ideal, but unrealistic. Most manuscripts have multiple authors, and very often, a single author is responsible for only a portion of the work being presented. For example, the manuscript may contain a crystal structure, determined by an expert crystallographer; spectral data, determined by an expert spectroscopist; kinetic data, determined by an expert kineticist; etc. In cases such as this, a single author cannot be held responsible for all of the results presented. A more realistic assessment of what authorship implies is that each author should have read the manuscript carefully and understood the findings, but the technical responsibility is only for the area in which a given author has the appropriate expertise. The responsibility of the corresponding author is to ensure that all authors have approved the manuscript before submission and for all subsequent revisions.

## What Went Before: Attribution and Context

Every scientific publication must include the proper attribution of the contributions of others by appropriate referencing and the placement of results within the context of the research field.

Referencing is a complex subject (see Chapter 14 of this volume). Every reference in the field cannot be cited, or the reference list would become intolerably long. However, important ideas and experiments must be cited. The introduction and discussion sections of a manuscript should be absolutely clear as to what the work of others has contributed to the research being reported. If data are presented that have been previously published, this should be clearly indicated. Direct quotations of more than a few words should be indicated by quotation marks and referenced. Paraphrases of quotations also should be referenced. Plagiarism—taking the writings or ideas of another and passing them off as one's own—of any type represents unethical conduct.

Occasionally, the attribution of an idea or fact may be to a "private communication" of a colleague or fellow scientist. In such cases, permission must be obtained from the individual in question before the citation is made. Reference to unpublished material should be avoided if possible because it generally will not be available to interested readers.

---

✐ **Reminder:** Every manuscript must reference the contributions of others and place results in the context of the research field.

---

The results and conclusions sections of a manuscript should be placed within the context of the research area. What was known before the research being presented? What has this research contributed that is new and significant? It should also be clear what conclusions are based on the work presented and which are speculations. It is appropriate to speculate—in fact, this is a stimulus to the field—as long as speculations are labeled as such. In this regard, the values and judgments of the authors and current thinking appropriately come into play.

Not all attributions to previous work cite supportive data. In some cases, results under discussion may differ from previous work, or authors may make critical comments about earlier research. Differences between the work reported and previous results must be discussed and reconciled. Criticism of previous work should be presented carefully and objectively, in terms of the facts only. This is part of normal scientific discourse. Criticism should never be directed at individuals or laboratories; it is essential to consider only the facts that have been presented.

Acknowledgments should be made to people who have assisted in the project, but not sufficiently for authorship, and to sponsoring agencies. It is also imperative to acknowledge potential conflicts of interest that may exist. For example, if the research being reported concerns drug XYZ and one of the authors has a substantial financial interest in a company that makes drug XYZ or is conducting clinical trials with drug XYZ, these facts should be explicitly stated.

## What Next: After Publication

An author's obligations do not stop with publication. If errors are found in the published work, they should be corrected with the publication of errata. If other investigators request more information or more complete data, the requests should be fulfilled without delay.

A trickier issue concerns the distribution of special materials used in the research. The rule of thumb is that the authors should be willing to provide others with a reasonable supply of special materials that have been used in the research. However, some common sense should be applied to this rule. For example, if two years have been spent cloning a specific protein and it will be used in future research, it is unreasonable to expect researchers to give this clone to competitors who are planning similar experiments. Similarly, if a complex substance has been synthesized and only a small supply is available, it would be unreasonable to expect the material to be given away. However, the publication should provide sufficient detail so that other researchers can develop the clone themselves or synthesize the compound in question. Although ethical behavior in this area is not always clear, the general rule is that all aspects of the research should be fully disclosed and reasonable assistance should be given to other researchers. Progress in science depends greatly on open communication and cooperation.

## Obligations of a Reviewer

Scientific discourse depends on critical review of manuscripts before publication. (Peer review—including ethical considerations—is discussed in greater detail in Chapter 6 of this volume.) The primary obligation of reviewers is to provide a rational, objective review of the science. This requires a careful reading of the manuscript and a careful preparation of the review. The review process is anonymous for most journals, but this does not mean that the reviewer has free rein to criticize. Any criticism must be logically and objectively delineated, and it should never be directed at the authors personally. Reviewers also should place the work within the context of the field: is it a major contribution, minor contribution, or an insufficient contribution to merit publication? Promptness in carrying out reviews is important and an ethical issue. Delaying a publication could be costly to an author, especially in a competitive field. The usual golden rule applies: review with the care and speed you expect for your own manuscripts. If a reviewer cannot meet a deadline, he or she should inform the publisher as soon as possible.

Manuscripts sent to reviewers are confidential documents. Unfortunately, a significant number of reviewers interpret the word "confidential" incorrectly. Confidential does not mean that reviewers can expand the scope of confidentiality, for example, within their research groups, by including a few colleagues, and so on. Confidential documents should not be shared or discussed with anybody without the explicit consent of the journal editor, the editorial board member handling the manuscript, or both. For example, senior investigators sometimes have graduate students or postdoctorals review manuscripts. This is acceptable only if the permission of the editor or editorial board member has been obtained. In some cases, a reviewer may discuss the results with a colleague; this also is forbidden if permission has not been obtained. Although breaches of confidentiality do not usually do any harm and are not intended to do so, they are unethical and should be avoided.

If reviewers have conflicts of interest with regard to a given manuscript, the manuscript should be returned as quickly as possible to the editor. Conflicts of interest vary. Perhaps similar research is being carried out in the reviewer's laboratory, or the reviewer may be privy to confidential information that conflicts with the results reported. Conflicts of interest can be more personal in nature: perhaps a reviewer has had personal difficulties with or is a close friend of one of the authors. When in doubt, the usual rule is not to review or read the manuscript. If you are unsure, ask the editor handling the manuscript. The editor may want your expert opinion even if some level of apparent conflict exists.

Finally, the results in a manuscript under review cannot be quoted or incorporated into a reviewer's own research program. After the work is published, a reviewer may use the ideas and data presented (with proper attribution), but the reviewer should not do so based on the review process. Such behavior is akin to insider trading in the purchase of stocks. Although a prison term is unlikely for this breach of conduct, the ethical principle is quite clear.

## Obligations as a Reader

Not all errors are found before publication by authors and reviewers; some are discovered by readers. If the errors involve serious misinterpretation or misquotation of the literature, the most straightforward procedure is to contact the author(s) directly. If this is awkward, the editor can be informed. It is not worthwhile, however, to create a fuss for nonsubstantive errors. Self-serving complaints, such as not quoting the reader's own work enough, seldom have much credibility.

In rare situations, a scientist may have evidence that published material contains falsification, fabrication, or plagiarism. It is the obligation of every scientist to report such cases immediately to the editor of the journal. Institutions receiving financial support from the National Institutes of Health and the National Science Foundation are required to have mechanisms in place to investigate such occurrences, and direct reporting to the appropriate institutional office may be more expedient. Accusations must be supported by fact, not suspicions, because academic misconduct is a serious matter with career-threatening implications. Unpleasant as this situation may be, it should not be ignored.

## For the Health of Research

This chapter has emphasized the global ethics of the publication process. Ethics are not complicated, and the practices and rules are mainly common sense. Adherence to ethical standards in research and publication is not optional; rather, it is essential for the health of scientific research.

➤➤➤➤➤

APPENDIX 1-1

# Ethical Guidelines to Publication of Chemical Research

The guidelines embodied in this document were revised by the Editors of the Publications Division of the American Chemical Society in January 2000.

## Preface

The American Chemical Society serves the chemistry profession and society at large in many ways, among them by publishing journals which present the results of scientific and engineering research. Every editor of a Society journal has the responsibility to establish and maintain guidelines for selecting and accepting papers submitted to that journal. In the main, these guidelines derive from the Society's definition of the scope of the journal and from the editor's perception of standards of quality for scientific work and its presentation.

An essential feature of a profession is the acceptance by its members of a code that outlines desirable behavior and specifies obligations of members to each other and to the public. Such a code derives from a desire to maximize perceived benefits to society and to the profession as a whole and to limit actions that might serve the narrow self-interests of individuals. The advancement of science requires the sharing of knowledge between individuals, even though doing so may sometimes entail forgoing some immediate personal advantage.

With these thoughts in mind, the editors of journals published by the American Chemical Society now present a set of ethical guidelines for persons engaged in the publication of chemical research, specifically, for editors, authors, and manuscript reviewers. These guidelines are offered not in the sense that there is any immediate crisis in ethical behavior, but rather from a conviction that the observance of high ethical standards is so vital to the whole scientific enterprise that a definition of those standards should be brought to the attention of all concerned.

We believe that most of the guidelines now offered are already understood and subscribed to by the majority of experienced research chemists. They may, however, be of substantial help to those who are relatively new to research. Even

The ethical guidelines are also available in their most recent version on the Web at https://paragon.acs.org.

well-established scientists may appreciate an opportunity to review matters so significant to the practice of science.

# Guidelines

## A. Ethical Obligations of Editors of Scientific Journals

1. An editor should give unbiased consideration to all manuscripts offered for publication, judging each on its merits without regard to race, religion, nationality, sex, seniority, or institutional affiliation of the author(s). An editor may, however, take into account relationships of a manuscript immediately under consideration to others previously or concurrently offered by the same author(s).
2. An editor should consider manuscripts submitted for publication with all reasonable speed.
3. The sole responsibility for acceptance or rejection of a manuscript rests with the editor. Responsible and prudent exercise of this duty normally requires that the editor seek advice from reviewers, chosen for their expertise and good judgment, as to the quality and reliability of manuscripts submitted for publication. However, manuscripts may be rejected without review if considered inappropriate for the journal.
4. The editor and members of the editor's staff should not disclose any information about a manuscript under consideration to anyone other than those from whom professional advice is sought. (However, an editor who solicits, or otherwise arranges beforehand, the submission of manuscripts may need to disclose to a prospective author the fact that a relevant manuscript by another author has been received or is in preparation.) After a decision has been made about a manuscript, the editor and members of the editor's staff may disclose or publish manuscript titles and authors' names of papers that have been accepted for publication, but no more than that unless the author's permission has been obtained.
5. An editor should respect the intellectual independence of authors.
6. Editorial responsibility and authority for any manuscript authored by an editor and submitted to the editor's journal should be delegated to some other qualified person, such as another editor of that journal or a member of its Editorial Advisory Board. Editorial consideration of the manuscript in any way or form by the author-editor would constitute a conflict of interest, and is therefore improper.
7. Unpublished information, arguments, or interpretations disclosed in a submitted manuscript should not be used in an editor's own research except with the consent of the author. However, if such information indicates that some of the editor's own research is unlikely to be profitable, the editor could

ethically discontinue the work. When a manuscript is so closely related to the current or past research of an editor as to create a conflict of interest, the editor should arrange for some other qualified person to take editorial responsibility for that manuscript. In some cases, it may be appropriate to tell an author about the editor's research and plans in that area.

8. If an editor is presented with convincing evidence that the main substance or conclusions of a report published in an editor's journal are erroneous, the editor should facilitate publication of an appropriate report pointing out the error and, if possible, correcting it. The report may be written by the person who discovered the error or by an original author.
9. An author may request that the editor not use certain reviewers in consideration of a manuscript. However, the editor may decide to use one or more of these reviewers, if the editor feels their opinions are important in the fair consideration of a manuscript. This might be the case, for example, when a manuscript seriously disagrees with the previous work of a potential reviewer.

## *B. Ethical Obligations of Authors*

1. An author's central obligation is to present an accurate account of the research performed as well as an objective discussion of its significance.
2. An author should recognize that journal space is a precious resource created at considerable cost. An author therefore has an obligation to use it wisely and economically.
3. A primary research report should contain sufficient detail and reference to public sources of information to permit the author's peers to repeat the work. When requested, the authors should make a reasonable effort to provide samples of unusual materials unavailable elsewhere, such as clones, microorganism strains, antibodies, etc., to other researchers, with appropriate material transfer agreements to restrict the field of use of the materials so as to protect the legitimate interests of the authors.
4. An author should cite those publications that have been influential in determining the nature of the reported work and that will guide the reader quickly to the earlier work that is essential for understanding the present investigation. Except in a review, citation of work that will not be referred to in the reported research should be minimized. An author is obligated to perform a literature search to find, and then cite, the original publications that describe closely related work. For critical materials used in the work, proper citation to sources should also be made when these were supplied by a nonauthor.
5. Any unusual hazards inherent in the chemicals, equipment, or procedures used in an investigation should be clearly identified in a manuscript reporting the work.
6. Fragmentation of research reports should be avoided. A scientist who has done extensive work on a system or group of related systems should organize

publication so that each report gives a well-rounded account of a particular aspect of the general study. Fragmentation consumes journal space excessively and unduly complicates literature searches. The convenience of readers is served if reports on related studies are published in the same journal, or in a small number of journals.

7. In submitting a manuscript for publication, an author should inform the editor of related manuscripts that the author has under editorial consideration or in press. Copies of those manuscripts should be supplied to the editor, and the relationships of such manuscripts to the one submitted should be indicated.
8. It is improper for an author to submit manuscripts describing essentially the same research to more than one journal of primary publication, unless it is a resubmission of a manuscript rejected for or withdrawn from publication. It is generally permissible to submit a manuscript for a full paper expanding on a previously published brief preliminary account (a "communication" or "letter") of the same work. However, at the time of submission, the editor should be made aware of the earlier communication, and the preliminary communication should be cited in the manuscript.
9. An author should identify the source of all information quoted or offered, except that which is common knowledge. Information obtained privately, as in conversation, correspondence, or discussion with third parties, should not be used or reported in the author's work without explicit permission from the investigator with whom the information originated. Information obtained in the course of confidential services, such as refereeing manuscripts or grant applications, should be treated similarly.
10. An experimental or theoretical study may sometimes justify criticism, even severe criticism, of the work of another scientist. When appropriate, such criticism may be offered in published papers. However, in no case is personal criticism considered to be appropriate.
11. The coauthors of a paper should be all those persons who have made significant scientific contributions to the work reported and who share responsibility and accountability for the results. Other contributions should be indicated in a footnote or an "Acknowledgments" section. An administrative relationship to the investigation does not of itself qualify a person for coauthorship (but occasionally it may be appropriate to acknowledge major administrative assistance). Deceased persons who meet the criterion for inclusion as coauthors should be so included, with a footnote reporting date of death. No fictitious name should be listed as an author or coauthor. The author who submits a manuscript for publication accepts the responsibility of having included as coauthors all persons appropriate and none inappropriate. The submitting author should have sent each living coauthor a draft copy of the manuscript and have obtained the coauthor's assent to coauthorship of it.

12. The authors should reveal to the editor any potential conflict of interest, e.g., a consulting or financial interest in a company, that might be affected by publication of the results contained in a manuscript. The authors should ensure that no contractual relations or proprietary considerations exist that would affect the publication of information in a submitted manuscript.

## C. Ethical Obligations of Reviewers of Manuscripts

1. Inasmuch as the reviewing of manuscripts is an essential step in the publication process, and therefore in the operation of the scientific method, every scientist has an obligation to do a fair share of reviewing.
2. A chosen reviewer who feels inadequately qualified to judge the research reported in a manuscript should return it promptly to the editor.
3. A reviewer (or referee) of a manuscript should judge objectively the quality of the manuscript, of its experimental and theoretical work, of its interpretations and its exposition, with due regard to the maintenance of high scientific and literary standards. A reviewer should respect the intellectual independence of the authors.
4. A reviewer should be sensitive to the appearance of a conflict of interest when the manuscript under review is closely related to the reviewer's work in progress or published. If in doubt, the reviewer should return the manuscript promptly without review, advising the editor of the conflict of interest or bias. Alternatively, the reviewer may wish to furnish a signed review stating the reviewer's interest in the work, with the understanding that it may, at the editor's discretion, be transmitted to the author.
5. A reviewer should not evaluate a manuscript authored or coauthored by a person with whom the reviewer has a personal or professional connection if the relationship would bias judgment of the manuscript.
6. A reviewer should treat a manuscript sent for review as a confidential document. It should neither be shown to nor discussed with others except, in special cases, to persons from whom specific advice may be sought; in that event, the identities of those consulted should be disclosed to the editor.
7. Reviewers should explain and support their judgments adequately so that editors and authors may understand the basis of their comments. Any statement that an observation, derivation, or argument had been previously reported should be accompanied by the relevant citation. Unsupported assertions by reviewers (or by authors in rebuttal) are of little value and should be avoided.
8. A reviewer should be alert to failure of authors to cite relevant work by other scientists, bearing in mind that complaints that the reviewer's own research was insufficiently cited may seem self-serving. A reviewer should call to the editor's attention any substantial similarity between the manuscript under

consideration and any published paper or any manuscript submitted concurrently to another journal.

9. A reviewer should act promptly, submitting a report in a timely manner. Should a reviewer receive a manuscript at a time when circumstances preclude prompt attention to it, the unreviewed manuscript should be returned immediately to the editor. Alternatively, the reviewer might notify the editor of probable delays and propose a revised review date.
10. Reviewers should not use or disclose unpublished information, arguments, or interpretations contained in a manuscript under consideration, except with the consent of the author. If this information indicates that some of the reviewer's work is unlikely to be profitable, the reviewer, however, could ethically discontinue the work. In some cases, it may be appropriate for the reviewer to write the author, with copy to the editor, about the reviewer's research and plans in that area.
11. The review of a submitted manuscript may sometimes justify criticism, even severe criticism, from a reviewer. When appropriate, such criticism may be offered in published papers. However, in no case is personal criticism of the author considered to be appropriate.

## *D. Ethical Obligations of Scientists Publishing outside the Scientific Literature*

1. A scientist publishing in the popular literature has the same basic obligation to be accurate in reporting observations and unbiased in interpreting them as when publishing in a scientific journal.
2. Inasmuch as laymen may not understand scientific terminology, the scientist may find it necessary to use common words of lesser precision to increase public comprehension. In view of the importance of scientists' communicating with the general public, some loss of accuracy in that sense can be condoned. The scientist should, however, strive to keep public writing, remarks, and interviews as accurate as possible consistent with effective communication.
3. A scientist should not proclaim a discovery to the public unless the experimental, statistical, or theoretical support for it is of strength sufficient to warrant publication in the scientific literature. An account of the experimental work and results that support a public pronouncement should be submitted as quickly as possible for publication in a scientific journal. Scientists should, however, be aware that disclosure of research results in the public press or in an electronic database or bulletin board might be considered by a journal editor as equivalent to a preliminary communication in the scientific literature.

➤➤➤➤➤ CHAPTER 2

# Scientific Papers

The chemistry community, like other scientific communities, depends on the communication of scientific results. Scientists communicate in a variety of ways, but much of the communication is through publication in books and journals. In this chapter, the different types of book and journal presentations are described, along with the components of the standard format for reporting original research.

## Types of Books

Books for the professional scientific community fall into one of three categories: proceedings volumes, monographs, and handbooks.

### *Proceedings Volumes*

Books based on meetings are called proceedings volumes. These are multiauthored volumes. The chapters in proceedings volumes may be accounts of original research or literature reviews. Generally, the chapters are developed and expanded from presentations given at symposia, but additional chapters may be written especially for the book to make sure that the coverage of the topic is complete. Proceedings volumes should contain at least one chapter that reviews the subject and also provides an overview of the book to unify the chapters into a coherent treatment of the subject. In a longer book that is divided into sections, each section may need a short overview chapter.

### *Monographs*

Monographs are books that examine a single topic in detail. They are written by one author or collaboratively by more than one author. Each chapter treats one subdivision of the broader topic.

### *Handbooks*

Handbooks are large, multiauthored volumes that discuss a field in depth. Generally, the individual submissions are short, about three or four pages. Each submission is written by one or two authors and provides a detailed discussion of a narrow topic within the scope of the book.

## Journal Presentations

There are four general types of presentations published in journals: articles, notes, communications, and reviews.

### *Articles*

Articles, also called full papers, are definitive accounts of significant, original studies. They present important new data or provide a fresh approach to an established subject. The organization and length of an article should be determined by the amount of new information to be presented and by space restrictions within the publication.

### *Notes*

Notes are concise accounts of original research of a limited scope. They may also be preliminary reports of special significance. The material reported must be definitive and may not be published again later. Appropriate subjects for notes include improved procedures of wide applicability or interest, accounts of novel observations or of compounds of special interest, and development of new techniques. Notes are subject to the same editorial appraisal as full-length articles.

### *Communications*

Communications, called "letters" or "correspondence" in some publications, are usually preliminary reports of special significance and urgency that are given expedited publication. They are accepted if the editor believes that their rapid publication will be a service to the scientific community. Communications are generally subject to strict length limitations; they must contain specific results to support their conclusions, but they may not contain nonessential experimental details.

The same rigorous standards of acceptance that apply to full-length articles also apply to communications. Like all types of presentations in journals, communications are submitted to review. In many cases, authors are expected to publish complete details (not necessarily in the same journal) after their communications have been published. Acceptance of a communication, however, does not guarantee acceptance of the detailed manuscript.

### *Reviews*

Reviews integrate, correlate, and evaluate results from published literature on a particular subject. They seldom report new experimental findings. Effective review articles have a well-defined theme, are usually critical, and may present novel theoretical interpretations. Ordinarily, reviews do not give experimental details, but in special cases (as when a technique is of central interest), experimental procedures may be included. An important function of reviews is to serve as a guide to the original literature; for this reason, accuracy and completeness of references cited are essential.

## Standard Format for Reporting Original Research

The main text of scientific papers presenting original research is generally organized into a standard format: abstract, introduction, experimental details or theoretical basis, results, discussion, and conclusions, although not necessarily in this order. This format has become standard because it is suitable for most reports of original research, it is basically logical, and it is easy to use. The reason it accommodates most reports of original research is that it parallels the scientific method of deductive reasoning: define the problem, create a hypothesis, devise an experiment to test the hypothesis, conduct the experiment, and draw conclusions. Furthermore, this format enables the reader to understand quickly what is being presented and to find specific information easily. This ability is crucial now more than ever because scientists, if not all professionals, must read much more material than in the past.

---

✎ **Reminder:** Journal articles and proceedings chapters are usually organized with an abstract, introduction, experimental details or theoretical basis, results, discussion, and conclusions.

---

Use the standard form for reports of original research whether the report is published in a journal or proceedings volume. Even if the information is more suited to one of the shorter types of presentations, the logic of the standard format applies, although some headings or sections may be omitted or other sections and subsections added. Manuscripts for monographs, handbooks, literature reviews, or theoretical papers generally do not follow the standard form. Consult author guidelines for information on how to organize these types of presentations or look at previously published work. Regardless of the type of presentation, be sure to present all parts of the paper as concisely as possible.

An extremely important step is to check the specific requirements of the publication targeted and follow them. Some publishers provide templates that help authors produce manuscripts in the requested format. Templates are also useful in making sure that the manuscript is not too long. Most editors require revisions of manuscripts that are not in their requested format. Thus, not following a publication's requirements can delay publication and make more work for authors.

## *Title*

The best time to determine the title is after the text is written, so that the title will reflect the paper's content and emphasis accurately and clearly. The title must be brief and grammatically correct but accurate and complete enough to stand alone. A two- or three-word title may be too vague, but a 14- or 15-word title is unnecessarily long. If the title is too long, consider breaking it into title and subtitle.

The title serves two main purposes: to attract the potential audience and to aid retrieval and indexing. Therefore, include several keywords. The title should provide the maximum information for a computerized title search.

➤ Choose terms that are as specific as the text permits, e.g., "a vanadium–iron alloy" rather than "a magnetic alloy". Avoid phrases such as "on the", "a study of", "research on", "report on", "regarding", and "use of". In most cases, omit "the" at the beginning of the title. Avoid nonquantitative, meaningless words such as "rapid" and "new".

➤ Spell out all terms in the title, and avoid jargon, symbols, formulas, and abbreviations. Whenever possible, use words rather than expressions containing superscripts, subscripts, or other special notations. Do not cite company names, specific trademarks, or brand names of chemicals, drugs, materials, or instruments.

➤ Series titles are of little value. Some publications do not permit them at all. If consecutive papers in a series are published simultaneously, a series title may be relevant, but in a long series, paper 42 probably bears so limited a relationship to paper 1 that they do not warrant a common title. In addition, an editor or reviewer seeing the same title repeatedly may reject it on the grounds that it is only one more publication on a general topic that has already been discussed at length.

### *Byline and Affiliation*

Include in the byline all those, and only those, who made substantial contributions to the work, even if the paper was actually written by only one person. Chapter 1 and Appendix 1-1 in this book are more explicit on this topic.

➤ Many ACS publications specifically request at least one full given name for each author, rather than only initials. Use your first name, initial, and surname (e.g., John R. Smith) or your first initial, second name, and surname (e.g., J. Robert Smith). Whatever byline is used, be consistent. Papers by John R. Smith, Jr., J. Smith, J. R. Smith, Jack Smith, and J. R. Smith, Jr., will not be indexed in the same manner; the bibliographic citations may be listed in five different locations, and ascribing the work to a single author will therefore be difficult if not impossible.

➤ Do not include professional, religious, or official titles or academic degrees.

➤ The affiliation is the institution (or institutions) at which the work was conducted. If the author has moved to another institution since the work was done, many publications include a footnote giving the current address. Contact the editor about this.

➤ If there is more than one author, use an asterisk or superscript (check the specific publication's style) to indicate the author or authors to whom correspondence should be addressed. Clarify all corresponding authors' addresses by accompanying footnotes if they are not apparent from the affiliation line. E-mail addresses may be included in corresponding author footnotes.

### *Abstract*

Most publications require an informative abstract for every paper, even if they do not publish abstracts. For a research paper, briefly state the problem or the purpose of the research, indicate the theoretical or experimental plan used, summarize the principal findings, and point out major conclusions. Include chemical safety information when applicable. Do not supplement or evaluate the conclusions in the abstract. For a review paper, the abstract describes the topic, scope, sources reviewed, and conclusions. Write the abstract last to be sure that it accurately reflects the content of the paper.

---

**Reminder:** The abstract allows the reader to determine the nature and scope of the paper and helps technical editors identify key features for indexing and retrieval.

---

➤ Although an abstract is not a substitute for the article itself, it must be concise, self-contained, and complete enough to appear separately in abstract publications. Often, authors' abstracts are used with little change in abstract publications. The optimal length is one paragraph, but it could be as short as two sentences. The length of the abstract depends on the subject matter and the length of the paper. Between 80 and 200 words is usually adequate.

➤ Do not cite references, tables, figures, or sections of the paper in the abstract. Do not include equations, schemes, or structures that require display on a line separate from the text.

➤ Use abbreviations and acronyms only when it is necessary to prevent awkward construction or needless repetition. Define abbreviations at first use in the abstract (and again at first use in the text).

### *Introduction*

A good introduction is a clear statement of the problem or project and the reasons for studying it. This information should be contained in the first few sentences. Give a concise and appropriate background discussion of the problem and the significance, scope, and limits of the work. Outline what has been done before by citing truly pertinent literature, but do not include a general survey of semirelevant literature. State how your work differs from or is related to work previously published. Demonstrate the continuity from the previous work to yours. The introduction can be one or two paragraphs long. Often, the heading "Introduction" is not used because it is superfluous; opening paragraphs are usually introductory.

### *Experimental Details or Theoretical Basis*

In research reports, this section can also be called "Experimental Methods", "Experimental Section", or "Materials and Methods". Be sure to check the specific publication for the correct title of this section. For experimental work, give sufficient detail about the materials and methods so that other experienced workers can repeat the work and obtain comparable results. When using a standard method, cite the appropriate literature and give only the details needed.

➤ Identify the materials used and give information on the degree of and criteria for purity, but do not reference standard laboratory reagents. Give the chemical names of all compounds and the chemical formulas of compounds that are new or uncommon. Use meaningful nomenclature; that is, use standard systematic nomenclature where specificity and complexity require, or use trivial nomenclature where it will adequately and unambiguously define a well-established compound.

➤ Describe apparatus only if it is not standard or not commercially available. Giving a company name and model number in parentheses is nondistracting and adequate to identify standard equipment.

➤ Avoid using trademarks and brand names of equipment and reagents. Use generic names; include the trademark in parentheses after the generic name only if the material or product used is somehow different from others. Remember that trademarks often are recognized and available as such only in the country of origin. In ACS publications, *do not use* trademark (™) and registered trademark (®) symbols.

➤ Describe the procedures used, unless they are established and standard.

➤ Note and emphasize any hazards, such as explosive or pyrophoric tendencies and toxicity, in a separate paragraph introduced by the heading "Caution:". Include precautionary handling procedures, special waste disposal procedures, and any other safety considerations in adequate detail so that workers repeating the experiments can take appropriate safety measures. Some ACS journals also indicate hazards as footnotes on their contents pages.

In theoretical reports, this section is called, for example, "Theoretical Basis" or "Theoretical Calculations" instead of "Experimental Details" and includes sufficient mathematical detail to enable other researchers to reproduce derivations and verify numerical results. Include all background data, equations, and formulas necessary to the arguments, but lengthy derivations are best presented as supporting information.

### Results

Summarize the data collected and their statistical treatment. Include only relevant data, but give sufficient detail to justify the conclusions. Use equations, figures, and tables only where necessary for clarity and brevity. Extensive but relevant data should be included in supporting information.

### Discussion

The purpose of the discussion is to interpret and compare the results. Be objective; point out the features and limitations of the work. Relate your results to current knowledge in the field and to the original purpose in undertaking the project: Was the problem resolved? What has been contributed? Briefly state the logical implications of the results. Suggest further study or applications if warranted.

Present the results and discussion either as two separate sections or as one combined section if it is more logical to do so. Do not repeat information given elsewhere in the manuscript.

### Conclusions

The purpose of the conclusions section is to put the interpretation into the context of the original problem. Do not repeat discussion points or include irrelevant material. Conclusions should be based on the evidence presented.

### Summary

A summary is unnecessary in most papers. In long papers, a summary of the main points can be helpful, but be sure to stick to the main points. If the summary itself is too long, its purpose is defeated.

### Acknowledgments

Generally, the last paragraph of the paper is the place to acknowledge people, organizations, and financing. As simply as possible, thank those persons, other than coauthors, who added substantially to the work, provided advice or technical assistance, or aided materially by providing equipment or supplies. Do not include their titles. If applicable, state grant numbers and sponsors here, as well as auspices under which the work was done, including permission to publish if appropriate.

Follow the publication's guidelines on what to include in the acknowledgments section. Some journals permit financial aid to be mentioned in acknowledgments, but not meeting references. Some journals put financial aid and meeting references together, but not in the acknowledgments section.

### References

In many books and journals, references are placed at the end of the article or chapter; in others, they are treated as footnotes. In any case, place the list of references at the end of the manuscript.

In ACS books and most journals, the style and content of references are standard regardless of where they are located. Follow the reference style presented in Chapter 14.

The accuracy of the references is the author's responsibility. Errors in references are one of the most common errors found in scientific publications and are a source of frustration to readers. Increasingly, hypertext links are automatically generated in Web-based publications, but this cannot be done for references containing errors. If citations are copied from another source, check the original reference for accuracy and appropriate content.

---

✎ **Reminder:** The accuracy of the references is the author's responsibility.

---

### Special Sections

This discussion on format applies to most manuscripts, but it is not a set of rigid rules and headings. If the paper is well organized, scientifically sound, and appropriate to the publication, adding other sections and subsections may be helpful to readers. For example, an appendix contains material that is not critical to understanding the text but provides important background information.

### Supporting Information

Material that may be essential to the specialized reader but not require elaboration in the paper itself is published as supporting information, usually on the journal's Web page. Examples of supporting information include large tables, extensive figures, lengthy experimental procedures, mathematical derivations, analytical and spectral characterization data, biological test data for a series, molecular modeling coordinates, modeling programs, crystallographic information files, instrument and circuit diagrams, and expanded discussions of peripheral findings.

More journals are encouraging this type of publishing to keep printed papers shorter. For ACS journals, supporting information is available immediately by linking to it from the citing paper on the Web. For example, for the article "Vanadium-Based, Extended Catalytic Lifetime Catechol Dioxygenases: Evidence for a Common Catalyst" by Cindy-Xing Yin and Richard G. Finke in *The Journal of the American Chemical Society* **2005,** *127,* 9003–9013, the supporting information consists of two files, ja051594esi20050517_053152.pdf (453 K) and ja051594erom20050320_064528.cif (24 K).

When including supporting information, place a statement to that effect at the end of the paper, using the format specified in the author instructions for the specific journal. For complete instructions on how to prepare this material for publication, check the author instructions for the publication.

### Web-Enhanced Objects

Some publishers, including ACS, have started exploring various Web-based technologies to enhance the way that information in a research article is conveyed. Selected papers in Web editions may contain Web-enhanced objects (WEOs) to supplement a reader's understanding of the research being reported. These types of files include color figures (including three-dimensional, rotatable figures), chemical structures, animations, spectra, video, and sound files. Links to WEOs will appear in the Web edition of the paper. These objects, although not essential to the understanding of the science, should help to augment a reader's understanding of the research being reported. The types of objects suitable for this form of publication should be viewable with commonly available plug-ins (e.g., Chime) or helper applications (e.g., WebLab Viewer, RasMol), which allow viewing and manipulating these objects within the HTML file itself or in a separate window. For example, a figure in the journal article "Orientation and Phase Transitions of Fat Crystals under Shear" by Gianfranco Mazzanti, Sarah E. Guthrie, Eric B. Sirota, Alejandro G. Marangoni, and Stefan H. J. Idziak, in

*Crystal Growth & Design* **2003,** *3,* 721–725, is supplemented by a movie WEO (in .mov format) depicting the time sequence of synchrotron X-ray diffraction patterns for the crystallization of cocoa butter in chocolate (see http://pubs.acs.org/isubscribe/journals/cgdefu/asap/objects/cg034048a/Mazzantivideouip.mov).

As with other types of special information, authors should check the author guidelines for the publication for instructions on how to prepare and submit WEOs.

➤➤➤➤➤ CHAPTER 3

# The Editorial Process

Publishing a manuscript, whether intended for a journal or a book, is a process. It has four stages: the draft manuscript, manuscript review, the final manuscript, and processing of accepted manuscripts. Along the way, responsibility for the different stages passes from the author, to the journal or book editor, back to the author, and finally to the technical editor. This chapter provides an overview of each of these stages as they evolve in scientific, technical, and medical (STM) publishing.

## The Draft Manuscript

### *Getting Started*

Before beginning to write, authors should review the ethical principles of scientific publication (see Chapter 1). The editorial process is supported by the ethical obligations of authors, editors, reviewers, and readers. Author integrity and adherence to the principles that guide scientific publications—such as deciding when it is the appropriate time to publish, determining who should author the manuscript, and providing the proper attribution and context for the research—are as integral to the success of scientific publication as providing science that is sound and of high quality.

Although there is no fixed set of "writing rules" to be followed like a cookbook recipe or an experimental procedure, some guidelines can be helpful. Start by considering the questions in Box 3-1; answering these questions will clarify your goals and make it easier to write the manuscript with the proper amount of detail. It will also make it easier for the book or journal editor to determine the

## Box 3-1. Questions for Drafting Your Manuscript

What is the function or purpose of this manuscript? Are you describing original and significant research results? Are you reviewing the literature? Are you providing an overview of the topic? Something else?

Who is the audience? Why would they want to read your manuscript? What will you need to tell them to help them understand your work?

How is your work different from that described in other reports on the same subject? (*Unless you are writing a review, be sure that your manuscript will make an original contribution. Most STM publishers, including ACS, do not publish previously published material.)*

What is the best format for publishing this manuscript—as a journal article, book, or book chapter? If you choose a journal article, which journal is most appropriate? *(Links to ACS journals can be found at http://pubs.acs.org/about.html.)*

manuscript's suitability for the publication. Writing is like so many other things: once the goal is identified, the details fall into place.

After you have determined the function of the manuscript and identified the audience, review your material for completeness or excess. Reports of original research, whether intended for a journal or a book, can be organized in the standard format: abstract, introduction, experimental details or theoretical basis, results, discussion, and conclusions. These sections are discussed in Chapter 2.

Keep in mind that scientific writing is not literary writing. Scientific writing serves a purpose completely different from that of literary writing, and it must therefore be precise and unambiguous. You and your colleagues probably have been discussing the project for months, so the words seem familiar, common, and clear to you. However, the readers will not have been part of these discussions. Many words are clear when speaking because you can amplify the meaning with gestures, expressions, and vocal inflections—but when these same words are written, they may be clear only to you. Chapter 4 presents strategies on how to write clearly and concisely as well as to select words that convey the meaning intended.

If English is not your first language, ask an English-speaking colleague—if possible, a native English speaker—for help with grammar and diction.

### *Publishers' Requirements*

An extremely important step is to check the specific requirements of the publication and to follow them. Journals often specify a format, the number of pages, what software packages or file formats are acceptable, how to cite references, and many other aspects of manuscript preparation. Requirements can vary from journal to journal even if the same publisher publishes them. Author guidelines for journals are generally posted on the Web at the journal's Web site, and they are also typically published in the first issue of each year. Book publishers also have author guidelines that need to be followed to expedite publication. Understanding the requirements for the manuscript cannot be overemphasized.

---

**Publishing with ACS:** The author guidelines for ACS journals can be seen at http://paragon.acs.org/paragon/index.jsp (see "Author Information"). The author guidelines for ACS books can be found at http://pubs.acs.org/books (see "Info for Authors").

---

Some publishers provide templates for authors to use when preparing their manuscripts. Use of a template makes it easier for authors to control margins, fonts, and paragraph styles, as well as the length of the manuscript. It also facilitates peer review by placing tabular and graphical material near the discussion in the text and providing journal and book editors with a single file to work with. Templates are generally available for Windows and Macintosh platforms, and they can be downloaded from a publisher's or journal's Web page.

---

**Publishing with ACS:** For ACS journals, templates can be accessed at http://paragon.acs.org/paragon/index.jsp (see "Download Manuscript Templates"). For ACS proceedings books, templates are available at http://pubs.acs.org/books/authorinfo.shtml (see "Request instructions on how to prepare your camera-ready manuscript").

---

## *Artwork*

As you write your draft manuscript, consider where structures, schemes, figures, and tables could be used appropriately to illustrate or support the material. Well-placed and well-designed artwork communicates information effectively, but too much artwork can be distracting.

Few scientists have access to graphic arts professionals. Consequently, chemical professionals need to know how to prepare art for manuscripts. Fortunately, software packages are available that can be easily mastered to produce good-looking graphs, charts, schemes, and structures. Chapters 15, 16, and 17 provide guidelines on when to use artwork and how to create figures, tables, or chemical structures and schemes that publishers can use effortlessly. These chapters also describe how to number figures, tables, structures, and schemes.

Sometimes you may wish to use artwork that has been previously published, whether from your own publications or from those of other authors. To use previously published artwork, you must get permission from the copyright holder, which is generally the publisher, even if you wrote the manuscript. Because it can take some time to secure reprint permission, it is a good idea to start obtaining permissions as you prepare your draft manuscript. If you wait until your manuscript is accepted for publication to initiate any permissions correspondence, publication of your manuscript may be delayed because publishers generally will not begin working on a manuscript when permissions are missing. Chapter 7 discusses how to get permission to reprint figures that have been previously published. Publishers' policies, and forms if required, are generally posted on their Web sites.

---

**Publishing with ACS:** Authors can reprint artwork previously published in ACS books and journals in other ACS publications without permission, provided that ACS is the original copyright holder. ACS's copyright policy and procedures can be found at http://pubs.acs.org/copyright_info.html.

---

Journals vary in their requirements about where tables and figures are placed in the manuscript. Some journals permit tables and figures to be inserted into the text for the draft but require that the tables and figures be submitted separately in the final manuscript. Other journals request that the tables and figures be embedded in the text. Some publishers accept figures prepared in a wide range of software packages, whereas others specify use of certain drawing programs. Check the specific requirements of the publication targeted before submitting the draft manuscript.

---

**Publishing with ACS:** Placement of artwork submitted to ACS journals depends on whether the manuscript is submitted through Paragon or the Paragon Plus environment. Be sure to check the author guidelines for the specific journal.

---

## *References*

References are an important component of every scholarly manuscript. Having complete and accurate references is the author's responsibility. Errors in references are one of the most common mistakes authors make. Although correct citations have always been important, the increasing number of hypertext links in Web-based publications makes correct citations more important than ever. Given the volume of manuscripts that publishers produce yearly, technical editors cannot verify each reference in each manuscript.

The citation of references in text is a subject that varies widely from journal to journal and publisher to publisher. There are three ways to cite references in text in ACS publications: superscript numbers, italic numbers in parentheses, or author name and year of publication. Authors are encouraged to check the author guidelines for a specific publication to find information on citing references. Chapter 14 explains how to cite references in ACS publications and how to format references from a variety of publications, in both print and electronic formats.

---

✐ **Reminder:** Although correct citations have always been important, the increasing number of hypertext links in Web-based publications makes correct citations more important than ever.

---

### *Revising the Draft Manuscript*

Once you have written your initial draft, the next step is a careful revision with an eye to organization, content, and editorial style, beginning with the questions in Box 3-2. Several chapters in this book are designed to help you communicate clearly. Chapter 9 reviews grammar, punctuation, and spelling. Chapter 10 provides guidelines on stylistic and editorial conventions, such as hyphenation and capitalization. Chapter 10 also includes a large appendix with abbreviations, acronyms, and symbols. Guidelines for using numbers, mathematics, and units of measure are given in Chapter 11. Two other chapters focus on more specific issues related to chemistry. Chapter 12 examines the use of proper chemical nomenclature. It provides rules for general chemistry nomenclature, as well as nomenclature in several specialized areas, such as polymer chemistry, biological chemistry, and combinatorial chemistry. Chapter 13 presents a quick reference guide for the use of typefaces, Greek letters, superscripts and subscripts, and special symbols that are commonly used in chemistry. Chapter 13 also includes an appendix containing symbols for commonly used physical quantities.

## Manuscript Review

When your draft manuscript is complete, check the journal or book author guidelines again for information on how and where to submit your draft. Some editors request that authors suggest possible reviewers. Some journals require that multiple copies of a draft manuscript be submitted and only accept manuscripts through the mail. Other journals request that the manuscript be submitted electronically via e-mail. Still others, like ACS, are using a Web-based system where authors submit a word-processing file or a PDF. For more information on submitting manuscripts using a Web-based system, see Chapter 5.

Once the editor has reviewed the manuscript and determined that it is appropriate for the publication, the peer-review process begins. Chapter 6 describes peer review and the responsibilities of reviewers and authors.

### Box 3-2. Questions for Revising Your Manuscript

Does your manuscript as it is written perform the function—new research, literature review, or topic overview—that you identified before you began your draft? Do you still think the format you selected—journal article, book, book chapter—is the best choice?

Have you explained terms, concepts, and procedures in a way that is appropriate to the audience you identified at the start?

Is your material presented in a logical fashion, so that a reader can easily follow your reasoning?

Is the manuscript too long? If so, what sections could be eliminated or possibly used as supporting information?

Do some sections need to be expanded to further clarify the material?

Are the sentences clear and unambiguous?

Are all the words spelled correctly and technical terms used appropriately?

Did you follow generally accepted conventions—such as those in this book—for communicating math and chemistry?

Could you use another opinion? You may find it helpful to ask a colleague, preferably one who is not closely involved with the research on which the manuscript is based, and preferably a native English speaker, to read and comment on your draft.

## The Final Manuscript

If your manuscript is accepted, the editor of the book or journal will return the peer-reviewed manuscript with a cover letter synthesizing the reviewers' comments and indicating what changes must be made for the final manuscript to be accepted. You, the author, then revise the manuscript accordingly. When you submit your final manuscript, include a cover letter indicating what changes you made. If you decide not to make some of the requested changes, you should write a rebuttal and send it with your final manuscript. For more information, see Chapter 6.

Authors are encouraged to submit all the paperwork with the final, revised manuscript. This includes all necessary permissions correspondence and, if required, a signed form transferring copyright from the author to the publisher. Chapter 7 gives a general introduction to copyright. If the manuscript is transmitted electronically, mail the forms separately.

**Reminder:** When you submit your final manuscript, you should include the final versions of text, tables, and illustrations, as well as any necessary permissions correspondence and a signed copyright transfer form.

Finally, keep a copy of the revised manuscript and all permissions correspondence. You will need the revised manuscript to check against the proofs that your publisher sends. Copies of the permissions correspondence can save you time and effort if the permissions correspondence gets lost or separated from your manuscript.

## Processing of Accepted Manuscripts

Journal editors and multiauthored book editors send accepted manuscripts directly to the publisher. Authors of monographs interact directly with the publisher, generally through an acquisitions editor. Accepted manuscripts go through three phases before publication: technical editing, proofing and review by the author, and correction by the publisher.

### *Technical Editing*

During the process of creating a book or journal issue, authors' electronic word-processing files are manipulated in a variety of ways. Files are tagged to identify data elements for print production and links for online products. Artwork is prepared for both publication media. The manuscript is copyedited to ensure consistency, clarity, and grammatical accuracy; changes are introduced to ensure the use of standard chemical conventions, graphics presentation, and tabular format. Copy editors often contact authors or query them at the proof stage for clarification of material.

### *Author's Proof*

One author, generally the author to whom correspondence should be addressed, receives a proof of the manuscript for final approval before publication. Papers are not generally released for printing until the author's proof or other approval has been received. Hence, proofs should be checked and returned promptly according to individual journal or book instructions.

**Publishing with ACS:** ACS journals request that proofs be returned within 48 hours of receipt.

Authors should check proofs very carefully and submit all of the corrections at one time; see Box 3-3 and Appendix 3-1 for information about reviewing

➤➤➤➤➤

## Box 3-3. Tips for Checking Proofs

➤ If you are instructed to return changes via the Web, list all corrections, revisions, and additions and clearly identify their location.

➤ If you are instructed to return changes in hard copy (paper printouts), mark corrections legibly in the margins of the proofs as instructed by the publisher. Do not erase or obliterate type; instead, strike one line through copy to be deleted and write the change in the margin.

➤ Clarify complicated corrections by rewriting the entire phrase or sentence.

➤ Check all text, data, and references against the original manuscript. Pay particular attention to equations; formulas; tables; captions; spelling of proper names; and numbering of equations, illustrations, tables and references.

➤ Answer explicitly all queries made by the technical editor.

Proofreader's marks and a sample of marked manuscript are given in Appendix 3-1.

proofs. Only corrections and necessary changes can be made to proofs. Although all authors may look at the proofs, only the corresponding author should submit corrections. Extensive changes may require editorial approval, which delays publication. Printer's errors are corrected at no cost to authors, but some publishers charge authors the cost of extensive production work made necessary by their own alterations.

## *Publication*

After you return your corrected proofs, the technical editor will review them and ensure that the corrections are made properly.

### ASAP (As Soon As Publishable) Articles

Many STM publishers, including ACS, publish journal articles on the Web before publishing them in print. Papers accepted for publication in ACS journals will be posted on the "Articles ASAP" page on the journal Web site as soon as they are ready for publication; that is, when the proofs are corrected and all author concerns are resolved. Publication on the Web usually occurs within four working days of receipt of proof corrections; this can be any time from three to six weeks before the date of the printed issue. Once a paper appears on the Web, ACS and the scientific community consider it published.

### Article Reprints and Complimentary Copies

Generally, authors receive a form for reprint orders with the author proof. Authors should follow the instructions on the form. Some publishers provide electronic reprints, as well as paper reprints. Customarily, there is a charge for paper reprints, and reprints with color artwork cost more.

Book authors sometimes also receive complimentary copies of the volume in which their chapter appears. On contributions with more than one author, the number of complimentary copies is generally limited; that is, not all authors will receive complimentary copies.

### Corrections to Published Manuscripts

Corrections of consequence to a paper that has already been published should be sent to the editor. Most journals publish corrections soon after they have been received. Some journals have a specific format for additions or corrections; check the author guidelines. In books, errata sheets will be printed and included in every book, and the book itself will be corrected before reprinting. However, additions and corrections generally reflect poorly on the authors, and careful manuscript preparation and attention to detail in the entire publication process can prevent the necessity for subsequent corrections.

➤➤➤➤➤

# APPENDIX 3-1
# Proofreaders' Marks

Publishers long ago established conventions for marking changes to manuscripts and proofs. These conventions, known as *proofreaders' marks,* evolved as an economical and precise shorthand for indicating on paper various types of changes.

Changes to double-spaced manuscript are marked one way (Figure 3A-1). Authors may be called on to interpret these proofreaders' marks if the publisher has copyediting done by hand, on paper. Common proofreaders' marks for manuscripts are presented in Figure 3A-2. Note that corrections to manuscripts are made in place (not in the margins) and usually need no additional explanation.

Changes to typeset proofs are marked somewhat differently (Figure 3A-3). Authors may be called on to use this proofreading method if the publisher supplies hard-copy (paper) proofs. Common proofreaders' marks for proofs (also called *galleys* or *page proofs*) are presented in Figure 3A-4. Note that corrections to proofs are made in two places: a minimal mark is made in the typeset text, to indicate where a change is being made, and an explanatory mark is made in the margin to describe the exact change. For example, a carat mark (^) in the typeset text indicates where new words are to be inserted; the words themselves are written in the margin. If there is more than one change to a typeset line, the changes in the margin are separated by slashes. (If you want to insert a slash, you should write out the word "slash" and circle it in the margin.) Two slashes in a row indicate that the first correction should be repeated. Try not to black out or obliterate the typeset characters. Avoid using arrows and lines to indicate where corrections go because more than one or two on a page breed confusion.

The photochemistry of α,β-unsaturated ketones has attracted much attention and is still a field ~~field~~ of current interest. 1 Numerous examples of such photochemical transformations are well-documented for cyclic enones and dienones, including both cycloaddition re actions and rearrangements. For example, cyclopentenones 1 and 2 readily rearrange to cyclopropyl ketenes upon irrandiation. Recently, the related cyclohexadienone-butadienyl ketene rearrangement has been shown to be a highly useful tool in the synthesis of natural products and macrocyclic lactones 2

Whereas cis trans isomerization, photodimerization, and [2 + 2] cycloadditions of acyclic α,β-undsaturated ketones are well-investigated photochemical transformations, comparatively little is documented concerning the photochemistry of such enones involving photodissociation, rearrangement or both. Clearly, the absence of ring strain lowers the reactivity twoard bond cleavage and renders an initial Norrish type I reaction unlikely. Introduction of radical stabilizing groups in the α-position of the enone may, however, be expected to change the reactivity of the enone in fagor ofthe photochemical α -cleavage and subsequent reactions derived from the resulting radical pairs.

**Figure 3A-1.** Sample of a manuscript copyedited by hand.

| | |
|---|---|
| ~~word~~ | Strike through to delete a word or words. |
| word | Triple underline to capitalize the "w". |
| Word | Slash to make the "w" lowercase. |
| wrod | Transpose two letters. |
| words two | Transpose two words. |
| Word | Double underline to make "ord" small capitals |
| word | Draw a wavy line to indicate bold face. |
| word | Underline to indicate italic type. |
| word1 | Draw an inverted carat to indicate superscript. |
| word2 | Draw a carat to indicate subscript. |
| keep ~~word~~ | Put dots or short dashes under copy that you wish to retain as it originally appeared. |

**Figure 3A-2.** Common proofreaders' marks for copyediting manuscripts.

The photochemistry of α,β unsaturated ketones has attracted much attention and is still a field ~~field~~ of current interest. 1 Numerous examples of such photochemical transformations are well-documented for cyclic enones and dienones, including both cycloaddition re actions and rearrangements. For example, cyclopentenones 1 and 2 readily rearrange to cyclopropyl ketenes upon irradiation. Recently, the related cyclohexadienone/butadienyl ketene rearrangement has been shown to be a highly useful tool in the synthesis of natural products and macrocyclic lactones 2

Whereas *cis/trans* isomerization, photodimerization, and [2 + 2] cycloadditions of acyclic α,β-undsaturated ketones are well-investigated photochemical transformations, comparatively little is documented concerning the photochemistry of such enones involving photodissociation, rearrangement or both. Clearly, the absence of ring strain lowers the reactivity twoard bond cleavage and renders an initial Norrish type I reaction unlikely. Introduction of radical stabilizing groups in the α-position of the enone may, however, be expected to change the reactivity of the enone in fador ofthe photochemical α-cleavage and subsequent reactions derived from the resulting radical pairs.

**Figure 3A-3.** Sample of a marked proof.

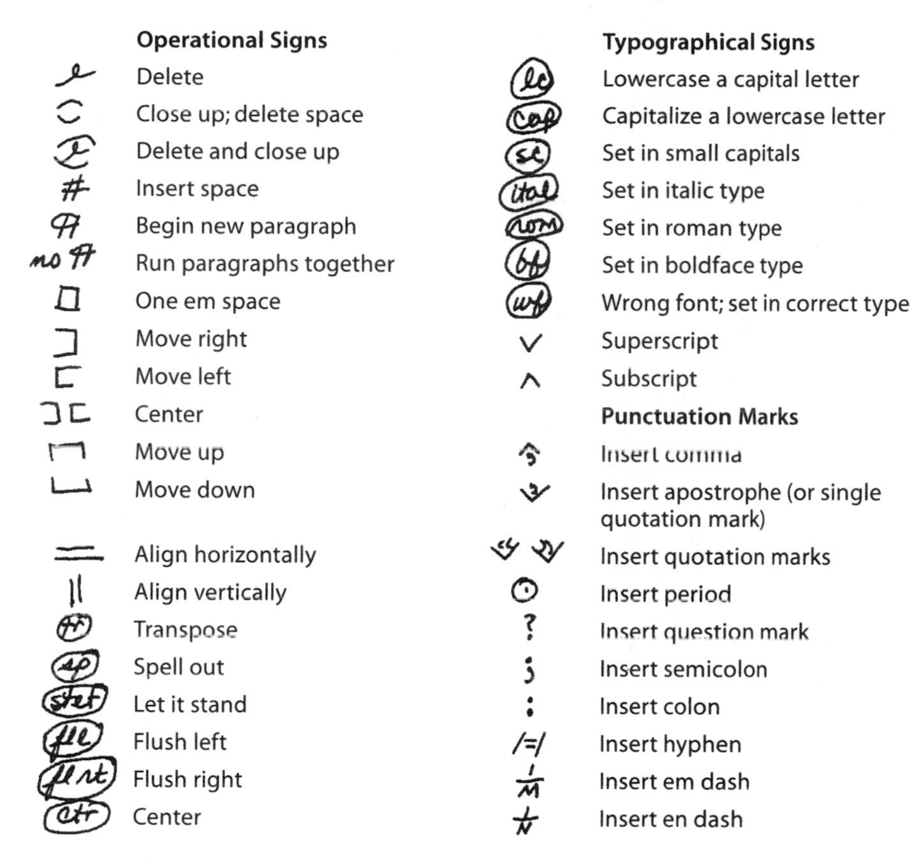


**Figure 3A-4.** Common proofreaders' marks for marking proofs.

➤➤➤➤➤ CHAPTER 4

# Writing Style and Word Usage

Every writer has a personal style, but all good writing tends to observe guidelines and conventions that communicate meaning clearly and exactly to readers. Scientific writing, in particular, must be precise and unambiguous to be effective.

This chapter presents guidelines for correct sentence structure and word usage. Other chapters of this book present topics also related to good writing style. Chapter 2 discusses the parts of a scientific paper; Chapter 3 presents an overview of the editorial process. Chapters in Part 2 address more specialized rules for usage; see, especially, Chapter 9 on grammar, punctuation, and spelling and Chapter 10 on editorial style.

## Correct Sentence Structure

Good organization (see Chapter 2) and sentence structure are an author's primary tools for conveying information in a logical, persuasive manner. When the words in a sentence are placed so that the reader follows easily from one fact or point to the next, then the reader is best able to comprehend the author's intended meaning. Poorly structured or ordered sentences create confusion for readers, who are then unable to understand accurately the author's meaning.

Short, simple *declarative* sentences—that is, sentences that make statements, rather than pose questions, issue commands, or exclaim—are the easiest to write and the easiest to read. They are also usually clear. However, too many short sentences in a row can sound abrupt or monotonous. They also can place too heavy a burden on the reader to connect the ideas from one sentence to the next. To add sentence variety and to enhance the flow of ideas, it is better to start with simple declarative sentences and then combine some of them, rather than to start with long rambling sentences and then try to shorten them. Two or more *simple* sentences (sentences with one independent clause and no subordinate clauses) can be combined to form a *compound* sentence. A *complex* sentence is created by adding one or more subordinate clauses to a simple sentence. A *clause* is a group of words that has a subject and a verb. If a clause can stand all by itself, it is an *independent clause.* If a clause cannot stand alone, it is a *dependent* or *subordinate* clause.

## Verbs

### Voice

A sentence is said to be in *active voice* when the subject of the sentence is the doer of the action indicated by the verb. The subject of an active verb is doing the action of the verb. In *passive voice,* the subject is the receiver of the action indicated by the verb.

➤ Use the active voice when it is less wordy and more direct than the passive.

POOR

> The fact that such processes are under strict stereoelectronic control is demonstrated by our work in this area.

BETTER

> Our work in this area demonstrates that such processes are under strict stereoelectronic control.

➤ Use the passive voice when the doer of the action is unknown or not important or when you would prefer not to specify the doer of the action.

> The solution is shaken until the precipitate forms.
>
> Melting points and boiling points have been approximated.
>
> Identity specifications and tests are not included in the monographs for reagent chemicals.

### Tense

Using the appropriate verb tense helps to orient the reader as to the nature of the information.

➤ Simple past tense is correct for stating what was done, either by others or by you.

> The solutions were heated to boiling.

Jones reviewed the literature and gathered much of this information.

We found that relativistic effects enhance the bond strength.

The structures were determined by neutron diffraction methods.

➤ Present tense is correct for statements of fact.

Absolute rate constants for a wide variety of reactions are available.

Hyperbranched compounds are macromolecular compounds that contain a branching point in each structural repeat unit.

➤ Present and simple past tenses may both be correct for results, discussion, and conclusions.

The characteristics of the voltammetric wave indicate that electron transfer occurs spontaneously.

The absence of substitution was confirmed by preparative-scale electrolysis.

IR spectroscopy shows that nitrates are adsorbed and are not removed by washing with distilled water.

However, the use of present or simple past tense for results, discussion, and conclusions should be consistent within a paper.

### Other Forms

➤ It is acceptable to use split infinitives to avoid awkwardness or ambiguity.

AWKWARD

The program is designed to assist financially the student who is considering a career in chemistry.

BETTER

The program is designed to financially assist the student who is considering a career in chemistry.

AMBIGUOUS

The bonded phases allowed us to investigate fully permanent gases.

BETTER

The bonded phases allowed us to fully investigate permanent gases.

## *Subjects and Subject–Verb Agreement*

➤ Use first person when it helps to keep your meaning clear and to express a purpose or a decision.

Jones reported xyz, but I (or we) found ....

> I (or we) present here a detailed study ....
>
> My (or our) recent work demonstrated ....
>
> To determine the effects of structure on photophysics, I (or we) ....

However, avoid clauses such as "we believe", "we feel", and "we can see", as well as personal opinions.

➤ Subjects and verbs must agree in person and number; this important point is discussed in detail in Chapter 9.

## *Sentence Modifiers*

Modifiers made up of phrases or dependent clauses can be added to simple sentences to indicate, for example, cause and effect, or time sequence, or comparison.

➤ A *restrictive* phrase or clause is one that is essential to the meaning of the sentence. Restrictive modifiers are not set off by commas.

> Only doctoral students who have completed their coursework may apply for this grant.
>
> Several systems that take advantage of this catalysis can be used to create new palladium compounds.

➤ A *nonrestrictive* phrase or clause is one that adds meaning to the sentence but is not essential; in other words, the meaning of the basic sentence would be the same without it. Nonrestrictive modifiers are set off by commas.

> Doctoral students, who often have completed their coursework, apply for this teaching fellowship.
>
> Several systems, which will be discussed below, take advantage of this catalytic reaction.

➤ A *misplaced modifier* is one that is placed next to the wrong word in the sentence, so it inadvertently misrepresents the author's intended meaning.

INCORRECT

> We commenced a new round of experiments unable to point to meaningful conclusions.

CORRECT

> Unable to point to meaningful conclusions, we commenced a new round of experiments.

➤ A *dangling modifier* is one that lacks a word in the sentence to modify in a logical or sensible way. It should not be confused with an *absolute construction*, which modifies an entire sentence. (See also the discussion of dangling modifiers in Chapter 9.)

INCORRECT

Adding 2 mL of indicator solution, the end point for the titration was reached.

CORRECT

Adding 2 mL of indicator solution, we reached the end point for the titration.

When we added 2 mL of indicator solution, the end point for the titration was reached.

## Sentence Construction and Word Order

➤ Use an affirmative sentence rather than a double negative.

| INSTEAD OF | CONSIDER USING |
| --- | --- |
| This reaction is not uncommon. | This reaction is common.<br>This reaction is not rare.<br>This reaction occurs about 40% of the time. |
| This transition was not unexpected. | This transition was expected.<br>We knew that such transitions were possible. |
| This strategy is not infrequently used. | This strategy is frequently used.<br>This strategy is occasionally used. |
| This result is not unlikely to occur. | This result is likely to occur.<br>This result is possible. |

➤ Watch the placement of the word "only". It has different meanings in different places in the sentence.

Only the largest group was injected with the test compound. (Meaning: and no other group)

The largest group was only injected with the test compound. (Meaning: and not given the compound in any other way)

The largest group was injected with only the test compound. (Meaning: and no other compounds)

The largest group was injected with the only test compound. (Meaning: there were no other test compounds)

➤ Be sure that the antecedents of pronouns are clear; in other words, when you use a pronoun (for example, "he", "she", "it", or "they"), the noun to which the pronoun refers should be obvious (for example, "Isaac Newton", "Marie Curie", "the compound", or "the research team"). This is particularly true for the pronouns "this" and "that". If there is a chance of ambiguity, use a noun to clarify your meaning.

AMBIGUOUS

> The photochemistry of transition-metal carbonyl complexes has been the focus of many investigations. This is due to the central role that metal carbonyl complexes play in various reactions.

UNAMBIGUOUS

> The photochemistry of transition-metal carbonyl complexes has been the focus of many investigations. This interest is due to the central role that metal carbonyl complexes play in various reactions.

➤ Use the proper subordinating conjunctions. (*Conjunctions* join parts of a sentence; *subordinating conjunctions* join subordinate clauses to the main sentence.) "While" and "since" have strong connotations of time. Do not use them where you mean "although", "because", or "whereas".

POOR

> Since solvent reorganization is a potential contributor, the selection of data is very important.

BETTER

> Because solvent reorganization is a potential contributor, the selection of data is very important.

POOR

> While the reactions of the anion were solvent-dependent, the corresponding reactions of the substituted derivatives were not.

BETTER

> Although the reactions of the anion were solvent-dependent, the corresponding reactions of the substituted derivatives were not.
>
> The reactions of the anion were solvent-dependent, but (or whereas) the corresponding reactions of the substituted derivatives were not.

## Parallelism

*Parallelism,* or *parallel construction,* is the use of words or groups of words of equal grammatical rank. *Equal grammatical rank* means that words are connected only to words, phrases only to phrases, subordinate clauses only to other subordinate clauses, and sentences only to other sentences. Establish parallel construction by using coordinating conjunctions, correlative conjunctions, and correlative constructions.

➤ A *coordinating conjunction* is a single word, such as "and", "but", "or", "nor", "yet", "for", and sometimes "so".

INCORRECT

Compound **12** was prepared analogously and by Lee's method (*5*).

CORRECT

Compound **12** was prepared in an analogous manner and by Lee's method (*5*).

➤ A *correlative conjunction* is a pairing of words, such as "either ... or"; "neither ... nor"; "both ... and"; "not only ... but also"; and "not ... but".

INCORRECT

The product was washed either with alcohol or acetone.

CORRECT

The product was washed with either alcohol or acetone.

The product was washed either with alcohol or with acetone.

INCORRECT

It is best to use alternative methods both because of the condensation reaction and because the amount of water in the solvent increases with time.

CORRECT

It is best to use alternative methods both because of the condensation reaction and because of the increase in the amount of water in the solvent with time.

INCORRECT

Not only was the NiH functionality active toward the C-donor derivatives but also toward the N donors.

CORRECT

The NiH functionality was active not only toward the C-donor derivatives but also toward the N donors.

The NiH functionality was not only active toward the C-donor derivatives but also active toward the N donors.

Not only was the NiH functionality active toward the C-donor derivatives, but it was also active toward the N donors.

➤ A *correlative construction* is a sentence structure that uses "as ... as" (for example, "as well as").

He performed the experiment as well as I could have done it.

➤ Do not try to use parallel construction around the word "but" when it is not used as a coordinating conjunction.

> Increasing the number of fluorine atoms on the adjacent boron atom decreases the chemical shift, but only by a small amount.
>
> The reaction proceeded readily, but with some decomposition of the product.

➤ Use parallel constructions in series and lists, including section headings and subheadings in text and tables and listings in figure captions.

### *Comparisons*

➤ Introductory phrases that imply comparisons should refer to the subject of the sentence and be followed by a comma.

INCORRECT

> Unlike alkali-metal or alkaline-earth-metal cations, hydrolysis of trivalent lanthanides proceeds significantly at this pH.

CORRECT

> Unlike that of alkali-metal or alkaline-earth-metal cations, hydrolysis of trivalent lanthanides proceeds significantly at this pH.
>
> Unlike alkali-metal or alkaline-earth-metal cations, trivalent lanthanides hydrolyze significantly at this pH.

INCORRECT

> In contrast to the bromide anion, there is strong distortion of the free fluoride anion on the vibrational spectroscopy time scale.

CORRECT

> In contrast to the bromide anion, the free fluoride anion is strongly distorted on the vibrational spectroscopy time scale.

➤ Use the verb "compare" followed by the preposition "to" when similarities are being noted. Use "compare" followed by the preposition "with" when differences are being noted. Only things of the same class should be compared.

> Compared to compound **3**, compound **4** shows an NMR spectrum with corresponding peaks.
>
> Compared with compound **3**, compound **4** shows a more complex NMR spectrum.

➤ Do not omit words needed to complete comparisons, and do not use confusing word order. The subordinating conjunction "than" is often used to introduce the second element in a comparison, following an adjective or adverb in the comparative degree.

INCORRECT

The alkyne stretching bands for the complexes are all lower than the uncoordinated alkyne ligands.

CORRECT

The alkyne stretching bands for the complexes are all lower than those for the uncoordinated alkyne ligands.

The alkyne stretching bands are all lower for the complexes than for the uncoordinated alkyne ligands.

INCORRECT

The decrease in isomer shift for compound **1** is greater in a given pressure increment than for compound **2**.

CORRECT

The decrease in isomer shift for compound **1** is greater in a given pressure increment than that for compound **2**.

The decrease in isomer shift in a given pressure increment is greater for compound **1** than for compound **2**.

➤ Idioms often used in comparisons are "different from", "similar to", "identical to", and "identical with". Generally these idioms should not be split.

INCORRECT

The complex shows a significantly different NMR resonance from that of compound **1**.

CORRECT

The complex shows an NMR resonance significantly different from that of compound **1**.

INCORRECT

Compound **5** does not catalyze hydrogenation under similar conditions to compound **6**.

CORRECT

Compound **5** does not catalyze hydrogenation under conditions similar to those for compound **6**.

EXCEPTION These idioms can be split if an intervening prepositional phrase modifies the first word in the idiom.

The single crystals are all similar in structure to the crystals of compound **7**.

Solution A is identical in appearance with solution B.

➤ Phrases such as "relative to", "as compared to", and "as compared with" and words such as "versus" are also used to introduce the second element in a comparison. The things being compared must be parallel.

> The greater acidity of nitric acid relative to nitrous acid is due to the initial-state charge distribution in the molecules.
>
> The lowering of the vibronic coupling constants for Ni as compared with Cu is due to configuration interaction.
>
> This behavior is analogous to the reduced Wittig-like reactivity in thiolate versus phenoxide complexes.

## Correct Word Usage

The words chosen by a writer are one of the defining characteristics of that author's style; however, word choice is not governed by style alone. The audience for a paper (as discussed in Chapter 3) must influence a writer's choice of words so that the writer can select words that are likely to be known to the audience and define the words that are not. The type of document also may influence a writer's word choices because some documents, such as scientific papers, journal articles, and books, tend to more formal word usage, whereas other documents, such as e-mails, allow less formality.

The choice of the correct word to express meaning begins with a good dictionary, but it also extends to understanding small differences in meaning between two words or phrases that are almost synonymous or that are spelled similarly but have significant differences in meaning. It is best to use words in their primary meanings and to avoid using a word to express a thought if such usage is uncommon, informal, or primarily literary. Many words are clear when you are speaking because you can amplify your meaning with gestures, expressions, and vocal inflections—but when these same words are written, they may be clear only to you.

This chapter presents only a few words and phrases that are commonly misused in scientific writing; consult a good reference on word usage for more comprehensive assistance. (Several such references are listed under the heading "References on Scientific Communication" in Chapter 18.)

### *Grouping and Comparison Words*

➤ Use "respectively" to relate two or more sequences in the same sentence.

> The excitation and emission were measured at 360 and 440 nm, respectively. (That is, the excitation was measured at 360 nm, and the emission was measured at 440 nm.)

➤ Use the more accurate terms "greater than" or "more than" rather than the imprecise "over" or "in excess of".

greater than 50% (*not* in excess of 50%)
more than 100 samples (*not* over 100 samples)
more than 25 mg (*not* in excess of 25 mg, *not* over 25 mg)

➤ Use "fewer" to refer to number; use "less" to refer to quantity.

fewer than 50 animals
fewer than 100 samples
less product
less time
less work

➤ However, use "less" with number and unit of measure combinations because they are regarded as singular.

less than 5 mg
less than 3 days

➤ Use "between" with two named objects; use "among" with three or more named or implied objects.

Communication between scientists and the public is essential.

Communication among scientists, educators, and the public is essential.

Communication among scientists is essential.

### Commonly Confused Words and Phrases

➤ Choose "myself" and "me" depending on your meaning. "Myself" is a reflexive pronoun that is used only in sentences in which "I" is the subject, whereas "me" is used as a direct or indirect object or the object of a preposition. "Myself" is never a substitute for "me".

Please give a copy of the agenda to Anne and me. (*not* to Anne and myself)

I myself checked the agenda.

Cheryl and I checked the agenda. (*not* Cheryl and myself)

The agenda was checked by Barbara and me. (*not* by Barbara and myself)

➤ Choose "due to", which means "attributable to", only to modify a noun or pronoun directly preceding it in the sentence or following a form of the verb "to be".

Cutbacks due to decreased funding have left us without basic reference books.

The accuracy of the prediction is due to a superior computer program.

➤ Choose “based on” and “on the basis of” depending on your meaning. Phrases starting with “based on” must modify a noun or pronoun that usually immediately precedes or follows the phrase. Use phrases starting with “on the basis of” to modify a verb.

> The doctors’ new methods in brain surgery were based on Ben Carson’s work.
>
> On the basis of the molecular orbital calculations, we propose a mechanism that can account for all the major features of alkali and alkaline earth catalyzed gasification reactions. (*not* Based on …)

➤ Choose “assure”, “ensure”, and “insure” depending on your meaning. To assure is to affirm; to ensure is to make certain; to insure is to indemnify for money.

> He assured me that the work had been completed.
>
> The procedure ensures that clear guidelines have been established.
>
> You cannot get a mortgage unless you insure your home.

➤ Choose “affect”, “effect”, and “impact” depending on your meaning. When “affect” is used as a verb, it means to influence, modify, or change. When “effect” is used as a verb, it means to bring about, but as a noun it means consequence, outcome, or result. “Impact” is a noun meaning a significant effect.

> The increased use of pesticides affects agricultural productivity.
>
> The use of polychlorinated benzenes has an effect on the cancer rate.
>
> The effect of the added acid was negligible.
>
> The new procedure effected a 50% increase in yield.
>
> The impact of pesticide use on health is felt throughout the world.
>
> The acid did not have a great impact on the reaction rate.

➤ Use “whether” to introduce at least two alternatives, either stated or implied.

> I am not sure whether I should repeat the experiment.
>
> I am not sure whether I should repeat the experiment or use a different statistical treatment.
>
> I am going to repeat the experiment whether the results are positive or negative.

Use “whether or not” to mean “regardless of whether”.

INCORRECT

> I am not sure whether or not to repeat the experiment.

CORRECT

I am not sure whether to repeat the experiment.

Whether or not the results are positive, I will repeat the experiment.

Whether or not I repeat the experiment, I will probably leave the laboratory late tonight.

➤ Use "to comprise" to mean "to contain" or "to consist of"; it is not a synonym for "to compose". The whole *comprises* the parts, or the whole *is composed of* the parts, but the whole is not comprised of the parts. Never use "is comprised of".

INCORRECT

A book is comprised of chapters.

CORRECT

A book comprises chapters.

A book is composed of chapters.

INCORRECT

Our research was comprised of three stages.

CORRECT

Our research comprised three stages.

## Use of "A" and "An"

➤ Choose the articles "a" and "an" according to the pronunciation of the words or abbreviations they precede. See pp 257 and 264 for the use of "a" and "an" with chemical elements and isotopes.

a nuclear magnetic resonance spectrometer
an NMR spectrometer

➤ Use "a" before an aspirated "h"; use "an" before the vowel sounds of a, e, i, o, "soft" or "short" u, and y.

a house, a history (*but* an hour, an honor)
a union, a U-$^{14}$C (*but* an ultimate)
a yard (*but* an ylide, an yttrium compound)

➤ Choose the proper article to precede B.A., B.S., M.A., M.S., and Ph.D., according to pronunciation of the first letter.

a B.S. degree
an M.S. degree
a Ph.D.

## *Words and Phrases To Avoid*

➤ Avoid slang and jargon.

➤ Be brief. Wordiness obscures your message and annoys your readers.

➤ Omit empty phrases such as

As already stated
It has been found that
It has long been known that
It is interesting to note that
It is worth mentioning at this point
It may be said that
It was demonstrated that

➤ Omit excess words.

| INSTEAD OF | CONSIDER USING |
|---|---|
| It is a procedure that is often used. | This procedure is often used. |
| There are seven steps that must be completed. | Seven steps must be completed. |
| This is a problem that is …. | This problem is …. |
| These results are preliminary in nature. | These results are preliminary. |

➤ Write economically (and usually more precisely) by using single words instead of phrases.

| INSTEAD OF | CONSIDER USING |
|---|---|
| a number of | many, several |
| a small number of | a few |
| are found to be | are |
| are in agreement | agree |
| are known to be | are |
| at present | now |
| at the present time | now |
| based on the fact that | because |
| by means of | by |
| despite the fact that | although |
| due to the fact that | because |
| during that time | while |
| fewer in number | fewer |
| for the reason that | because |
| has been shown to be | is |
| if it is assumed that | if |
| in color, e.g., red in color | just state the color, e.g., red |
| in consequence of this fact | therefore, consequently |
| in length | long |
| in order to | to |
| in shape, e.g., round in shape | just state the shape, e.g., round |
| in size, e.g., small in size | just state the size, e.g., small |

| INSTEAD OF | CONSIDER USING |
|---|---|
| in spite of the fact that | although |
| in the case of … | in …, for … |
| in the near future | soon |
| in view of the fact that | because |
| is known to be | is |
| it appears that | apparently |
| it is clear that | clearly |
| it is likely that | likely |
| it is possible that | possibly |
| it would appear that | apparently |
| of great importance | important |
| on the order of | about |
| owing to the fact that | because |
| prior to | before |
| reported in the literature | reported |
| subsequent to | after |

➤ Do not use contractions in scientific papers.

INCORRECT

The identification wasn't confirmed by mass spectrometry.

CORRECT

The identification was not confirmed by mass spectrometry.

➤ Do not use the word "plus" or the plus sign as a synonym for "and".

INCORRECT

Two bacterial enzymes were used in a linked-enzyme assay for heroin plus metabolites.

CORRECT

Two bacterial enzymes were used in a linked-enzyme assay for heroin and its metabolites.

➤ Do not use "respectively" when you mean "separately" or "independently".

INCORRECT

The electrochemical oxidations of chromium and tungsten tricarbonyl complexes, respectively, were studied.

CORRECT

The electrochemical oxidations of chromium and tungsten tricarbonyl complexes were studied separately.

➤ Avoid misuse of prepositional phrases introduced by "with".

POOR

Nine deaths from leukemia occurred, with six expected.

BETTER

Nine deaths from leukemia occurred, and six had been expected.

POOR

Of the 20 compounds tested, 12 gave positive reactions, with three being greater than 75%.

BETTER

Of the 20 compounds tested, 12 gave positive reactions; three of these were greater than 75%.

POOR

Two weeks later, six more animals died, with the total rising to 25.

BETTER

Two weeks later, six more animals died, and the total was then 25.

➤ Do not use a slash to mean "and" or "or".

INCORRECT

Hot/cold extremes will damage the samples.

CORRECT

Hot and cold extremes will damage the samples.

➤ Replace "and/or" with either "and" or "or", depending on your meaning.

INCORRECT

Our goal was to confirm the presence of the alkaloid in the leaves and/or roots.

CORRECT

Our goal was to confirm the presence of the alkaloid in the leaves and roots.

Our goal was to confirm the presence of the alkaloid in either the leaves or the roots.

Our goal was to confirm the presence of the alkaloid in the leaves, the roots, or both.

➤ If you have already presented your results at a symposium or other meeting and are now writing the paper for publication in a book or journal, delete all references to the meeting or symposium, such as "Good afternoon, ladies and gentlemen", "This morning we heard", "in this symposium", "at this meeting", and

"I am pleased to be here". Such phrases would be appropriate only if you were asked to provide an exact transcript of a speech.

➤ Avoid using the word "recently". Your article or book may be available for a long time. This word will make it look dated in little time.

POOR

It was recently found that these effects enhance the bond strength.

BETTER

Harris and Harris (2006) found that these effects enhance the bond strength.

## Gender-Neutral Language

The U.S. government and many publishers have gone to great effort to encourage the use of gender-neutral language in their publications. Gender-neutral language is also a goal of many chemists. Recent style guides and writing guides urge copy editors and writers to choose terms that do not reinforce outdated sex roles. Gender-neutral language can be accurate and unbiased and not necessarily awkward.

The most problematic words are the noun "man" and the pronouns "he" and "his", but there are usually several satisfactory gender-neutral alternatives for these words. Choose an alternative carefully and keep it consistent with the context.

➤ Instead of "man", use "people", "humans", "human beings", or "human species", depending on your meaning.

OUTDATED

The effects of compounds **I–X** were studied in rats and man.

GENDER-NEUTRAL

The effects of compounds **I–X** were studied in rats and humans.

OUTDATED

Men working in hazardous environments are often unaware of their rights and responsibilities.

GENDER-NEUTRAL

People working in hazardous environments are often unaware of their rights and responsibilities.

OUTDATED

Man's search for beauty and truth has resulted in some of his greatest accomplishments.

GENDER-NEUTRAL

The search for beauty and truth has resulted in some of our greatest accomplishments.

➤ Instead of “manpower”, use “workers”, “staff”, “work force”, “labor”, “crew”, “employees”, or “personnel”, depending on your meaning.

➤ Instead of “man-made”, use “synthetic”, “artificial”, “built”, “constructed”, “manufactured”, or even “factory-made”.

➤ Instead of “he” and “his”, change the construction to a plural form (“they” and “theirs”) or first person (“we”, “us”, and “ours”). Alternatively, delete “his” and replace it with “a”, “the”, or nothing at all. “His or her”, if not overused, is also acceptable. Using passive voice or second person (“you”, “your”, and “yours”) also works sometimes.

OUTDATED

The principal investigator should place an asterisk after his name.

GENDER-NEUTRAL

Principal investigators should place asterisks after their names.

If you are the principal investigator, place an asterisk after your name.

The name of the principal investigator should be followed by an asterisk.

➤ Do not use a plural pronoun with a singular antecedent.

INCORRECT

The principal investigator should place an asterisk after their name.

CORRECT

The principal investigators should place asterisks after their names.

➤ Instead of “wife”, use “family” or “spouse” where appropriate.

OUTDATED

The work of professionals such as chemists and doctors is often so time-consuming that their wives are neglected.

GENDER-NEUTRAL

The work of professionals such as chemists and doctors is often so time-consuming that their families are neglected.

OUTDATED

the society member and his wife

GENDER-NEUTRAL

the society member and spouse

➤➤➤➤➤ CHAPTER 5

# Electronic Submission of Manuscripts Using Web-Based Systems

Sarah C. Blendermann

Electronic submission of manuscripts to journals is undergoing significant change as publishers respond to mounting pressure to publish faster, better, and more efficiently. The use of the Web and e-mail enables the peer-review process to move more rapidly, speeding review and decision cycles. And although scientific research has always been an international activity, journal publishing is increasingly global, with authors, reviewers, and editors contributing from numerous countries and all participants benefiting from electronic communications.

---

✐ **Reminder:** Good practices and appropriate file creation start early in the manuscript preparation process. If author source files are not of adequate quality for production, then the publication process will be delayed. Authors should be mindful of publishers' requirements early in the writing process.

---

This chapter covers the major systems used by dominant commercial publishers and professional societies to manage the submission, review, and acceptance of scholarly manuscripts, and it endeavors to guide authors through the routine tasks associated with submitting a manuscript online. It is important to acknowledge that publishers will revise their systems and add new features quickly, so the information that follows will likely age rapidly. However, authors can rely on this chapter to guide them through the general process of online submission.

Proprietary and commercial editorial systems are frequently designed to accommodate both electronic and paper publication processes. Capabilities vary, but many systems have become similar as publishers establish parallel mechanisms for managing the peer-review process. Leading commercial software packages currently include Bench>Press from HighWire Press, Editorial Manager from Aries Systems, EJPress from eJournal Press, ScholarOne Manuscript Central, and Rapid Review from Cadmus Systems. Other Web-based commercial products include EdiKit from Berkeley Electronic Press; myICAAP, EPRESS, ESPERE, Fontisworks, and Open Journal Systems from University of British Columbia; PublishASAP; Temple Peer Review Manager; and Xpress Track. Each package provides authors, reviewers, and editors with submission acknowledgments, decision letters, and review documents transmitted using e-mail; no paper correspondence is needed.

Several publishers have opted to develop proprietary programs rather than purchase services from a third party. Examples include the American Chemical Society's ACS Paragon System and the American Institute of Physics' Peer X-Press. These packages offer features and workflow similar to those contained in commercial editorial packages.

Appendix 5-1 matches selected scientific publishers and research grant agencies with the manuscript submission software they use and manuscript submission sites.

## Preparing Materials

The author guidelines for each journal contain generic and journal-specific instructions concerning manuscript preparation. They indicate the types of components that are required for online submissions, such as the cover letter, abstract, manuscript document, supporting information, figures, and proposed reviewers. Publishers require these items to be submitted in common word-processing or graphics formats, and author guidelines provide the technical details for preparing these manuscript components. Assembling and organizing electronic components of a manuscript in advance will streamline the process and reduce the possibility of errors in submission.

The author guidelines will indicate if there are word-processing templates available for authors to download and will specify if there are requirements about their use. Template files contain all the necessary formatting for a particular journal and are provided by the publisher from its Web site. Templates make it easier to format submissions to meet the publisher's specifications.

It is important to understand why authors are asked to provide manuscripts as both word-processing and PDF files. These versions of the manuscript are used in different ways; the PDF version of a manuscript is more suitable for peer review because it is easily transferred among authors, editors, and reviewers and is readable with the ubiquitous free Adobe Acrobat Reader. PDF files are portable across computer platforms, whereas original word-processing files may not be compatible with the system of an editor or reviewer. Publishers may require that authors submit a PDF file, or they may automatically generate one as part of the online submission process. A PDF file assists in maintaining control of versions of the manuscript because editors, authors, and reviewers can insert comments and notes for correction without altering the original text. Most publishers will require word-processing (or TeX or its derivatives) files for production purposes. These versions of the manuscript are manipulated during the final electronic creation of a journal for publication online or in print.

---

**Reminder:** Most publishers require both PDF and word-processing versions of a manuscript to be used for peer review and publication production, respectively.

---

Another critical element of your submission is the figures associated with the manuscript. Graphics should be high-resolution and of good quality to ensure clarity and accuracy in the final published copy and to facilitate any needed reduction required in the print publication (see Chapter 15 for more detail on preparing illustrations). Many software programs—including PowerPoint, Word, Excel, and WordPerfect—do not create high-resolution images suitable for publication. For that reason, it is recommended that authors create graphics using applications that can prepare figures in TIFF or EPS formats. Figures should have clear, sharp lines, should be clearly labeled, and should be at a high enough resolution that they can be used to compose the print journal.

Appendix 5-2 presents the text and image formats accepted by seven major manuscript peer-review and submission systems.

## Beginning Your Submission

Depending on the publisher, authors may be asked to e-mail the prepared manuscript to the editor, to upload it via an FTP server, or to upload files through the publisher's Web browser to a secure Web site using HTTP protocol. It is extremely important to submit manuscripts in the method designated for that particular journal because submission mechanisms may vary, even within one publisher's journals. If the journal requires online manuscript submission, submit all files to that Web site. Similarly, if the journal uses FTP or e-mail, follow the instructions carefully and send all manuscript components via the same route, unless indicated otherwise.

When authors are asked to e-mail the manuscript directly to an editor, pay particular attention to the type of files authors are asked to provide, and note if the manuscript should be provided in the body of the e-mail or as an attachment. Publishers offering an FTP server for author use will have detailed instructions on how to access the FTP server. Authors may be asked to compress files when transmitting manuscripts by e-mail, FTP, or as part of an online submission system. If so, the publisher's Web site will detail the acceptable formats for compressed files. These may include zipped, tarred, uuencoded, or BinHex files. Under some circumstances, Web uploads will automatically invoke a compression process.

Leading online submission packages Bench>Press, Editorial Manager, EJPress, Manuscript Central, and Rapid Review, as well as the ACS Paragon System and the American Institute of Physics' Peer X-Press all use the author home page. This simple online profile contains the author's contact details and establishes a user name and password. The e-mail address entered here must be correct; a typo could delay important e-mail notifications about acceptance or requests for revision. If the author has already established an account, future submissions can begin with the author login.

Several publishers have adopted the concept of a submitting agent or a second author if someone other than the author is submitting the manuscript on the author's behalf. This person is responsible for the tasks associated with submission, but correspondence and requests for further changes are made to the party designated as the corresponding or primary author of the manuscript. Submitting agent accounts are created similarly to author accounts, but they require information about the primary author also to be added to the user account profile.

Passwords to either author accounts or submitting agent accounts should not be shared. Doing so leaves the account holder at risk of incomplete submissions being changed or new submissions being created under his or her name.

## The Author Home Page

After logging in, the author is presented with his or her home page. Typically, the author home page is divided into several areas; from this location, authors can begin a new submission, check the status of a previous submission, continue a submission begun earlier, or submit a revised manuscript. The author home page also shows the progress of accepted manuscripts through the production cycle to publication. As publishers increasingly recognize the value of this author home page, new features are likely to be added. In the case of the Cadmus Rapid Review software, the author home page has already evolved to include a scientist's activities as both a referee and an author.

---

✐ **Reminder:** If an author home page is available, use it to submit new articles and to track the status of current submissions.

---

## Submitting Your Manuscript

Preparing materials in advance allows the authors either to complete their full submissions in one sitting or to partially complete the process, depending on individual preferences and the requirements of the publisher. E-mail and FTP submissions require the submission to be completed at one time. However, more publishers with online submission sites allow authors to interrupt their online submissions and to complete them later. During a partial submission, each step must be completed for the information pertaining to that submission step to be saved. The author can complete the submission by logging into the journal site and accessing his or her author home page. After selecting the link for that manuscript, the submission process can be resumed, beginning with the first incomplete step. Before the manuscript is submitted, authors are asked to review all the component parts and make any final changes.

Supplemental information should also be reviewed and validated. Crystallographic information files (CIFs) in particular can be verified using the free CheckCIF utility, which is available from the International Union of Crystallography.

Usually in the cases of e-mail, FTP, or online submission, no additional changes to the manuscript or associated documents are permitted after submission unless an editor requests a revision or the publisher contacts the author to fix a problem.

## Submitted Manuscripts

After the submission is complete, the author typically receives an e-mail acknowledging the submission from either the editor or the submission system. If the publisher offers an author home page, the status of the manuscript can be tracked from that site. These sites also notify authors by e-mail as the status of the manuscript changes. If the publisher makes use of an online system for submission, the editors for that publisher may be able to use features available only to them to shepherd the manuscript through the peer-review process. In these cases, the system notifies the editor of newly arrived manuscripts, allows the editor to view all associated files and details of the submission, and allows for the manuscript to be assigned to the appropriate editors and reviewers. Many systems generate the correspondence that accompanies the notification of editors, reviewers, and authors. As decisions are made and correspondence is sent, the corresponding status changes for the manuscript are displayed on the author home page. Additional details about the peer-review process can be found in Chapter 6.

## Stops Along the Way (Revisions)

When a revision of the paper is requested, an e-mail from the editor will detail the necessary changes. Where FTP or direct-to-editor e-mail is the preferred submission mechanism, the author will send revised files directly to the indicated e-mail address or FTP location. For publishers with author home pages, authors can view the status of submitted manuscripts and determine that a revision has been requested. From the author home page, authors access the submission and upload revised manuscript documents, images, and associated files. During the submission of the revision, authors may also provide rebuttal information or revision details to clarify how the manuscript has been altered. If an author determines that a revision is necessary before a request is made by the editor, authors using Web-based submission systems must contact the appropriate editor and request that the status of the manuscript be changed within the system to indicate that a revision is needed. This step will allow the author to adjust and replace files and then to submit a revision. Once the revision is received, an acknowledgment is sent by e-mail, and the author will not be able to make further changes to the paper unless the editor requests another revision to the manuscript.

## Acceptance

Authors can expect to be notified of their manuscripts' acceptance by e-mail. Accepted manuscripts are copyedited and formatted according to specific journal style, using the original word-processing files from the author's online submission. As the manuscript progresses through the various stages of production, the status on the author home page changes to reflect each new step.

➤➤➤➤➤

# APPENDIX 5-1

# Online Submission at Selected Scientific Publishers and Research Grant Agencies

This appendix contains a list of scientific publishers and research grant agencies, along with the software they use for online manuscript submission and the manuscript submission site, if there is one.

**Table 5A-1.** Scientific Publishers

| *Publisher* | *Software* | *URL* |
|---|---|---|
| American Academy of Forensic Science | Information and forms only | No electronic submission |
| American Association for Clinical Chemistry | Bench>Press | http://submit.clinchem.org/?ctst=y |
| American Association of Pharmaceutical Scientists | Editorial Manager | *Pharmaceutical Research* submissions at https://www.editorialmanager.com/pharmres/ |
| American Chemical Society | ACS Paragon System | https://paragon.acs.org/paragon/index.jsp |
| American Geophysical Union | Geophysical Electronic Manuscript System | http://gcubed-submit.agu.org/ |
| American Institute for Chemical Engineers | Rapid Review | https://www.rapidreview.com/AIChE2/CALogon.jsp |
| American Institute of Physics | Some journals use Peer X-Press | http://www.aipservices.org/peerxpress/index.html After acceptance, you will be asked to use their FTP or e-mail site at http://www.aip.org/epub/submittext.html. |
| American Mathematical Society | Information and forms only | No electronic submission |
| American Peptide Society | ScholarOne Manuscript Central | http://bip-pep-wiley.manuscriptcentral.com/ |
| American Pharmacists Association | ScholarOne Manuscript Central | *Journal of Pharmaceutical Sciences* submissions at http://jpharmsci-wiley.manuscriptcentral.com/ |
| American Physical Society | — | http://publish.aps.org/ESUB/ |
| American Society for Biochemistry and Molecular Biology | — | *The Journal of Biological Chemistry* submissions at http://osrs.jbc.org/asbmb/osrs.nsf/StartSubmission?OpenForm |

*Continued on next page*

**Table 5A-1.** Scientific Publishers—Continued

| *Publisher* | *Software* | *URL* |
|---|---|---|
| American Society for Cell Biology | EJPress | http://www.mbcpapers.org/ |
| American Society for Mass Spectrometry | Elsevier author gateway | http://authors.elsevier.com/JournalDetail.html?PubID=505727&Precis=DESC |
| American Society for Microbiology | Rapid Review | https://www.rapidreview.com/ASM2/CALogon.jsp |
| Biophysical Society | Bench>Press | http://submit.biophysj.org/?ctst=y |
| Blackwell Publishing | Information and forms only | No electronic submission |
| Cambridge University Press | Information and forms only | No electronic submission |
| Electrochemical Society | Peer X-Press | http://jes.peerx-press.org/cgi-bin/main.plex |
| Elsevier | Elsevier author gateway | Each journal has its own instructions for submitting work. Go to the home page of your journal of interest at http://authors.elsevier.com/. |
| Institute of Electrical & Electronics Engineers (IEEE) | — | Each journal has its own instructions for submitting work. Go to the home page of your journal of interest at http://www.ieee.org/organizations/pubs/guide.html. |
| Institute of Food Technologists | ScholarOne Manuscript Central | http://ift.manuscriptcentral.com/ |
| Materials Research Society | ScholarOne Manuscript Central | http://jmr.manuscriptcentral.com/ |
| Nature | EJPress | http://npg.nature.com/npg/servlet/Content?data=xml/05_sub.xml&style=xml/05_sub.xsl |
| Oxford University Press | Some journals use ScholarOne Manuscript Central | Each journal has its own instructions for submitting work. Go to the home page of your journal of interest at http://www3.oup.co.uk/jnls/. |
| Royal Society of Chemistry | ReSourCe | http://chemistry.rsc.org/Publishing/ReSourCe/ |
| Science | Submit to Science | http://www.submit2science.org/ws/menu.asp |
| Society of Plastics Engineers | Information and forms only | No electronic submission |
| Society of Toxicology | ScholarOne Manuscript Central | http://toxsci.manuscriptcentral.com/ |
| Taylor & Francis | Information and forms only | No electronic submission |
| Wiley-VCH | Some journals use ManuscriptXpress | Each journal has its own instructions for submitting work. Go to the home page of your journal of interest. |
| Wolters Kluwer/Springer | ScholarOne Manuscript Central | Each journal has its own instructions for submitting work. Go to the home page of your journal of interest at http://www.kluweronline.com. |
| World Scientific | | Each journal has its own instructions for submitting work. Go to the home page of your journal of interest at http://www.worldscinet.com/subject.shtml. |

**Table 5A-2.** Research Grant Agencies

| *Research Grant Agency* | *Software* | *URL* |
|---|---|---|
| Alexander von Humboldt Stiftung | Information and forms only | No electronic submission http://www.humboldt-foundation.de/en/programme/bewerbung.htm |
| American Chemical Society Petroleum Research Fund | — | Electronic submission only. http://www.chemistry.org/prf |
| American Heart Association | — | Electronic submission only. http://www.americanheart.org/presenter.jhtml?identifier=270 https://home.heart.org/resch/ |
| Australian Research Council | Grant Application Management System (GAMS) | Electronic submission only. https://gams.arc.gov.au/ |
| Camille and Henry Dreyfus Foundation | Limited online submission | http://www.dreyfus.org/index.shtml |
| Ford Foundation Diversity Fellowships | NRCOnline | http://nrc58.nas.edu/nrconline/ford/login/Login.asp |
| NASA | NASA Solicitation and Proposal Integrated Review and Evaluation System (NSPIRES) | http://nspires.nasaprs.com/external/ |
| The National Academies, Research Associateship Programs | NRC WebRAP | http://nrc58.nas.edu/nrcwebrap/rap/login/Register.asp |
| National Institutes of Health, Office of Extramural Research | Information and forms only | No electronic submission http://grants1.nih.gov/grants/oer.htm |
| National Science Foundation | FastLane | https://www.fastlane.nsf.gov/fastlane.jsp |
| Office of Naval Research | Information only | No electronic submission http://www.onr.navy.mil/02/how_to.asp |
| proposalCENTRAL | Research and Management System (RAMS) | https://v2.ramscompany.com/ |
| U.S. Air Force Office of Scientific Research | Information and forms only | No electronic submission http://www.afosr.af.mil/ |
| U.S. Department of Energy, Office of Science | PureEdge | http://www.science.doe.gov/grants/ |
| Welch Foundation | Information and forms only | No electronic submission http://www.welch1.org/ |
| Wellcome Trust | eGrants | http://www.wellcome.ac.uk/node2110.html or http://www.wellcome.ac.uk/doc_WTD004053.html |

➤➤➤➤➤

# APPENDIX 5-2

# Key Features of Selected Online Submission Systems

The following systems are entirely Internet-based and incorporate all elements of submission, review, and manuscript management, unless otherwise noted. They operate with standard browsers and require that authors, editors, and reviewers have Adobe Acrobat Reader to view PDF documents. Authors, reviewers, and editors are offered a secure, password-protected login. Each system incorporates an author home page for the submission, revision, and tracking of manuscripts. Similarly, each system allows referees to submit reviews of manuscripts online.

Publishers modify these systems in varying degrees to reflect different workflows. For this reason, journals using the same software may have slightly different features. This list indicates acceptable submission file types for each system and highlights the system's distinctive features. As technology changes and new features become available, publishers will amend submission sites and systems. Therefore, authors should check regularly for current requirements with the journal before beginning a manuscript submission.

Table 5A-3 catalogs a variety of scientific publishers and the systems they currently use to accept manuscripts online. This table is not all-encompassing, but it is intended to be representative of publishing organizations that contribute to the peer-review literature used by ACS members.

**Table 5A-3.** Text and Image Formats Acceptable to Different Web-Based Manuscript Submission Systems

| *Format* | *ACS Paragon System* | *Bench> Press* | *Editorial Manager* | *EJPress* | *Manuscript Central* | *Peer X-Press* | *Rapid Review* |
|---|---|---|---|---|---|---|---|
| ***Text Formats*** | | | | | | | |
| MS Word (.DOC) | Yes | Yes | Yes | Yes | Yes | Yes | — |
| WordPerfect (.WPD) | — | Yes | Yes | Yes | Yes | Yes | — |
| Encapsulated PostScript (.EPS) | Yes | — | Yes | Yes | Yes | Yes | — |
| PostScript (.PS) | Yes | — | Yes | Yes | Yes | Yes | — |
| Text—ASCII (.TXT) | — | — | Yes | — | Yes | — | — |
| Rich text format (.RTF) | Yes | — | Yes | Yes | Yes | Yes | — |
| TeX | Yes | — | Yes | — | Yes | — | — |
| LaTeX | Yes | — | Yes | — | Yes | — | — |
| Portable document format (.PDF) | Yes | Yes | Yes | Yes | Yes | Yes | — |
| ***Image Formats*** | | | | | | | |
| Graphics interchange format (.GIF) | Yes | Yes | Yes | Yes | Yes | Yes | — |
| Joint Photographic Experts Group (.JPEG) | Yes | Yes | Yes | Yes | Yes | Yes | |
| Tagged image file (.TIF) | Yes | Yes | Yes | Yes | Yes | Yes | Yes |
| Encapsulated PostScript (.EPS) | — | Yes | Yes | Yes | Yes | Yes | Yes |
| PostScript (.PS) | — | — | — | Yes | Yes | Yes | — |
| MS PowerPoint (.PPT) | — | — | — | — | — | — | Yes |
| ***Other Acceptable Formats*** | *a* | *b* | *b* | *a* | *b* | *a* | *b* |
| Chemical markup language (.CML) | Yes | Yes | Yes | — | Yes | — | Yes |
| Chemical structures, such as those created by ChemDraw, ChemSketch, ISIS/Draw | Yes | Yes | Yes | — | Yes | — | Yes |
| DrawIt (formerly Chemwindow) | Yes | Yes | Yes | — | Yes | — | Yes |
| Crystallographic information file (.CIF) | Yes | Yes | Yes | — | Yes | — | Yes |
| Executable (.EXE) | Yes | Yes | Yes | — | Yes | — | Yes |
| MS Excel (.XLS) | Yes | Yes | Yes | — | Yes | — | Yes |
| MOL | Yes | Yes | Yes | — | Yes | — | Yes |
| Video files, such as QuickTime (.AVI, .MPEG) | Yes | Yes | Yes | — | Yes | — | Yes |
| Protein Data Bank (.PDB) | Yes | Yes | Yes | — | Yes | — | Yes |
| Windows metafile (.WMF) | Yes | Yes | Yes | — | Yes | — | Yes |
| Winzip file (.ZIP) | Yes | Yes | Yes | — | Yes | — | Yes |

*Note:* The systems and their developers are as follows: ACS Paragon System, American Chemical Society; Bench>Press, HighWire Press; Editorial Manager, Aries Systems; EJPress, eJournal Press; Manuscript Central, ScholarOne; Peer X-Press, American Institute of Physics; and Rapid Review, Cadmus Systems.

[a]These systems allow a variety of designated file types to be uploaded as supplemental information.

[b]These systems allow any file type to be uploaded as supplemental information.

CHAPTER 6

# Peer Review

Barbara A. Booth

Peer review is a process used by scientific publications to assist editors in evaluating manuscripts, particularly for scientific merit. It is not the only system used to evaluate manuscripts, and it is not perfect. Editors of peer-reviewed books and journals send manuscripts to several reviewers and request their opinions on originality and scientific importance of the topic, the quality of the work performed, and the appropriateness for the specific journal. Reviewers may also comment on language usage, clarity of figures and tables, manuscript length, and anything that they find relevant to effective communication. Although not every editor uses peer review in the same manner, this chapter presents a summary of many common attributes of the peer-review process. (For a broader discussion of peer review, see "Peer Review and the Acceptance of New Scientific Ideas" at http://www.senseaboutscience.org.uk/PDF/peerReview.pdf.)

## Purpose of Peer Review

When a manuscript is submitted for consideration, peer review provides the editor with advice on whether to accept the manuscript for publication. Reviewers also provide suggestions for improving the manuscript. The decision on whether to accept the manuscript for publication rests solely with the editor. Reviewers provide additional expertise and have perspectives that may complement that of the editor. Customarily, peer review is anonymous: the identities of reviewers are not revealed by editors. Some journals also hold authors' names and affiliations in confidence, a double-blind review approach. Occasionally, reviewers request that the editor disclose their names to authors; this is allowed at the discretion of the editor, based on the policy of the individual publication. In one variation of the peer-review process, review is not anonymous; all reviewers are identified.

Peer review is also intended to help authors. External review can help improve the presentation and interpretation of data alike, and ultimately, the research. Clearly and succinctly describing a scientific study is challenging, and reviewers provide valuable feedback. Few manuscripts are so well written that they are accepted without revision. Data and interpretation that seem clear to authors are not always comprehensible to readers. Scientific research is both competitive and cooperative; at its best, peer review is a part of the cooperative process. For example, after evaluating data, a reviewer may suggest an alternative explanation or additional experiments that trigger ideas for further research. Also, a reviewer pointing out an error can save an author the embarrassment of subsequently publishing a correction.

## The Peer-Review Process

When the editor receives a manuscript, he or she examines it and determines whether the manuscript fits within the scope of the journal, whether it meets the specific requirements of the journal, and whether it is of sufficient scientific merit for consideration. Not all manuscripts are transmitted to reviewers. In some cases, the editor decides to reject a manuscript without review, or rarely, to accept it for publication. Manuscripts are rejected without review for various reasons, including the following: the topic is inappropriate for the journal; the concept or the data are not novel; the format is incorrect; the writing is so poor that the manuscript is unreadable; or the authors have previously published overly similar papers. If a manuscript is rejected without review, the editor will usually briefly inform the author of the reason. Some editors offer suggestions to help authors with future submissions.

If the editor decides to send the manuscript for peer review, customarily two to four individuals with appropriate expertise—training or research experience—are asked to review the manuscript. The editor may identify reviewers in a number of ways. Many editors ask authors to recommend reviewers; some do not. Author-recommended reviewers may or may not be used. Most editors will not send manuscripts to specific reviewers if an author so requests. Other potential reviewers may be authors cited in the manuscript, acknowledged experts in the field, or other active researchers in the field. Editors often use scientific search services (such as SciFinder or SciFinder Scholar for chemists) to identify qualified potential reviewers. Reviewers may or may not be known personally to the editor. Most journals maintain records on thousands of reviewers, including their expertise, manuscripts they have reviewed, performance, and so on.

Usually, reviewers are asked whether they are willing to review a manuscript. If they agree, the manuscript is provided in either hard copy or electronically, often with an accompanying review form. Editors generally ask reviewers to submit their reviews in two or three weeks. When a review is overdue, the editor usually sends a reminder to the reviewer.

Once reviews are returned, the editor reads the reviews in conjunction with the relevant manuscript, evaluating both the manuscript and the reviews, and then makes a decision whether to accept the manuscript, request revisions, reject the manuscript, or send it for additional review. In some cases of conflicting advice or opinions of reviewers, editors may seek advice from others. Editors are not obligated to follow the recommendations of reviewers. Reviewer ratings are not averaged; often, a single cogent negative review leads to rejection of a manuscript.

## Responsibilities of Reviewers

Peer review is a critical component of formal scientific communication, and every scientist has an obligation to do a fair share of reviewing.

---

**Reminder:** Manuscripts should be reviewed in a timely and balanced manner and should be kept confidential until publication.

---

When a manuscript is under review, it is a confidential document and it should not be discussed or shown to others. After reading a manuscript, a reviewer may conclude that a better review could be accomplished with assistance from a colleague. In this circumstance, the reviewer should inform the editor before engaging the colleague. Reviewers are expected to provide reviews that are thorough and unbiased. A reviewer in direct competition with the authors of a manuscript should inform the editor that there is a potential conflict of interest and discuss the issue with the editor. In addition, if reviewers are asked to review the work of someone at the same institution or the work of a previous student or co-worker, the reviewer should inform the editor. In some cases, the editor will excuse the reviewer from doing the review; in other cases, the editor will consider the relationship when evaluating the review.

Reviewers are expected to submit their reviews on time. Most reviewers are also authors and expect reviews of their manuscripts to be handled expeditiously. If circumstances arise that prevent or delay a review, the editor should be informed as soon as possible.

The entire manuscript should be read carefully and critically. Most reviewers read a manuscript more than once. Manuscripts should be rated on technical quality, significance of the work, importance to the research field, and adequacy of expression. Often a standard form is provided for this portion of the review.

## Box 6-1. Suggested Topics for a Peer Review

Are the methods (experimental section) adequately described and referenced?

Are there any unsupported conclusions?

Is there anything that is confusing or ambiguous?

Do figures and tables appropriately illustrate the data?

Is the introduction clear and informative?

Is either the introduction or discussion longer than necessary, and do they make sense in relation to the subject and the data?

Although the discussion is the appropriate place for speculation, is it excessive?

Are the appropriate references cited? Are the references accurate?

Is English usage and grammar adequate? Some reviewers may be inclined to edit or annotate the manuscript. However, reviews are more valuable to editors if reviewers mention that there are problems with the English and concentrate on evaluating the data and its interpretation.

Is the length of the manuscript unwarranted? Suggestions on how a manuscript can be shortened are appreciated by editors.

Is the use of color warranted? Printing color is a significant expense for the publisher.

Reviewers should feel free to comment on the suitability of the manuscript for the particular publication. Sometimes first-rate manuscripts are submitted to an inappropriate publication. In addition to the actual review, some editors allow reviewers to submit confidential comments about the manuscript. These are not forwarded to the author. If suspicions of abuse, plagiarism, or fraud arise, the editor should be informed immediately.

Many reviewers divide their reviews into general comments and specific, detailed comments. In the general section, reviewers should draw attention to both the strong and the weak points of the manuscript, the concepts, the objectives, and the methods. Like an author writing a manuscript, reviewers should write reviews in a comprehensive but concise manner, addressing the questions presented in Box 6-1.

Reviews should be written in a helpful, tactful manner. Editors generally edit or do not pass discourteous comments to authors. Abusive reviews lack credibility and also reflect poorly on the reviewer.

## Responsibilities of Authors

Authors should read the current instructions to authors for the publication to which they intend to submit a manuscript. For journals, these instructions are typically published in the first issue of the year and can often be found on the Web. Author guidelines for books are available from the publisher, and sometimes they are posted on the publisher's Web site. Many submissions are rejected without review because the manuscript does not conform to the publication guidelines.

---

**Reminder:** Authors should read the current instructions to authors for any publication to which they intend to submit a manuscript.

---

If English is not the author's native language, it is a good idea to ask a native English speaker to edit the manuscript before submission. Alternatively, authors can employ a technical editing service. Such editing services can be found through recommendations from colleagues, by searching the Web, or advertised in scientific publications. However, it is the author's responsibility to ensure that the writing is accurate.

In suggesting potential reviewers, authors should not recommend members of their own institution or current or recent collaborators (including students or postdoctoral associates). Many authors suggest the names of eminent scientists in their fields. These individuals are frequently asked to review a large number of manuscripts and, constrained by time, may not be available to review additional manuscripts. Therefore, it is more helpful to suggest the names of highly qualified, less well known researchers. Many book and journal editors also allow authors to request that certain individuals, such as those in direct competition, not be selected to review. Editors usually comply with these requests.

Authors should read reviews carefully and dispassionately, with the expectation that reviewers' comments will improve their manuscripts. When an editor requests a revision, authors should respond to all comments and answer all questions. Editors generally reject perfunctory revisions. When it is apparent that a reviewer has misread or misinterpreted the text, authors should provide a direct but tactful reply. Revisions should be submitted in a timely manner. Generally, editors expect revisions within a few weeks, not months. If there is long delay in submitting a revision, some editors will consider the revision to be a new manuscript. Sometimes an editor or reviewer will ask authors to perform additional experiments or to reanalyze data, necessitating a delay in submitting a revision. Authors should ask the editor for an extension, if necessary, providing an explanation.

When a manuscript is rejected, authors should read the editor's letter and reviews carefully to determine the reasons for the rejection. Rejection usually means that another publication should be considered for submission of the manuscript. Some manuscripts should be abandoned. Although some editors will consider an appeal of a rejection, successful appeals are rare. Only if authors are confident that with the help of the reviews, the research and its presentation can be improved, should they resubmit the manuscript to the same publication. A cover letter accompanying the revision of the rejected manuscript should describe all changes to the manuscript, reply to all reviewer comments, and respond to all questions. Ultimately, whether the manuscript is submitted to a different publication, revised and resubmitted to the same publication, or further experiments are performed to broaden the scope of the investigation, critical assessment from reviewers can aid both research and its communication.

## Encouragement to New Investigators

The peer-review process may seem daunting and at times unfair to new investigators. On occasion, reviewers will be unduly critical and unhelpful—and infrequently, outright nasty and rude. Sometimes reviewers can be exasperatingly slow. Good manuscripts are rejected, in particular by prestigious journals, which receive many more high-quality manuscripts than they can publish. However, many reviewers and editors will be solicitous and encouraging to investigators in the early stages of their careers. Despite its shortcomings, peer review is regarded by the scientific community as an essential component to high-quality, effective communication that further advances science.

---

**Reminder:** Frequently consult the "Ethical Guidelines to Publication of Chemical Research" (Appendix 1-1 in this book, or on the Web at https://paragon.acs.org). The guidelines are reviewed regularly to ensure their clarity.

---

➤➤➤➤➤ CHAPTER 7

# Copyright Basics

Karen S. Buehler, C. Arleen Courtney, and Eric S. Slater

Copyright law is a cornerstone of intellectual and scientific exchange. This chapter is intended to introduce a complex and constantly changing legal area; we do not intend to provide legal advice. The first section is an overview of U.S. copyright law: what materials are subject to copyright, who owns copyright, and when copyrighted materials may be used by others. The second part of the chapter presents the methods of obtaining permission to use someone else's copyrighted work. The third part briefly explains how copyright is transferred from the author to the publisher.

## U.S. Copyright Law

### *What Can Be Copyrighted*

*Copyright* is a doctrine of federal law (Title 17, U.S. Code) and is defined as a form of intellectual property law that protects original works of authorship fixed in a tangible medium of expression for a specified period of time. A *tangible medium of expression* is something that can be seen, touched, or heard (see the examples below). Copyright applies to all media of expression, including print, the Internet, CD-ROM, and videotape. Copyright allows authors and creators certain rights to protect their original works. Copyright law protects both published and unpublished works.

The authors wish to thank Barbara F. Polansky and William J. Cook for their time in reviewing and commenting on this chapter.

Copyright applies to a variety of works including the following:

- literary works (including scientific works and computer programs);
- musical works (including lyrics);
- dramatic works (including accompanying music);
- pantomimes and choreographic works;
- pictorial, graphic, and sculptural works;
- motion pictures and other audiovisual works;
- sound recordings;
- architectural works; and
- compilations and databases to the extent that they reflect originality in the selection and arrangement of elements.

### What Cannot Be Copyrighted

Copyright does *not* protect the following:

- Works not fixed in a tangible form of expression. Copyright does not protect ideas, only the fixed expression of ideas. Thus a thought, not written down in any way, is not protected.
- Titles, names, short phrases, slogans, familiar symbols, or designs. (These items may be protected under trademark or service mark laws.)
- Lists of ingredients, contents, or facts.
- Ideas, procedures, methods, systems, processes, concepts, principles, discoveries, and devices. (These items may be protected under patent law.)
- Standard calendars, rulers, lists, or tables taken from the public domain and other works containing no original authorship.

### Who Owns the Copyright

The copyright owner is the author or creator of the original work. An original work can be in the form of an article, photograph, illustration, figure, table, etc. Copyright does not protect ideas, only the actual expression of the ideas. Therefore, the copyright owner is the person who drew a figure or table on a computer and saved it, or took a photograph (the subject of the picture is not the copyright owner), or wrote a narrative on a piece of paper.

There are two exceptions when the author or creator of the original work is not the copyright owner: (1) when the copyright was transferred in writing to another person or entity (usually via a copyright status form) and (2) when the work was created as a work-made-for-hire. In a work-made-for-hire situation, employees create the work within the scope of their employment; therefore, the copyright owner is the employer. The employer may be an individual, corporation, or university. The above two examples represent typical scenarios when publishing scientific or scholarly works. Work-made-for-hire can also cover, among other specified categories, independent contractor situations under certain conditions and if the parties have agreed in writing.

## Box 7-1. Copyright at ACS

Each division of the ACS is responsible for copyright issues related to material that appears in its publications. The ACS Copyright Office can assist ACS authors and editors with copyright issues pertaining to materials published by the ACS Publications Division. The e-mail address is copyright@acs.org. The mailing address is American Chemical Society, Copyright Office, 1155 Sixteenth Street, NW, Washington, DC 20036. The fax number is 202-776-8112, and the telephone number is 202-872-4368 or 4367. Contact other ACS divisions directly for permission to use their material. For specific advice regarding copyright matters, it is best to seek legal counsel.

### *Rights of the Copyright Owner*

The copyright owner can be described as controlling a "bundle" of stated rights. The copyright owner (whether an individual, a corporation, or a publisher) retains certain exclusive rights in the work. The bundle of rights includes

- reproducing copies of the work;
- distributing copies of the work to the public;
- creating derivative works based on the work;
- performing the work publicly (for certain types of works) and, in the case of sound recordings, by digital audio transmission; and
- displaying the work publicly (for certain types of works).

Most scientific and scholarly publishers require authors (or their employers) to transfer copyright as part of their publication agreement. The ACS Publications Division requires authors (or their employers) to transfer copyright to the American Chemical Society, although some rights are transferred back to the authors (or their employers). For specific information regarding transferring copyright to ACS and authors' rights, see http://pubs.acs.org/copyright_info.html or contact the ACS Copyright Office (see Box 7-1).

### *Copyright Notice*

A work is protected by copyright even if it does not contain a formal copyright notice (the word "copyright", abbreviation "copr", or symbol © with the year of first publication and name of copyright owner). It is strongly recommended to place a copyright notice in or on a work, thereby giving notice that someone controls the rights to that work. Registration of a work with the U.S. Copyright Office is not necessary to obtain copyright protection, although there are substantial benefits to doing so. For more information about copyright registration, contact the U.S. Copyright Office at 202-707-5959, or at http://www.copyright.gov/.

The U.S. copyright law also provides remedies for infringement. Authors should contact the appropriate book or journal editor in cases where they believe their work has been infringed upon. If a copyright owner (in the case of an ACS publication, the ACS) believes that his or her work has been infringed upon, it is the responsibility of the copyright owner to bring action.

## *Fair Use*

The U.S. copyright law contains limitations on rights granted to the copyright owner. Probably one of most misunderstood of these limitations is fair use. Essentially, fair use permits certain actions that might otherwise infringe upon an exclusive right of the copyright owner. It has been our experience that the interpretation of what constitutes fair use can vary widely. For instance, a copyright owner's interpretation may be very narrow; a party looking to take advantage of fair use will interpret it broadly.

Using excerpts from works for the following purposes may qualify as fair use:

- criticism or comment;
- news reporting;
- teaching (but only in a spontaneous situation. If one knows in advance that copyrighted material will be used in a class, permission must be obtained.); and
- scholarship or research. (Private researchers and nonprofit employees may make one photocopy of an article for their own research purposes only.)

These uses may not constitute copyright infringement. The determination of whether a use is fair depends on the facts in each case. Even if the facts are virtually identical, one degree of minutiae can cut against fair use.

The U.S. copyright law provides four factors that must be considered in determining whether any particular use is fair use. Courts weigh the four factors against each other, and no one factor is determinative in every case. The four factors are the following:

1. the purpose and character of the use, including whether it is of a commercial nature or for nonprofit educational purposes;
2. the nature of the work;
3. the amount and substantiality of the portion used in relation to the work as a whole; and
4. the effect of the use in question on the potential market for or value of the work.

There are no numerical guidelines or percentages that can be used for each situation where fair use might be involved. Even one percent of a work used in a manuscript may be determined to be against fair use if the material is considered the most important part or the heart of that work. Fair use is a defense, and a strong argument must be made that the use is actually fair. Never assume that a use is a fair use; it is always best to seek permission to avoid potential and sometimes costly problems.

---

**Reminder:** Never assume that a use is a fair use; it is always best to seek permission to avoid potential problems.

---

## *Public Domain and U.S. Government Works*

Works that are in the public domain are not protected by copyright. The following are situations in which a work can have public domain status:

1. works published before 1923 whose copyright term has expired (see the section called "Duration of Copyright"), or works for which copyright was not renewed;
2. works published before 1978 without a proper copyright notice;
3. works dedicated to the public domain by their creator, e.g., so-called shareware; and
4. works authored by U.S. government employees within the scope of their employment.

Generally, it is not necessary to seek anyone's permission to use public domain works. However, when using material that appeared in a U.S. government publication, check the caption or reference section to see if credit is given to a source for which permission is required.

---

**Reminder:** Even though material in a U.S. government publication generally does not require permission to reproduce it, sometimes government publications include material that is protected by copyright. Always check the caption or reference section to determine whether you need to obtain permission to reproduce it.

---

## *Common Misconceptions about Copyright*

One misconception about copyright status involves material found on Internet Web sites. Someone does own copyright to the material on a Web site, and it is necessary to obtain permission to reuse it. The lack of a copyright notice does not mean that the work is not protected, nor does it mean that a work is in the public domain or that the author of the work has waived his or her rights.

Other misconceptions involve photographs. The individuals appearing in photographs are not the copyright owners. The photographer, or employer if the photo is a work-made-for-hire, is the copyright owner, and permission to reuse a photograph must come from the copyright owner. (Permission of people appearing in photographs may also be required by a publisher, but this is not a copyright issue. It is a privacy/right of publicity concern.)

Another problem involves the use of photographs of old works of art. It may be true that the art itself is in the public domain (i.e., a van Gogh or Michelangelo painting). However, the *photograph* of the artwork is protected by copyright. In these cases, the copyright owner may not be an individual photographer but a museum or a historical society.

### *Duration of Copyright*

Under current U.S. copyright law, length of copyright is life of the author plus 70 years. For works-made-for-hire, length of copyright is 95 years from the date of first publication, or 120 years from the date of creation, whichever is shorter.

### *Copyright in the Electronic Age*

Several recent court cases have involved copyright infringement in electronic media. The legal disputes centered on which party controlled specific rights under the contractual arrangements authors had with publishers. A common thread to these cases is whether a publisher controls all rights to publish content in any format.

Publishers and authors need to be aware of the rights they hold, either explicitly under the U.S. copyright law, or via a licensing arrangement. ACS requires complete transfer of copyright in all formats, thus allowing ACS to publish material in different formats.

In cases where authors do not transfer copyright, publishers generally require a *nonexclusive* licensing arrangement. Here, the author retains copyright and grants certain rights to the publisher. Placing material on the Internet without authorization from the copyright owner is a copyright infringement, even if the publisher had permission to use the material in print.

It is important to have a working knowledge of copyright law, and it is helpful to have some familiarity with contracts. Publishers and authors need to read contracts and copyright transfer forms thoroughly before signing and be certain that they understand what they are agreeing to.

There are numerous Web sites containing copyright information. We recommend the following Web sites for accurate and up-to-date information, although this is by no means a complete list:

- U.S. Copyright Office, http://www.copyright.gov/
- Association of American Publishers, http://www.publishers.org/
- American Chemical Society Copyright Learning Module, http://pubs.acs.org/copyright/learning_module/module.html

## Obtaining Permission To Reproduce Material

Whenever an author wishes to include a figure, a table, or a substantial portion of text that has already been published elsewhere (in print or on the Internet), the author must obtain permission from the copyright holder to reproduce the material. It is the author's responsibility to

1. identify material in a paper or manuscript that has been previously published in print or on the Internet;
2. determine whether that material is subject to copyright protection;
3. if so, identify the copyright owner and request permission in writing; and
4. ensure that permission is granted and forwarded to the author's publisher.

The first two points are discussed at the beginning of this chapter. The second two points will be considered next in a general way applicable to requesting permission from any scientific or scholarly publisher. For information on how to reproduce materials from books, journals, or magazines published by the ACS Publications Division, see Box 7-2.

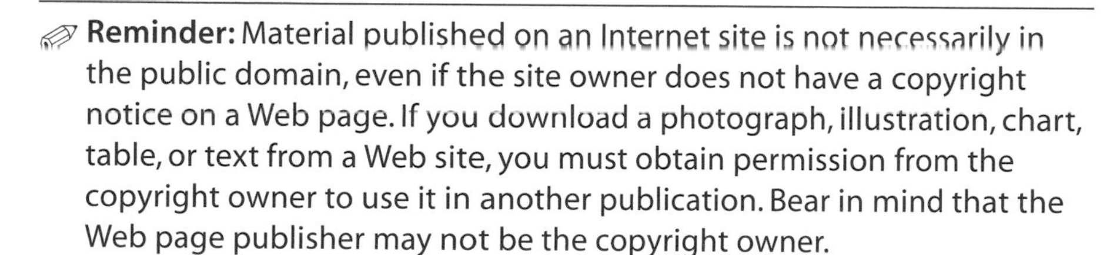

**Reminder:** Material published on an Internet site is not necessarily in the public domain, even if the site owner does not have a copyright notice on a Web page. If you download a photograph, illustration, chart, table, or text from a Web site, you must obtain permission from the copyright owner to use it in another publication. Bear in mind that the Web page publisher may not be the copyright owner.

### *Writing To Request Permission*

Once an author has identified the tables, figures, and text that require permission, the next step is for the author to write to the copyright owner to request permission. This is not a difficult task, but it requires some organization and attention to detail, and it may take several months for permission to be granted. Fortunately, the Internet has made the task less burdensome than it used to be because most publishers post their permission policy, forms, and contact information on the Web. Some publishers now have Web-based permission systems.

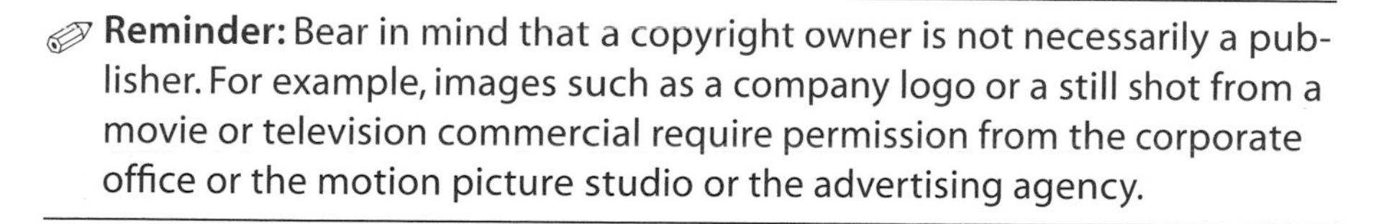

**Reminder:** Bear in mind that a copyright owner is not necessarily a publisher. For example, images such as a company logo or a still shot from a movie or television commercial require permission from the corporate office or the motion picture studio or the advertising agency.

➤ Begin by organizing your materials requiring permission according to publisher and source. This way, you can list all materials from a single publisher and source on one form, reducing paperwork for yourself and the publisher.

## Box 7-2. Permissions at ACS

Permission is needed to reproduce tables, figures, or text from books, journals, or magazines published by the ACS Publications Division. Permission is also needed if adapting or using part of a figure or table. Permission is not needed if using data from the text to create an original figure or table. When using a figure that appeared in an ACS journal and ACS owns copyright to that figure, permission is not needed to use it in a paper that will appear in the same or another ACS journal, although a credit line is required.

The ACS permission policy and forms for ACS journals and magazines are available at http://pubs.acs.org/copyright_info.html and for ACS books, at http://pubs.acs.org/books/forms.shtml. An interactive form can be found at http://pubs.acs.org/cgi-bin/display-copyright?bichaw.

Permission requests to reproduce materials from the books, journals, and magazines of the ACS Publications Division should be directed to the ACS Copyright Office. All permission requests must be in writing. The ACS Copyright Office accepts permission requests via e-mail, mail, and fax. Permission is granted only by fax or by mail, so it is important to include a fax number.

The Copyright Office e-mail address is copyright@acs.org. The mailing address is American Chemical Society, Copyright Office, 1155 Sixteenth Street, NW, Washington DC 20036. The fax number is 202-776-8112, and the telephone number is 202-872-4368 or 4367.

➤ Determine whether the publisher has permission-request information on the Web and, if so, download or print out the appropriate request forms. (Even if the publisher has a Web-based permission system, you might want to print out the blank form, so you can be sure that you have collected the necessary information when you fill it out online.)

➤ If the publisher does not have forms on the Web, then draft a letter requesting permission. Be sure to include the elements listed in Box 7-3.

➤ List all the material for which you need permission from a given publisher on one form or on the form with an attachment. Unless the publisher specifically directs differently, avoid submitting one form for each item requiring permission.

➤ Include your specific deadline date by which you wish to have the permission granted.

➤➤➤➤➤

## Box 7-3. Components of a Permission Request

Your request for permission to reproduce material will be processed more quickly by a publisher if you provide all the necessary information on the publisher's form or in a request letter. A permission request should provide the following information:

1. A list of the original figure or table numbers for each figure or table that you wish to reproduce from a given source, along with a complete reference citation for the source.
   - The reference citation for a journal article should include the name(s) of the author(s); the name of the journal; the month, day, and year of publication; volume and issue numbers; and inclusive page numbers for the article.
   - The reference citation for a book should include the title of the book (and series name and number if applicable); the name(s) of the author(s) or editor(s); the year of publication; and the page numbers on which the original figures or tables appear. If appropriate, also list the title of the chapter, the name(s) of the chapter author(s), and the inclusive page numbers for that chapter.
2. A description of where the material will be published. Include the title of the forthcoming publication, the type of publication (e.g., journal, book, magazine, or Web journal), and the name of your publisher.
3. A list of the formats in which the requested material will appear, such as print, online, CD-ROM, or proceedings. Unless all rights are requested, only print rights will be granted.
4. Your complete mailing address, e-mail address, and telephone and fax numbers.
5. A specific deadline (calendar date).

Many permission request letters also include a space for the original publisher to specify the publisher's preferred credit line.

➤ Include complete contact information for yourself, including a mailing address, e-mail address, and telephone and fax numbers.

➤ If you do not have a response from the copyright owner in a reasonable amount of time, contact the publisher again and try to determine the reasons for the delay; perhaps your request was not received or was missing information. Do not assume that you have permission unless you have heard explicitly from the publisher, even if you set a deadline.

### What Sort of Credit Line Is Appropriate?

➤ Once the publisher has granted permission to reproduce materials, incorporate the required copyright credit line into your manuscript. This credit line notifies readers that the publisher owns copyright to the material. Each publisher has its own style for credit lines. For example, the credit line required by the ACS is shown below:

> Reprinted with permission from REFERENCE CITATION. Copyright YEAR American Chemical Society. *(Insert the appropriate information in place of the capitalized words.)*

➤ Include the credit line on the first page where copyrighted text appears and under each copyrighted table or figure.

➤ If you are adapting or modifying copyrighted material or if parts of copyrighted material are being used, the words "Adapted" or "Reprinted in part" should replace "Reprinted" in the credit line.

## Transferring Copyright

Most scientific and scholarly publishers require the transfer of copyright ownership as part of the publication agreement between the publisher and authors. When the publisher owns copyright, it removes the burden from authors to grant permission and it assures them that only legitimate requests for their material are approved. Similarly, if the publisher owns copyright, it is easier for authors who are seeking permission to come to one source that handles requests from multiple references and processes them quickly. If individual authors owned copyright, it might be difficult and cumbersome to contact authors for permission to use their works or portions of their works. If you are in doubt about a publisher's requirements, contact the publisher or seek legal advice.

Each publisher has forms for transferring copyright to the publisher, and the submission of a completed copyright transfer form is nearly always necessary before a publisher will schedule publication of or begin production work on a manuscript. When authors publish in an ACS publication, they assign copyright interest via the ACS Copyright Status Form. This form is available on the ACS Publications Division Web site at http://pubs.acs.org/cgi-bin/display-copyright?bichaw and in the January issues of ACS journals. Authors publishing books with ACS will find a form at http://pubs.acs.org/books/forms.shtml. Completed Copyright Status Forms should be submitted to the appropriate book or journal editor, not the ACS Copyright Office.

➤➤➤➤➤ CHAPTER 8

# Markup Languages and the Datument

Peter Murray-Rust and Henry S. Rzepa

A style guide is presumed in its basic objective as guiding authors in the task of producing a document describing their work. In a conventional sense, style is traditionally presentational in objective and is aimed at producing a (visually) homogeneous form suitable for aggregation into a journal issue or a book. At the outset of this chapter, we should pin our flag firmly to a rather different mast. This chapter is less about the application of style to the presentation of a document and more about the general principles involved in its application to data to achieve a homogeneous form capable of being reused. It is not so much about how to create a document, but rather about how to create a more data-focused entity we call a *datument* (*1*), that is, a container for data and its associated descriptions. The datument can have conventional attributes of authorship, affiliation, and other familiar structural components, such as sections, tables, figures, and bibliography, but it extends this by also carrying data. This differs from, for instance, tabulated or quoted data typical of that found in most chemical documents. To illustrate this most important difference, consider the following assertion:

> The melting point of aspirin is 135°, and its molecular ion has the formula $C_9H_8O_4^{+\cdot}$.

As a human trained in chemistry, you probably understand much of the semantics, but consider how much potential ambiguity and implied meaning is contained in this admittedly concise statement.

- You understand what is meant by the term *melting point*, but you might have to seek a librarian's help to locate a relevant dictionary of chemical terms where you can check a more precise definition should the need arise.

- After a little thinking, you conclude that the glyph ° indicates that the number preceding it is a *temperature* expressed in *Celsius units.* You probably also recognize that this number is probably only accurate to ± 1 °C, only because you have made such measurements yourself. At the back of your mind is probably the knowledge that a value of 135° is reasonable for an *organic compound* (and that 1135° would not be), that this implies that this substance is a *solid* at *room temperature* and that this value may be used as an approximate indicator of *chemical purity.*
- You certainly recognize the term *aspirin* as a trivial, unsystematic but commonly used description of a nevertheless well-defined *molecule* (the structure, or more accurately the *connection table,* of which you may need to ascertain).
- The term *molecular ion* is a term that tends to be used in a particular branch of *spectroscopy* known as *mass spectrometry.*
- You have little difficulty in recognizing the *molecular formula* as by convention listing the number of carbon atoms as the subscript to the initial C, then followed by the number of hydrogen and then other atoms in alphabetical order. The final suffix indicates the charge on the system and reconciles with the use of the term *ion,* and you will notice that the suffix reveals the presence of an *unpaired electron.*
- You probably are aware of rules that would allow you to check the validity of this formula (i.e., are the $^{+}$ and the $^{\cdot}$ consistent with each other; is the formula physically possible within the constraints of valence theory, etc.).
- You might want to derive a *molecular mass* from the formula, in which case you need to know about *atomic weights, isotopes,* and other concepts.
- You might want to relate the formula to a *two-dimensional* structural representation, indicating perhaps where the charge might reside, or how the ion might *fragment,* or a *three-dimensional* representation for *molecular modeling.*
- As a human, you will infer properties such as *aromaticity* or the probable presence of a *hydrogen bond.*

A knowledgeable chemist could probably make quite a few more inferences from the above, but the purpose here is to illustrate how much implicit (i.e., undeclared) information there is in such a brief statement (all well-defined chemical terms are shown above in italics to emphasize this). The purpose in writing it all down is to reinforce the idea that although such a statement (as might be found in a document) contains much data, only a human can really make significant use of it. Well, that would be true only if the human were to give undivided attention to the contents of such a document; the human certainly could not cope well with more than a few such documents, and not at all with, say, millions of documents. A normal response to such issues of scale would be to say that the essential data and semantics of the above phrase need to be abstracted (with inevitable loss of some data and information) into a database representation and then suitably queried if needed; after all, agencies such as Chemical Abstracts Service exist to provide such a service. However, consider that the process of converting even a brief statement such as the above into a true chemical abstract requires an expert (and it must be said, error-prone) human. Even then, the knowledge required to construct the list above would have to be acquired from other sources.

Could instead much, perhaps all, of this work be handled by a computer? The answer is a clear "yes", but only if the ground is well prepared for such a task. The purpose of this chapter is to outline some of the basic principles (but not the technical details) of how this could be done and to set out the grander vision that creating an infrastructure that adopts such principles would be the first step toward what has been described as a *semantic web* of information and knowledge. Not only humans but machines could roam, on vast scales if need be, on such a web, and by doing so discover connections between data and concepts which in days past might have been described as the art of scientific serendipity (*2*).

## Markup Languages and the World Wide Web

The World Wide Web arose from the need for high-energy physicists at Conseil Européen pour la Recherche Nucléaire (CERN) to communicate and exchange data and information within a large dispersed community. The basic design, which has been well documented, involved the creation of structured documents containing identifiable information components linked by uniform resource identifiers using markup tags that could be recognized by machines and used as formatting or styling instructions rather than being part of the actual content. Using the HTML version of the previous example:

```
<p>The melting point of aspirin is 135&deg;, and its
molecular ion has the formula C<sub>9</sub>H<sub>8
</sub>O<sub>4</sub><sup>+.</sup>.</p>
```

The "tags" in the angle brackets are recognized by the processor as markup and are used as instructions rather than content to produce the previously rendered sentence. Although such examples will be familiar to many readers, we emphasize it here because it illustrates the critical importance of separating content from style (or form). The `<p>` tags precisely define a paragraph, a unit for structuring the document. A machine could now easily count the paragraphs in a document and the number of characters (but not individual words, which are separated not by tags, but by spaces) in each. HTML provides a flexible (perhaps rather too flexible) document structure for text (paragraphs, headers, tables, lists), embedded images (and other multimedia objects), human interactivity (through forms), programs (through scripts, applets, and plug-ins), styling (some degree of formatting and screen layout), and metadata (essentially descriptions of data).

Whereas this is a substantial list, its success has generated many problems, which HTML in its original form cannot solve:

- HTML can only support a fixed tagset (for example 59 in the latest specification for XHTML 2.0), and even this number is regarded as close to unmanageable (no software yet implements this full set consistently, accurately, and completely). Any other tags that might be present (e.g., <molecule>) are simply ignored (strictly speaking, they should be marked as invalid HTML, although they may be valid for other languages).
- Much of the behavior (semantics) is undefined. This lack has led to specific disciplines creating their proprietary methods of supporting functionality (e.g., through scripting languages, plug-ins, applets, and other software).
- HTML was designed to be error-tolerant, in recognition that it would be authored (and viewed) mostly by humans. Browsers may try to recover from nonconforming documents and may do so in different ways. Humans are good at recognizing and often correcting errors in HTML (missing links, broken formatting, incomplete text). Machines cannot normally manage broken HTML other than in a "fuzzy" manner.
- Author-provided metadata is often entirely absent. If present, it will likely adhere to a general form (the so-called Dublin Core schema, http://dublincore.org/), of limited utility in scientific, technical, and medical (STM) areas.
- The emphasis on presentation in many of the original tags (such as fonts, colors, and layout) muddled the separation of content from style. The World Wide Web Consortium soon developed technologies (CSS, or cascading style sheets, and XSL, or extensible stylesheet language; more information on this and other XML issues is available at http://www.w3c.org/) to help overcome this problem, but as with HTML, CSS is variably implemented in most browsers. Most commercial tools for authoring HTML emphasize presentation or interactivity (to capture the reader's attention), and in such HTML, the content is subservient to the style.

## *Examples of XML and of Chemical Markup Language*

These conventional markup approaches (HTML, CSS, and XSL) are inadequate for datuments because there is usually no domain-specific support. XML, or extensible markup language, was introduced as a solution to this problem. XML was designed to be simple, easy to use, and small; it is a fully conforming subset of the older SGML (essentially "SGML lite"). It allows new markup languages to be defined through an XML schema formalism (see http://www.w3c.org/XML/Schema). A schema specifies a set of rules (syntax, structure, and vocabulary) to which a document must conform; those that do are said to be "valid". Schemas allow more precise constraints, allow the definition of datatypes, and enhance the potential for machine processing.

We start by explaining the terms used in the XML language, illustrated with a small and simple example (kept brief for simplicity and hence not relating to a real molecule) (Scheme 8-1).

```
<molecule>
   <identifier convention="CAS-RN" >150-78-2 </identifier>
   <identifier version="1.0" convention="InChI" >1.0/C9H8O4/c1-6(10)
13-8-5-3-2-4-7(8)9(11)12/...</identifier>
   <atomArray>
      <atom id="a1" elementType="C" x2="-5.4753" y2="5.0867"/>
      <atom id="a2" elementType="C" x2="-5.4753" y2="3.5466"/>
      </atomArray>
   <bondArray>
      <bond atomRefs2="a1 a2" order="2"/>
   </bondArray>
</molecule>
```

**Scheme 8-1.** The basic features of an XML document.

The core of the language consists of a set of data containers, or more formally elements (not to be confused with the chemical elements), the enumeration of which is ideally defined by a schema. In this example, the elements are <molecule>, <identifier>, <atomArray>, <atom>, <bondArray>, and <bond>. These have a clearly defined relationship to one another (illustrated above by indentation of the text). Thus the element <atomArray> is said to be the parent of a child element termed <atom>, and both are children of the top-level element <molecule>, which can also be called the document root element. This hierarchy among elements is precisely defined and must carry no ambiguity.

Elements can specify data or information in two ways. First, data can be contained between the start and end of any particular element, such as <identifier> and </identifier> in the example. Such content can of course be other (child) elements, but it can also be character or numeric data, as in the example above, which uses both the CAS Registry Number, a unique identifier assigned to chemical structures by CAS (see Appendix 12-3), and a unique canonical molecule identifier known as InChI (International Chemical Identifier) (for more on InChI, see Appendix 8-1).

Second, data can also occur as the value of an attribute to the element. In the example above, the <identifier> element has two attributes, version="1.0" and convention="InChI". Both the name of the attribute and its value can be enumerated if needed by the schema; if the attribute is unknown, or its value is outside defined limits, the entire document or datument can be flagged as invalid by suitable software. Thus <atom> has attributes elementType="C" and x2="-5.4753". For the former, a value of "C" is allowed (because it is recognized as the standard symbol for the chemical element carbon), but a value of say "CX" would not be allowed. The second attribute is defined (in the schema) as the *x* coordinate of a set of two-dimensional molecular coordinates. As such, its presence implies that it should be paired with a *y2* coordinate. One can specify in the schema what kind of behavior to impose if, say, *y*2 were to be missing. One might decide that its presence would be inferred and that its value should be `y2="0.0"`, although in practice that would be a dangerous assumption, and it would be better to flag its absence as an error. Decisions also have to be made regarding the value of this attribute. With two-dimensional coordinates, no assumptions can really be made about the units in which the coordinates are specified, and it would be up to any software to process the values in a sensible manner. Whereas a human might think that, e.g., `x2="-54753.0"` looks unreasonable, it may still be internally consistent with the other coordinates. Such software would probably also be expected to trap conditions such as two atoms with identical coordinates, or truly unreasonable values. However, one can be a little more specific about, e.g., `x3="-5.4753"`. This would be interpreted as the *x* coordinate of a three-dimensional set, and as such a reasonable implicit behavior would be to treat this value as corresponding to Angstrom units unless otherwise specified. It is also worth noting that elements which specify data in the form of attributes need not enclose any further data; thus `<atom />` in this case represents both the start and the end of the element (in other words it is an empty container).

The preceding discussion has been fairly precise and meticulous, if only to illustrate how XML can be used to impose well-defined structures and relationships on data. We emphasize, however, that it would not normally be a human who has to cope with such levels of detail and precision; the design is such that in fact software will carry almost all of the burden of producing the XML in the first place and then validating and using it subsequently. The preceding argument served only to illustrate how such software can be made to safely operate without the need for human intervention in the process.

The second example (Scheme 8-2) is an elaboration of the first fragment, but formalized below as CML (chemical markup language) (*3*). We emphasize that this chapter is not meant to be an instructional manual for any given markup language, with CML here serving only to illustrate the general principles involved. Many other scientific applications of XML have been developed (*3, 4*), and syntactically, either of these examples could be replaced by other such modularized markup languages.

This more extensive example illustrates how a wider range of properties can be defined and also contains a new feature called a *namespace.* The purpose of this namespace is to enable this entire XML fragment to be combined or aggregated with other XML languages so that no conflict between the names used for the elements can arise. This aggregation is achieved by prefacing each element with a unique (to the document) short string: `<cml:molecule>`. An attribute `xmlns:cml` is now used to define what is called a URI (uniform resource identifier), which stamps a globally unique identifier on the meaning of the `cml:` prefix. This uniqueness will allow this datument to coexist with other XML languages without conflict (an example of which is described later in this chapter).

```
<cml:molecule id="m01" title="aspirin" xmlns:cml="http://www.
xml-cml.org/schema/CML2/Core">
  <cml:metadata name="dc:identifier" content="InChI"/>
  <cml:identifier version="1.0" convention="InChI">
1.0/C9H8O4/c1-6(10)13-8-5-3-2-4-7(8)9(11)12/h1H3,2-
5H,(H,11,12)
  </cml:identifier>
  <cml:metadata name="dc:identifier" content="CAS-RN"/>
  <cml:identifier convention="CAS-RN">150-78-2</cml:identifier>
  <cml:atomArray atomID="a1 a2 a3 a4 a5 a6 a7 a8 a9 a10
a11 a12 a13" elementType="C C C C C C C C O O O O"
formalCharge="0 0 0 0 0 0 0 0 0 0 0 0" hydrogenCount="1 1 1
1 0 0 0 0 3 0 1 0 0"
  x2="-5.475336 -5.475336 -4.141667 -2.807998 -4.141667 -
4.141667 -2.807998 -1.075737 -1.075737 -5.629193 -2.807988 -
1.474318 0.464263"
  y2="5.086684 3.546650 2.776633 3.546650 5.856700 7.396700
5.086684 7.344209 8.884209 7.795282 8.166700 5.856684 7.344209"
  />
  <cml:bondArray atomRef1="a5 a1 a2 a3 a4 a5 a5 a6 a6 a7 a12
a8 a8" atomRef2="a1 a2 a3 a4 a7 a7 a6 a11 a10 a12 a8 a13 a9"
  order="1 2 1 2 1 2 1 1 2 1 1 2 1" />
  <cml.propertyList>
     <cml:property dictRef="chem:mpt" title="melting point"
xmlns:chem="http://www.xml-cml.org/dict/core" >
      <cml:scalar dataType="xsd:decimal" errorValue="1.0"
dictRef="chem:mpt" units="unit:c">136</cml:scalar>
     </cml:property>
  </cml:propertyList>
</cml:molecule>
```

**Scheme 8-2.** A CML datument describing a property of aspirin.

The molecule element in this example contains five child elements: cml:metadata, cml:identifier, cml:atomArray, cml:bondArray, and cml:propertyList. Of these, cml:metadata, cml:atomArray, and cml:bondArray have no children and are empty containers, defining only attribute/value pairs, whereas cml:propertyList has one child, cml:property. The latter itself has a child: cml:scalar. As well as the namespace, the cml:molecule element itself has two other attributes, id and title.

The </cml:property> and </cml:scalar> elements reference namespaces other than CML. This is done to facilitate aggregation with other XML components, such as dictionaries, and by this means to reduce what has been called "tag soup". For example, <cml:property dictRef="chem:mpt"/> defines a namespace for a dictionary reference called chem:mpt. Any processing software that might need to process a melting point property would be directed to this dictionary for further information on the semantics of this term. Similarly, the attribute units="unit:c" would handle the conversion of scientific units, and dataType="xsd:decimal" would handle the basic datatype (i.e., the definition of a decimal number) itself. This mechanism avoids overburdening CML itself with the need to specify such semantics. No other elements or attributes in this example have XML-defined semantics; all other semantics are imposed by CML itself. Thus, the CML schema defines an enumeration (list) of allowed elementTypes and defines their meaning, use, and boundaries. These aspects are discussed in more detail below.

### *The Use of Identifiers*

The examples of XML shown in Schemes 8-1 and 8-2 illustrate the use of two types of identifiers. The identifier attributes seen in, e.g., <cml:atomArray atomID="a1 a2 a3 a4 a5 ..."> are used internally to enable specification of, e.g., <cml:bondArray atomRef1="a5 a1 a2 ..." atomRef2="a1 a2 a3 ..."> and should be unique within the datument (but not necessarily globally) to ensure that this XML document is well-formed and valid. The second type is an element containing identifiers. Various identifiers could be used, such as SMILES, CAS-RN, and the InChI canonical identifier (as shown here), precisely derived from the molecule connection table and used to establish molecular global uniqueness. Any two datuments that contain the same InChI identifier (in this example, C9H8O4/c1-6(10)13-8-5-3-2-4-7(8)9(11)12/h1H3,2-5H,(H,11,12)) should be presumed to refer to the same molecule (in the sense of a connection table, but not necessarily other properties, such as 3-D coordinates, for example). The whole aspect of identifiable data is pivotal to the concepts used here. The CAS-RN is widely used and offers comprehensive coverage from simple molecules to polymers to Markush structures. The InChI can be derived from the structure, but it is still a fairly new standard. It does not yet have wide support and does not yet offer complete coverage of all materials.

### *Display of XML and Specific XML Languages*

The default way of "displaying" or "browsing" any XML-compliant language, such as CML, is as a so-called tree view, outlining the structure of the document (Figure 8-1) but to which no style has been applied. Most modern Web browsers will support this feature (we recommend Firefox).

Of more utility is to associate a specific style or transform with this datument. A technology known as XSLT is essentially a specification of how an XML-based datument might be transformed into a different representation (or subset) of the data. Four examples of how this might be used to transform this datument are listed below:

- Extraction of atom two- or three-dimensional coordinates and rewrapping with appropriate syntax for interactive display on screen using appropriate

**(A)**

```
+<cml:molecule id="m01" title="aspirin"></cml:molecule>
```

**(B)**

```
-<cml:molecule id="m01" title="aspirin">
   <cml:metadata name="dc:identifer" content="InChI"/>
 + <cml:identifier version="1.0" convention="InChI"></cml:identifier>
   <cml:metadata name="dc:identifier" content="CAS-RN"/>
      <cml:identifier convention="CAS-RN">150-78-2</cml:identifier>
   <cml:atomArray atomID="a1 a2 a3 a4 a5 a6 a7 a8 a9 a10 a11 a12 a13"
   elementType="C C C C C C C C C O O O O" formalCharge="0 0 0 0 0 0
   0 0 0 0 0 0 0" hydrogenCount="1 1 1 1 0 0 0 0 3 0 1 0 0"
   x2="-5.475336 -5.475336 -4.141667 -2.807998 -4.141667 -4.141667
   -2.807988 -1.075737 -1.075737 -5.629193 -2.807988 -1.474318
   0.464263" y2="5.086684 3.546650 2.776633 3.546650 5.856700
   7.396700 5.086684 7.344209 8.884209 7.795282 8.166700 5.856684
   7.344209/>
   <cml:bondArray atomRef1="a5 a1 a2 a3 a4 a5 a5 a6 a6 a7 a12 a8 a8"
   atomRef2="a1 a2 a3 a4 a7 a7 a6 a11 a10 a12 a8 a13 a9"
   order="1 2 1 2 1 2 1 1 2 1 1 2 1"/>
 + <cml:propertyList></cml:propertyList>
</cml:molecule>
```

**(C)**

```
-<cml:molecule id="m01" title="aspirin">
   <cml:metadata name="dc:identifer" content="InChI"/>
 + <cml:identifier version="1.0" convention="InChI">
     1.0/c9h8O4/c1-6(10)13-8-5-3-2-4-7(8)9(11)12/h1H3, 2-5H,(H,11,12)
   </cml:identifier>
   <cml:metadata name="dc:identifier" content="CAS-RN"/>
      <cml:identifier convention="CAS-RN">150-78-2</cml:identifier>
   <cml:atomArray atomID="a1 a2 a3 a4 a5 a6 a7 a8 a9 a10 a11 a12 a13"
   elementType="C C C C C C C C C O O O O" formalCharge="0 0 0 0 0 0
   0 0 0 0 0 0 0" hydrogenCount="1 1 1 1 0 0 0 0 3 0 1 0 0"
   x2="-5.475336 -5.475336 -4.141667 -2.807998 -4.141667 -4.141667
   -2.807988 -1.075737 -1.075737 -5.629193 -2.807988 -1.474318
   0.464263" y2="5.086684 3.546650 2.776633 3.546650 5.856700
   7.396700 5.086684 7.344209 8.884209 7.795282 8.166700 5.856684
   7.344209/>
   <cml:bondArray atomRef1="a5 a1 a2 a3 a4 a5 a5 a6 a6 a7 a12 a8 a8"
   atomRef2="a1 a2 a3 a4 a7 a7 a6 a11 a10 a12 a8 a13 a9"
   order="1 2 1 2 1 2 1 1 2 1 1 2 1"/>
 - <cml:propertyList>
     <cml:property dictRef="chem:mpt" title="melting point"
       <cml:scalar dataType="xsd:decimal" errorValue="1.0"
       dictRef="chem:mpt" units="unit:c">136</cml:scalar>
     </cml:property>
   </cml:propertyList>
</cml:molecule>
```

**Figure 8-1.** Viewing the XML datument described in Scheme 8-2 in a Web browser, shown as (A) a collapsed tree view, (B) partially expanded, and (C) fully expanded.

software. This use was first demonstrated in an unusual series of articles about themselves, in which the theme was to demonstrate how they could be dynamically transformed into other representations of the data contained within (*5–7*).
- Use of the atom array to calculate a molecular formula and weight.
- Calculation of molecular properties via invocation of molecular modeling algorithms using, e.g., Web services.
- Use as a database query ("Is aspirin in this database?").

Only the first of these uses really corresponds to the conventional sense of a document (whether it be an HTML page on the Web or a journal article in PDF format). The last three uses would be true applications of the datument as part of a semantic web. Another difference is that the semantics in such a datument cannot be unambiguously deduced from inspecting such examples but must be formally defined (e.g., in an XML schema or similar tool). Thus, CML2 defines that, e.g., the `atomRef1` and `atomRef2` attributes (of the element `bondArray`) contains references to `id` attributes on `atomArray` elements and that the `order="1"` attribute relates to a single bond. How a single bond is handled must in turn be specified by the software; a display program may render the bond in a particular width, color, taper, etc., whereas a database query may handle this information quite differently, or indeed not need it at all.

## Datument Validation

Publishers often provide human-readable guidelines for authors (also known as style guides) for document preparation, which sometimes can extend to entire books. Humans are also quite prone to noncompliance or imperfect compliance because they are often busy or perhaps simply readily bored. Guidelines for data preparation, if they exist, are often to be found in optional categories, such as supporting information. If an author deposits data in such a form, how does the publisher know that it is correct? A key aspect of XML is that documents (and of course datuments) can be validated. For publishing purposes, validation implies a contract between the author and the publisher, which is machine-enforceable. A schema formalizes the syntax, vocabulary, document structure, and some of the semantics. It comprises a set of machine-based rules to which a datument must conform. If it does not, it is the author's responsibility to edit it until it does. If it conforms, it is assumed that the author has complied with the publisher's requirements.

Validation guarantees that the datument conforms to rules. The more powerful the rules, the more invalid data can be detected. Thus, schemas can allow the detection of some disallowed data, particularly with a controlled vocabulary. An atom in CML is not allowed an elementType of "CO" (arising perhaps as a misprint for the element "Co"), or a hydrogenCount of –1. It is, however, allowed a formalCharge of "+10". This might be corrupted data or a legitimate description of a highly ionized atom. Individual schema-based rules (e.g., for different journals) could allow discrimination between these possibilities.

# Datument Vocabularies

The construction of a schema immediately emphasizes the need for a communal vocabulary. An element such as `<molecule>` must be processed in the same consistent manner regardless of the author, the reader, or the processing software. We emphasize "processing"; the implementor must adhere to the same software specifications, and the software must behave in a predictable manner. For many scientists, this implementation will require a change in their thinking, and we emphasize the consequences here.

In its strictest form, this attitude is a controlled vocabulary. Only certain terms may be used, and their meaning is specified by a trusted authority. An example is IUPAC "gold book" vocabulary of chemical terms and definitions (at http://www.chemsoc.org/chembytes/goldbook/). Controlled vocabularies are widely used to enforce the mapping of a discipline such as chemistry onto a generally agreed or mandated vocabulary. They often require substantial formal guidelines or training sessions to ensure consistency of interpretation. Markup languages require us to use absolute precision in syntax and structure. It is highly desirable to have additional precision in semantics (the meaning and behavior of documents). The attachment of semantics to documents is not generally appreciated but is a critical process, and we must have a formal means of attaching semantics to every XML element and attribute and their content. At present, this can be achieved in four principal ways, ordered below in terms of machine-understandable rigor.

### A Human-Readable Prose Description

This description can be as simple as a definition in a dictionary, which may or may not give an indication as to how it might be used. An example from the CIF (crystallographic) dictionary is `_chemical_compound_source`, which is defined as "description of the source of the compound under study, or of the parent molecule if a simple derivative is studied. This includes the place of discovery for minerals or the actual source of a natural product." This description formalizes the concept but (deliberately) gives wide latitude in its implementation and content.

### A Human-Readable Set of Instructions for Machine Implementation

Another CIF entry (abbreviated) for `_atom_site_U_iso_or_equiv` might specify carefully how the concept must be implemented and indicate constraints, such as datatype, enumeration range, and units. The constraints are all machine-processable, and the definition includes an implementable algorithmic constraint.

Because CIF predates XML, this definition is not machine-processable (i.e., it acts as a specification for a human programmer, but it cannot be used to generate software automatically). XML schemas provide mechanisms to overcome this.

### Definitions by Software

Many elements of a controlled vocabulary are effectively defined by software implementation. Thus, the description of the HTML language requires certain elements to have specified behavior. For example, `<img>` supports the display of raster images, but the precise look may vary between implementations and file types. Implementation through software is useful and powerful where authors, publishers, readers, and processors all use the same system. Because STM publishing is increasingly multidisciplinary, this implementation becomes problematic. Often a reader may have to download specialist software that is idiosyncratic and that may not have enough functionality, especially the export of semantically rich data. Moreover, the semantic rules are often buried deep in the software and difficult to understand precisely.

### Formal Semantics

We believe that the chemical community should move toward the adoption of formal rules for expressing semantics and ontology (semantics is the branch of semiotics, the philosophy or study of signs, that deals with meaning. Ontology is defined as a description, such as a formal specification of a program, of the concepts and relationships that can exist for an agent or a community of agents). Our central message is that we need carefully constructed and curated machine-processable ontologies. We believe that scientific and scholarly organizations have a major role to play and that openness and free access to ontologies is critical.

## Authoring and Editing Tools

At present, most chemical publications are created by authors in a publisher-specific manner. Each publisher requires a particular document structure, often a particular technology (e.g., formats of text, images, references, and domain-specific data) and a particular (usually implicit) ontology. The author has to change each of these according to the publisher's requirements and independently of the content. The publisher (or author) then has to make significant technical edits, often as a result of author noncompliance. Original data are transformed into text-oriented formatting languages for rendering to human-readable output, either paper or e-paper, and during this process the machine-processability is lost. Supporting data are often prepared in a variety of (ill-suited) formats; thus, spectral information is frequently merely a scanned image corresponding to the original printed spectrum (itself formatted for human convenience rather than processability).

XML has the potential to revolutionize this situation. With agreed XML-based markup languages, authors can have a single environment independent of the publishers' requirements. Publishers can transform the XML into their in-house systems, but quite independently other "reusers" can do so to their own (possibly quite different) requirements. The original datument, which contains all the "supporting data", can be archived along with the semantics and ontology, all in XML. To achieve all of this consistency, a major change will be required in authoring tools. Instead of proprietary text-based tools, with little useful support for semantics of either text or data, we will require XML-based tools with domain-specific XML components.

An XML editor display environment contains generic mechanisms to manage any domain-specific schema and therefore ensures that a resulting datument is valid. It will also contain mechanisms for supporting domain-specific software, such as editors and browsers (e.g., for molecules, spectra, etc.). Much high-quality chemistry software capable of being used in this context is already available. One commercial example based on XML, which encapsulates this concept, has already appeared. The Publicon tool, for example, provides a comprehensive technical authoring environment capable of handling mathematics and expressing the result in XML; interestingly, its target audience includes chemists and bioscientists. Because all XML-conforming markup languages have essentially the same kinds of structures, syntax, and rules, it becomes much easier to write appropriate, often generic, software to handle it.

## The Datument as a Component of a Scientific Grid

The use of XML has the potential to create savings in certain areas (time for authors and staff costs for technical editors), but a major benefit is that the collected XML datuments, together with the ontologies, would effectively create a machine-processable knowledge base for chemistry. At present, primary publications do not create knowledge without a lot of additional human actions to maximize the knowledge created by primary publications: abstracting, collating, and validating are needed. If this knowledge could be captured and tagged within the primary publication itself, without simply transferring that human action to a different point in the process, a significant proportion of the knowledge base could be extracted by machine. If the metadata, structure, datatypes, ontology, semantics, and processing behavior of a piece of information are determined, it essentially becomes a self-describing *information component.* These information components—which might be implemented by a mixture of XML protocols and object-oriented code—can be regarded as stand-alone, self-describing parts of a knowledge base. Protocols such as XML Query are able to search a heterogeneous aggregate of such components, and RDF (resource description framework, a way to use metadata so that, for example, search engines can locate it) will be able to make deductions from their metadata. By combination of different markup languages, all information, even at a fine level, can be captured without loss. Any part of it can be retrieved, and hence a collection of marked up XML publications would constitute a knowledge base.

If each datument has sufficient high-quality metadata, there may be no essential need for a knowledge base to be centralized. By collecting those publications of interest, any reader or group can create their own personal base. In turn, such metadata can be exported to a wider community using new mechanisms such as RDF or RSS (RDF Site Summary or Rich Site Summary, a way of collecting metadata of interest to an individual reader).

With such exciting technologies in the offing, authors, funders, editors, publishers, and readers have a unique opportunity to start experimenting with creation and dissemination of machine-understandable information and data as an integral part of the process of scientific publishing.

## References


1. Murray-Rust, P.; Rzepa, H. S. *J. Digital Inf.* **2004,** *5,* article 248, 2004-03-18.
2. Murray-Rust, P.; Rzepa, H. S.; Tyrrell, S. M.; Zhang, Y. Y. *Org. Biomol. Chem.* **2004,** *2,* 3192–3203.
3. See Murray-Rust, P.; Rzepa, H. S. *J. Chem. Inf. Comput. Sci.* **2003,** *43,* 757–772 and references cited therein. For online information, including schemas, see http://cml.sourceforge.net/.
4. A full review and listing of all scientifically based XML languages is beyond the scope of this short chapter. Other than CML, the best documented is ThermoML, which is used for thermophysical and thermochemical data. Frenkel, M.; Chirico, R. D.; Diky, V. V.; Marsh, K. N.; Dymond, J. H.; Wakeham, W. A. *J. Chem. Eng. Data* **2004,** *49,* 381–393 and references cited therein.
5. Murray-Rust, P.; Rzepa, H. S.; Wright, M.; Zara, S. *Chem. Commun. (Cambridge, U.K.)* **2000,** 1471–1472.
6. Murray-Rust, P.; Rzepa, H. S.; Wright, M. *New J. Chem.* **2001,** 618–634.
7. Gkoutos, G. K.; Murray-Rust, P.; Rzepa, H. S.; Wright, M. *J. Chem. Inf. Comput. Sci.* **2001,** *41,* 1124–1130.

➤➤➤➤➤

# APPENDIX 8-1

# The IUPAC International Chemical Identifier, InChI

Stephen R. Heller and Alan D. McNaught

The ability to represent uniquely a chemical compound is a fundamental requirement for storage or transmission of chemical information. We define compounds by their molecular structure, as shown in two-dimensional diagrams or stored in computers. Pronounceable names have been developed for oral and written communication, ranging from the trivial, containing no structural information, to completely systematic names, which can be decoded to yield the original structure. However, the application of systematic nomenclature to complicated structures requires expert knowledge of elaborate systems of nomenclature rules. The use of systematic nomenclature to convey information about the increasingly complex molecular systems handled by today's chemists is both laborious and inefficient.

Over the past decade, with ever-increasing reliance by chemists on computer processing, the International Union of Pure and Applied Chemistry (IUPAC) recognized a need to develop methods of nomenclature that can be interpreted by computers, or more precisely, by computer algorithms. A new program was initiated, aimed at creating a method to generate a freely available, nonproprietary identifier for chemical substances that could be used in printed and electronic data sources. The technical development was carried out primarily at the U.S. National Institute of Standards and Technology, and the product is referred to as the IUPAC International Chemical Identifier (InChI).

InChI is not a registry system. It does not depend on the existence of a database of unique substance records to establish the next available sequence number for any new chemical substance being assigned an identifier. Instead, InChI transforms the chemical structure of a compound into a string of characters that uniquely identify that compound. This conversion of a graphical representation of a chemical substance into the unique InChI label can be carried out automatically by any organization, and the facility can be built into any chemical structure drawing program. InChI labels are completely transferable and can be created from existing collections of chemical structures.

Whereas the theory needed for conversion of a structure to a unique string of characters has been known for a long time, when work on InChI began there were no freely available unique representations for compound identification, nor was their development being actively discussed. Thus, before active development could proceed, a precise specification of requirements was wanted, and the following five characteristics were specified as needed for such an identifier:

1. The structure of the compound can be drawn using common conventions.
2. The identifier is derived directly from the structure by an algorithm.
3. Exactly one identifier is associated with a given structure, that is, different structures give different identifiers.
4. The identifier works for a large fraction of all "drawable" chemical substances.
5. The identifier must be openly available.

To be as precise and broadly applicable as desired, InChI uses a layered format to represent all available structural information relevant to compound identity. Each layer in an InChI representation contains a specific type of structural information. These layers, automatically extracted from the input structure, are designed so that each successive layer adds additional detail. The specific layers generated depend on the level of structural detail available and whether tautomerism is allowed. Any ambiguities or uncertainties in the original structure will remain in the InChI. The InChI layers are formula (standard Hill sorted); connectivity (no formal bond orders), including disconnected and connected metals; isotopes; stereochemistry, including double bond (*Z*/*E*) and tetrahedral ($sp^3$); and tautomers (on or off). Charges are not part of the basic InChI, but rather are added at the end of the InChI string.

An example of an InChI representation is given in Figure 8A-1. The acronym InChI and version number are regarded as part of the InChI string (InChI=1 in this case). It is important to recognize, however, that InChI strings are intended for use by computers, and end users need not understand any of their details. In fact, the open nature of InChI and its flexibility of representation, after implementation into software systems, may allow chemists to be even less concerned with the details of structure representation by computers. Source code and an executable version of the structure-to-InChI conversion algorithm are freely available from the IUPAC InChI Web site at http://www.iupac.org/inchi.

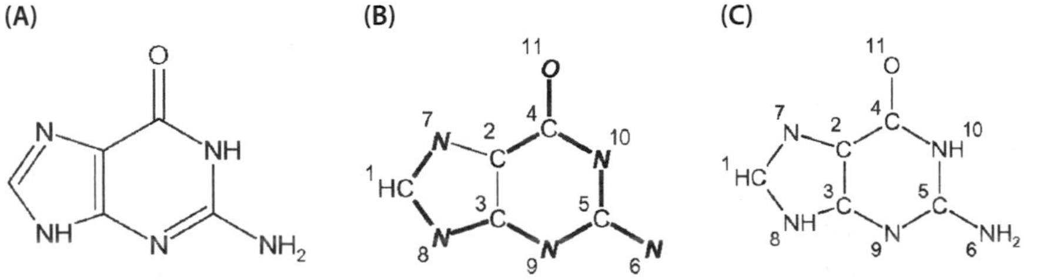

(A) (B) (C)

InChI=1/C5H5N5O/c6-5-9-3-2(4(11)10-5)7-1-8-3/h1H,(H4,6,7,8,9,10,11)/f/h8,10H,6H2

**Figure 8A-1.** InChI for guanine: (A) input structure; (B) mobile H canonical numbering, with the attachment points of four mobile H and changeable bonds indicated in bold; and (C) with fixed H canonical numbering.

PART 2

➤➤➤➤➤

# Style Guidelines

➤➤➤➤➤ CHAPTER 9

# Grammar, Punctuation, and Spelling

This chapter presents grammatical points that cover most situations. It does not attempt to discuss all the rules of grammar; many excellent grammar texts are available for that purpose, such as those given in the selected bibliography, Chapter 18. Writing style and word usage are discussed in Chapter 4. Punctuation, spelling, and word usage are also discussed in Chapter 11 with respect to numbers, mathematics, and units of measure and in Chapter 12 with respect to chemical names.

## Grammar

### *Subject–Verb Agreement*

Everyone knows that a subject and its verb must agree in number. Nevertheless, errors in subject–verb agreement are quite common. The primary cause is confusion about the number of the subject.

➤ The number of the subject can be obscured when one or more prepositional phrases come between the subject and the verb.

> Application of this technique to studies on the phytoplankton biomass and its environments is described. (The subject is "application", which is singular.)

➤ The number of the subject can be obscured when the sentence is constructed in the order prepositional phrase, verb, subject.

> To the mixture were added KCl, HEPES, and water.

> To the solution was added the parent compound.

➤ Two singular subjects joined by "and" require a plural verb.

Growth and isolation of M13 virus were described.

EXCEPTION A subject that is plural in form but singular in effect takes a singular verb. Here a compound subject functions as a single entity.

Research and development is attracting a growing number of young scientists.

Its inventor and chief practitioner is a native son of Boston, Robert Coles.

Much inconsistency and confusion exists with technical documentation.

➤ When two or more subjects are joined by "or", the verb takes the number of the closer or closest subject.

All of the pH values or the median pH value was used.

The median pH value or all of the pH values were used.

➤ Collective nouns take a singular verb when the group as a whole is meant; in that case, they are often preceded by the word "the". Collective nouns take a plural verb when individuals of the group are meant; in that case, they are often preceded by the word "a".

| | | |
|---|---|---|
| contents | majority | range |
| couple | number | series |
| dozen | pair | variety |
| group | | |

The number of metal amides synthesized was the largest to date. (Refers to the number as a unit.)

A number of metal amides were synthesized. (Refers to each amide.)

The series of compounds was prepared to test the hypothesis. (Refers to the series as a unit.)

A series of compounds were tested. (Refers to each compound.)

The variety of materials tested was sufficient for comparative analysis. (Refers to variety as a unit.)

A variety of materials were tested for selective removal of $^{90}Sr$ from nuclear waste solutions. (Refers to the materials individually.)

This group of workers is well aware of its responsibilities. (Refers to the group as a unit.)

This group of workers are willing to sign their names. (Refers to the individuals.)

➤ "Data" can be a singular or plural noun.

After the data is printed and distributed, we can meet to discuss it. (Refers to the whole collection of data as one unit.)

Experimental data that we obtained are compared with previously reported results. (Refers to the data as individual results.)

➤ Units of measure are treated as collective nouns that take a singular verb.

The mixture was stirred, and 5 mL of diluent was added.

Five grams of NaCl was added to the solution.

Three weeks is needed to complete the experiment.

To the mixture was added 5 g of compound **B**.

Under high pressure, 5 volumes of solution A was added.

➤ Nouns ending in "ics" and denoting a scientific discipline are usually singular.

| | |
|---|---|
| dynamics | mechanics |
| kinetics | physics |
| mathematics | thermodynamics |

Mechanics involves the application of Newton's three laws of motion.

The kinetics of electron transfer to and from photogenerated radicals was examined by laser flash photolysis.

The thermodynamics is governed by the positions of the valence and conduction bands.

➤ Compound subjects containing the words "each", "every", and "everybody" take singular verbs.

Each flask and each holder was sterilized before use.

Every rat injected and every rat dosed orally was included.

Everybody in the group and every visitor is assigned a different journal each month.

➤ Sometimes, one of these words is implicit; such cases take a singular verb.

Each name and address is entered into the database.

➤ If both components of the compound subject do not contain, explicitly or implicitly, one of the words "each", "every", or "everybody", the verb must be plural.

Each student and all the professors were invited.

➤ Indefinite pronouns themselves (or adjectives combined with the indefinite pronoun "one") can be the subject of the sentence.

- Those that take a singular verb are "each", "either", "neither", "no one", "every one", "anyone", "someone", "everyone", "anybody", "somebody", and "everybody".

Each was evaluated for its effect on metabolism.

Neither disrupts the cell membrane.

Regarding compounds **1–10**, every one reacts with the control agent.

Someone measures the volume every day.

- Those that take a plural verb are "several", "few", "both", and "many".

  Several were evaluated for their effects on metabolism.

  Few disrupt the cell membrane.

  Regarding compounds **1** and **2**, both react with the control agent.

  Many were chosen to be part of the study.

- Those that take either a singular or a plural verb, depending on context, are "some", "any", "none", "all", and "most". The number of the object of the preposition determines the number of the indefinite pronoun related to it.

  All of the money was stolen.

  Most of the books were lost.

  Not all the disks are here; some were lost.

➤ When a fraction is the subject of the sentence, the number of the attendant object of the preposition determines the number of the subject.

One-third of the precipitate was dissolved.

One-fourth of the electrons were excited.

➤ When a subject and its predicate noun disagree in number, the verb takes the number of the subject. (A *predicate noun* is the "complement" of a form of the verb "to be"; it refers to the same person or thing as the subject.)

The preparation and structure determination [plural subject] of these three compounds are the topic [singular predicate noun] of this paper.

The topic of this paper [singular subject] is the preparation and structure determination [plural predicate noun] of these three compounds.

## Awkward Omissions of Verbs and Auxiliary Verbs

➤ Each subject in a compound sentence must have the proper verb and auxiliary verb.

INCORRECT

The eluant was added to the column, and the samples collected in 10 mL increments.

CORRECT

The eluant was added to the column, and the samples were collected in 10 mL increments.

### *Restrictive and Nonrestrictive Expressions*

➤ A phrase or clause is *restrictive* when it is necessary to the sense of the sentence; that is, the sentence would become pointless without the phrase or clause. Restrictive clauses are best introduced by "that", not "which".

> It was necessary to find a blocking group that would react with the amino group but not with the hydroxyl group.
>
> Comparison will be restricted to acetylene compounds that have the same functional end groups.

If the clauses beginning with "that" were deleted, the sentences would not convey the information intended. Therefore, the clauses are restrictive.

Phrases can also be restrictive.

> Reactions leading to the desired products are shown in Scheme 1.

If the phrase "leading to the desired products" were deleted, the sentence would not convey the information intended.

➤ A phrase or clause is *nonrestrictive* if it adds information but is not essential; that is, the sentence does not lose its meaning if the phrase or clause is deleted. Nonrestrictive phrases and clauses are set off by commas. Nonrestrictive clauses may be introduced by "who" or "which" but not by "that".

> Squalene, a precursor of cholesterol, is a 30-carbon isoprenoid.
>
> This highly readable book, written in nontechnical language, surveys the field of chemistry by describing the contributions of chemistry to everyday life.
>
> Moore, working at the Rockefeller Institute, developed methods for the quantitative determination of amino acids.
>
> The current–voltage curves, which are shown in Figure 6, clearly demonstrate the reversibility of all four processes.
>
> Several hazardous waste disposal sites are located along the shores of the Niagara River, which is a major water source.
>
> Melvin Calvin, who won the Nobel Prize in 1961, elucidated the biochemical pathways in photosynthesis.
>
> James Aberdeen, professor emeritus of Central State University, which has provided significant scholarship support to minority students over the years, made a generous contribution to the school's building fund.

### *Dangling Modifiers*

A *dangling modifier* is a modifying word or phrase that does not clearly and logically modify another word in the sentence. In scientific writing, the passive voice is often necessary ("the solutions were heated"; "melting points were determined"), but its use can lead to dangling modifiers.

➤ If a modifier precedes the subject of a sentence, it must modify that subject and be separated from it by a comma. Otherwise, it is a dangling modifier.

INCORRECT

Splitting the atom, many new elements were discovered by Seaborg.

CORRECT

Splitting the atom, Seaborg discovered many new elements.

INCORRECT

Upon splitting the atom, many new elements were discovered by Seaborg.

CORRECT

Upon splitting the atom, Seaborg discovered many new elements.

INCORRECT

When confronted with these limitations, the experiments were discontinued.

CORRECT

When confronted with these limitations, we discontinued the experiments.

In light of these limitations, the experiments were discontinued.

INCORRECT

Understanding the effect of substituents on the parent molecules, the ortho hydrogens could be assigned to the high-frequency peak.

CORRECT

Understanding the effect of substituents on the parent molecules, we could assign the ortho hydrogens to the high-frequency peak.

INCORRECT

Using the procedure described previously, the partition function can be evaluated.

CORRECT

Using the procedure described previously, we can evaluate the partition function.

➤ In some cases, the passive voice can be used to correct a dangling modifier.

INCORRECT

After combining the reactants, the reaction mixture was stirred at room temperature for 3 h.

CORRECT

After the reactants were combined, the reaction mixture was stirred at room temperature for 3 h.

INCORRECT

After stirring the mixture, 5 mg of compound **2** was added.

CORRECT

After the mixture was stirred, 5 mg of compound **2** was added.

➤ Phrases starting with "based on" must modify a noun or pronoun that usually immediately precedes or follows the phrase. Use phrases starting with "on the basis of" to modify a verb.

INCORRECT

Based on resonance enhancement and frequency shifts, changes in the inter-ring separation were calculated.

CORRECT

On the basis of resonance enhancement and frequency shifts, changes in the inter-ring separation were calculated.

INCORRECT

Based on extensive study, this genetic deficiency was attributed to the loss of one isozyme. ("Based on extensive study" modifies the noun "deficiency", but this is not the meaning.)

CORRECT

On the basis of extensive study, this genetic deficiency was attributed to the loss of one isozyme. ("On the basis of extensive study" modifies the verb "was attributed".)

Style guidelines based on authoritative sources are included in this book. ("Based on authoritative sources" modifies the noun "guidelines".)

➤ "Due to" means "attributable to"; use it only to modify a noun or pronoun directly preceding it in the sentence or following a form of the verb "to be".

INCORRECT

Delays resulted due to equipment failure.

CORRECT

Delays due to equipment failure resulted.

The delays were due to equipment failure.

INCORRECT

This high value resulted due to the high conversion efficiencies of the enzymatic reactor.

CORRECT

This high value is due to the high conversion efficiencies of the enzymatic reactor.

This high value resulted from the high conversion efficiencies of the enzymatic reactor.

INCORRECT

Due to exposure to low levels of lead, children can be at risk for developmental problems.

CORRECT

Because of exposure to low levels of lead, children can be at risk for developmental problems.

Children can be at risk for developmental problems because of exposure to low levels of lead.

➤ *Absolute constructions* are words, phrases, or clauses that are grammatically unconnected with the rest of the sentence in which they appear. They are sometimes called "sentence modifiers" because they qualify the rest of the sentence. They may occur anywhere in the sentence, and they are always set off by commas. They are not dangling modifiers.

Contrary to the excited-state situation, metal–metal bonding interactions in the ground states are weak.

The conclusions were premature, considering the lack of available data.

Judging from the spectral changes, exhaustive photolysis of compound **4** had occurred.

The conformations about the Re–Re bond, in addition, are different for all three complexes.

When necessary, the solutions were deaerated by bubbling nitrogen.

Clearly, alternative synthetic methods are possible.

The instructor having made her point, the discussion continued.

Absolute constructions often begin with one of the following words:

concerning
considering
failing
given
judging
provided
providing
regarding

- In mathematical papers, absolute phrases beginning with the words "assuming" and "taking" are often used as sentence modifiers.

Assuming that distance *d* is induced by the norm, *M* is a symmetrical and positively defined matrix.

Taking this value as an upper limit, the two shortest distances are sometimes too long for incipient hydrogen bonds.

- A subordinate or elliptical clause may be used as a sentence modifier.

  The compound is stable in air, as we concluded from the experimental evidence.

  The Mo 5s orbitals, as expected, interact strongly with the ligands.

- An introductory infinitive or infinitive phrase may be a sentence modifier.

  To prepare compound **2**, the method of Garner was followed.

### *Reflexive Pronouns*

➤ Use the reflexive pronouns "myself", "yourself", "himself", "herself", "itself", "ourselves", and "themselves" only to refer back to a noun or another pronoun in the same sentence.

INCORRECT

Please send your manuscript to the associate editor or myself.

CORRECT

Please send your manuscript to the associate editor or me.

The associate editor herself will review your manuscript.

INCORRECT

My collaborators and myself will evaluate the results.

CORRECT

My collaborators and I will evaluate the results.

I will evaluate the results myself.

I myself will evaluate the results.

## Punctuation

### *Comma*

➤ Use a comma before, but not after, the coordinating conjunctions "and", "or", "nor", "but", "yet", "for", and "so" connecting two or more main clauses (complete thoughts).

Toluene and hexane were purified by standard procedures, and benzene was redistilled from calcium hydride.

The role of organic templates in zeolite synthesis has been studied extensively, but no general principles have been delineated.

Supported metals are among the most important industrial catalysts, yet only a few have been studied thoroughly.

No dielectric constants are available for concentrated acids, so it is difficult to give a quantitative explanation for the results.

➤ Use a comma after a subordinate clause that precedes the main clause in a complex sentence.

> Although 40 different P450 enzymes have been identified, only six are responsible for the processing of carcinogens.
>
> Since the institute opened, plant breeders have developed three new prototypes.
>
> Because the gene and the molecular marker are so close on the chromosome, they segregate together in the progeny.

➤ Use a comma after most introductory words and phrases.

> However, the public is being inundated with stories about cancer-causing chemicals.
>
> Therefore, the type of organic solvent used is an important factor in lipase-catalyzed enzymatic synthesis.
>
> After 3 months, the plants grown under phosphorus-deficient conditions were evaluated.
>
> Thus, their motion is the result of the rotation of ferromagnetic domains.
>
> On cooling, a crystalline phase may develop in coexistence with an amorphous phase.

➤ Use a comma before the coordinating conjunction in a series of words, phrases, or clauses of equal rank containing three or more items. (This comma is called the *serial comma.*)

> Water, sodium hydroxide, and ammonia were the solvents.
>
> The red needles were collected, washed with toluene, and dried in a vacuum desiccator.
>
> The compound does not add bromine, undergo polymerization by the Diels–Alder reaction, or react with electrophiles.

➤ In compound sentences containing coordinating conjunctions, the clause following the conjunction is punctuated as if it were alone.

> The reaction proceeds smoothly, and by use of appropriate reagents, the yields will be enhanced.
>
> The compounds were separated, and after the filters had been washed, the experiments were completed.

➤ Do not use a comma to separate a verb from its subject, its object, or its predicate noun.

INCORRECT

> The addition of substituted silanes to carbon–carbon double bonds, has been studied extensively.

CORRECT

The addition of substituted silanes to carbon–carbon double bonds has been studied extensively.

INCORRECT

The disciplines described in the brochure include, materials science, biotechnology, and environmental chemistry.

CORRECT

The disciplines described in the brochure include materials science, biotechnology, and environmental chemistry.

INCORRECT

The solvents used in this study were, cyclohexane, methanol, *n*-pentane, and toluene.

CORRECT

The solvents used in this study were cyclohexane, methanol, *n*-pentane, and toluene.

➤ Do not use a comma before the conjunction joining a compound predicate consisting of only two parts.

INCORRECT

The product distribution results were obtained in sodium hydroxide, and are listed in Table 10.

CORRECT

The product distribution results were obtained in sodium hydroxide and are listed in Table 10.

➤ Use commas to separate items in a series that contains another series in parentheses already separated by commas.

The structure was confirmed with spectroscopy ($^{1}$H NMR, UV, and IR), high-resolution mass spectrometry, and elemental analysis.

➤ Use a comma between two or more adjectives preceding a noun only if you can reverse the order of the adjectives without losing meaning. If you can insert the word "and", the comma is correct.

The intense, broad signals of the two groups confirmed their location.
The broad, intense signals of the two groups confirmed their location.

Sample preparation is a repetitious, labor-intensive task.
Sample preparation is a labor-intensive, repetitious task.

A powerful, versatile tool for particle sizing is quasi-elastic light scattering.
A versatile, powerful tool for particle sizing is quasi-elastic light scattering.

But:

> Polyethylene is an important industrial polymer.
>
> The rapid intramolecular reaction course leads to ring formation.
>
> The backbone dihedral angles were characterized by *J* couplings.
>
> The local structural environment of the Mn cluster was determined.

➤ Use a comma before, but not after, the subordinating conjunction in a nonrestrictive clause.

INCORRECT

> The bryopyran ring system is a unique requirement for anticancer activity whereas, the ester substituents influence the degree of cytotoxicity.

CORRECT

> The bryopyran ring system is a unique requirement for anticancer activity, whereas the ester substituents influence the degree of cytotoxicity.

➤ Use commas to set off nonrestrictive phrases or clauses.

> The products, which were produced at high temperatures, were unstable.

➤ Phrases introduced by "such as" or "including" can be restrictive (and thus not set off by commas) or nonrestrictive (and thus set off by commas).

> Potassium compounds such as KCl are strong electrolytes; other potassium compounds are weak electrolytes.
>
> Previously, we described a mathematical model including a description of chlorophyll degradation in foods.
>
> Divalent metal ions, such as magnesium(II) and zinc(II), are located in the catalytic active sites of the enzymes.
>
> Hydrogen-bonded complexes, including proton-bound dimers, are well-known species.

In the first two sentences, the phrases are restrictive because the sentences do not make their points without the phrases. In the third and fourth sentences, the phrases are nonrestrictive because the sentences can make their points without the phrases.

➤ An appositive is a noun that follows another noun and identifies or explains the meaning of the first noun.

> My wife, Jeanne, is a biochemist at the National Institutes of Health.
>
> My son James plays baseball, and my son John plays soccer.

An appositive is nonrestrictive (and therefore set off by commas) when it names the only possibility. In the first sentence, Jeanne is a nonrestrictive appositive. An appositive is restrictive (and therefore not set off by commas) when it points out one of two or more possibilities. In the second sentence, the names of the two sons are restrictive appositives.

➤ Use commas to set off the words "that is", "namely", and "for example" when they are followed by a word or list of words and not a clause. Also use a comma after the item or items being named. Use a comma after "i.e." and "e.g." in parenthetical expressions.

> The new derivatives obtained with the simpler procedure, that is, reaction with organocuprates, were evaluated for antitumor activity.
>
> Alkali metal derivatives of organic compounds exist as aggregates of ion pairs, namely, dimers, trimers, and tetramers, in solvents of low polarity.
>
> Many antibiotics, for example, penicillins, cephalosporins, and vancomycin, interfere with bacterial peptidoglycan construction.
>
> These oxides are more stable in organic solvents (e.g., ketones, esters, and ethers) than previously believed.

➤ Use commas to separate two reference citation numbers, but use an en dash ( ) to express a range of three or more in sequence, whether they are superscripts or are on the line in parentheses. When they are superscripts, do not use a space after the comma.

> Experimental investigations[10,14,18–25] concerned the relative importance of field and electronegativity effects.
>
> Certain complexes of cobalt were reported (*10, 11*) to have catalytic effects on hydrolysis reactions.
>
> Flash photolysis studies (*3–7*) demonstrated the formation of transient intermediate products such as triplet states.

➤ Use a comma before Jr. and Sr., but treat II and III according to the person's preference. Within a sentence, always use a comma after Jr. and Sr., but use a comma after II and III only if they are preceded by a comma.

> William M. Delaney, Jr.
> Charles J. Smith, III
> John J. Alden II
>
> William M. Delaney, Jr., was elected to the governing board.
>
> Charles J. Smith, III, received a majority of the votes.
>
> John J. Alden II did not run for office this year.

➤ Do not use a comma preceding "et al." unless commas are needed for other reasons.

Saltzman et al.
Saltzman, M. J., et al.
Saltzman, Brown, et al.

➤ In dates, use a comma after the day, but not after the month when the day is not given.

June 15, 1996
June 1996

When giving a complete date within a sentence, use a comma after the year as well.

On August 18, 1984, an extraordinary person was born.

➤ When a geographical location is named within a sentence and the name includes a comma, use a comma at the end of the name as well.

Iona College, in New Rochelle, New York, is the CEO's alma mater.

The lead researcher, who obtained her education at the University of Calgary, Alberta, Canada, addressed the reporters' questions.

➤ Use a comma to introduce quotations.

In the words of Pasteur, "Chance favors the prepared mind."

Pasteur said, "Chance favors the prepared mind."

➤ Do not use a comma after a quotation that is the subject of the sentence.

"Chance favors the prepared mind" is a translation from the French. (The quotation is the subject of the sentence.)

## *Period*

➤ Use a period at the end of a declarative sentence, but never in combination with any other punctuation marks.

He said, "Watch out!"

She asked, "May I go?"

➤ Do not use periods after most abbreviated units of measure, except when the abbreviation could be confused with a word (in. for inches, at. for atomic, no. for number).

➤ If a sentence ends with an abbreviation that includes a period, do not add another period.

She will return at 3 a.m.

➤ Use periods and spaces after initials in persons' names.

J.-L. Gay Lussac J. E. Lennard-Jones M. S. Newman

EXCEPTION Use periods but no spaces when referring to authors of a paper in the acknowledgment paragraph of the paper.

R.C.McD. and C.R. thank Dr. Rose Allan for carefully reading the manuscript.

C.-C.Y., L.B.-P., N.-h.X., and S.Zh.O. are grateful for generous support from the university.

➤ Do not use periods in abbreviations or acronyms of institution or organization names.

ACS CNRS NASA NIH

## *Semicolon*

➤ Use a semicolon to separate independent clauses that are not joined by a conjunction.

All solvents were distilled from an appropriate drying agent; tetrahydrofuran and diethyl ether were also pretreated with activity I alumina.

➤ Use semicolons between items in a series of words, phrases, or data strings if one or more of the items already contain commas.

We thank Zachary Axelrod, University of Michigan, for spectral data; Caroline Fleissner, Harvard University, for helpful discussions; and the National Science Foundation for financial support (Grant XYZ 123456).

The product was dried under vacuum to give compound **2**: yield 68%; IR 1991 m, 1896 s, sh, 1865 s $cm^{-1}$; $^{1}H$ NMR 0.36 ppm; $^{13}C$ NMR 221.3, 8.1 ppm.

Figure 1. Cyclic voltammograms in dichloromethane: (a) compound **1,** 23 °C; (b) compound **2,** –40 °C; (c) compound **4,** 23 °C.

Figure 6. Ru–H stretches in the IR spectrum of compound **5:** ×, 298 K; +, 90 K.

This rule holds even if the only group containing the commas is the last in the series.

The compounds studied were methyl ethyl ketone; sodium benzoate; and acetic, benzoic, and cinnamic acids.

➤ Use a semicolon between independent clauses joined by conjunctive adverbs or transitional phrases such as "that is", "however", "therefore", "hence", "indeed", "accordingly", "besides", and "thus".

The rate at which bleaching occurred was dependent on cluster size; that is, the degradation of the mononuclear cluster was about 5 times faster than that of the tetranuclear cluster.

Many kinetic models have been investigated; however, the first-order reactions were studied most extensively.

The proposed intermediate is not easily accessible; therefore, the final product is observed initially.

The restriction of the rotational motions of the *tert*-butyl group gives rise to large entropy changes for the association reaction; hence, the covalent form is relatively easy to identify.

The efficiency of the cross-coupling depends on the nature of X in RX; thus, the reaction is performed at room temperature by slow addition of the ester.

➤ Do not use a semicolon between dependent and independent clauses.

INCORRECT

The activity on bromopyruvate was decreased; whereas, the activity on pyruvate was enhanced.

CORRECT

The activity on bromopyruvate was decreased, whereas the activity on pyruvate was enhanced.

## Colon

➤ Use a colon to introduce a word, a phrase, a complete sentence, or several complete sentences that illustrate, clarify, or expand the information that precedes it. Capitalize the first word after a colon only if the colon introduces more than one complete sentence, a quotation, or a formal statement.

The electron density was studied for the ground state of three groups of molecules: (1) methane–methanol–carbon dioxide, (2) water–hydrogen peroxide, and (3) ferrous oxide–ferric oxide.

We now report a preliminary finding: no chemical shift changes were detected in the concentration range 0.1–10 M.

The following are our conclusions: Large-angle X-ray scattering studies give us an accurate picture of structures up to 9 Å. They do not allow the specification of defects, such as random ruptures of the chains. The structural models defined are strongly supported by magnetic measurements.

➤ In figure captions, use a colon to introduce explanations of symbols or other aspects of the figure.

Figure 1. Variable-temperature $^{1}$H NMR spectra of compound **12:** top, 403 K; middle, 353 K; bottom, 298 K.

Figure 3. Brønsted-type plots for aminolysis in 1 M KCl at 25 °C: ○, 2-nitrophenyl acetate; □, 3-chlorobenzoic acid; ◇, 2,6-dinitrobenzoic acid.

➤ Do not use a colon (or any punctuation) between a verb and its object or complement or between a preposition and its object.

INCORRECT

The rate constants for the reaction in increasing concentrations of sodium hydroxide are: 3.9, 4.1, 4.4, 4.6, and 4.9.

CORRECT

The rate constants for the reaction in increasing concentrations of sodium hydroxide are 3.9, 4.1, 4.4, 4.6, and 4.9.

INCORRECT

The thermal decomposition was investigated with: gas chromatography, BET surface areas, and X-ray powder diffraction.

CORRECT

The thermal decomposition was investigated with gas chromatography, BET surface areas, and X-ray powder diffraction.

INCORRECT

Transition-metal nitrides have many properties that make them suitable for industrial applications, including: high wear resistance, high decomposition temperature, and high microhardness.

CORRECT

Transition-metal nitrides have many properties that make them suitable for industrial applications, including high wear resistance, high decomposition temperature, and high microhardness.

➤ Use either a colon or a slash to represent a ratio, but not an en dash. Use either a slash or an en dash between components of a mixture, but not a colon.

dissolved in 5:1 glycerin/water
dissolved in 5:1 glycerin–water

the metal/ligand (1:1) reaction mixture
the metal–ligand (1:1) reaction mixture
the metal–ligand (1/1) reaction mixture

the methane/oxygen/argon (1/50/450) matrix
the methane/oxygen/argon (1:50:450) matrix

## *Quotation Marks*

Location of closing quotation marks with respect to other punctuation is a style point in which ACS differs from other authorities. In 1978, ACS questioned the traditional practice and recommended a deviation: logical placement. Thus, if the punctuation is part of the quotation, then it should be within the quotation marks; if the punctuation is not part of the quotation, the writer should not mislead the reader by implying that it is.

➤ Place closing quotation marks before all punctuation that is not part of the original quotation. Place them after all punctuation that is part of the quotation.

> The sample solution was stirred briefly with a magnetic "flea".
>
> Ralph Waldo Emerson said, "The reward of a thing well done is to have done it."

➤ Use quotation marks around words used in a new sense or words not used literally, but only the first time they appear in text.

> Plastocyanin is a soluble "blue" copper protein.
>
> The integrated intensity of each diagonal in the spectrum is proportional to a "mixing coefficient".
>
> The "electron-deficient" cations are, in fact, well-established intermediates.

➤ Use quotation marks to enclose the titles of uniquely named parts and sections of a book or a paper.

> A complete description of the oils is given in the section "Flavonoids in Citrus Peel Oils", and other references are listed in the bibliography.

But:

> The preface describes the complexity of the problem.

➤ Use quotation marks to enclose short direct quotations (up to three sentences).

> In the book *Megatrends*, Naisbitt concludes, "We are moving from the specialist who is soon obsolete to the generalist who can adapt."

➤ Use a narrower column width (that is, indented on both sides) for longer quotations (extracts) of 50 words or more. Do not use quotation marks.

> Everything is made of atoms. That is the key hypothesis. The most important hypothesis in all of biology, for example, is that everything that animals do, atoms do. In other words, there is nothing that living things do that cannot be understood from the point of view that they are made of atoms acting according to the laws of physics.
>
> —Richard Phillips Feynman

However, this convention does not apply in an article quoting someone who has been interviewed. In such cases, quoted text need not be differentiated by column width, and quotation marks should be used.

➤ Use single quotation marks only when they are within double quotation marks.

> He said, "You should read the article 'Fullerenes Gain Nobel Stature' in the January 6, 1997, issue of *Chemical & Engineering News*."

## *Parentheses*

Parenthetical expressions contain information that is subsidiary to the point of the sentence. The sentence does not depend on the information within the parentheses.

➤ Use parentheses for parenthetical expressions that clarify, identify, or illustrate and that direct the reader.

> The total amount (10 mg) was recovered by modification of the procedure.
>
> The final step (washing) also was performed under a hood.
>
> The curve (Figure 2) obeys the Beer–Lambert law.
>
> The results (Table 1) were consistently positive.
>
> Only 15 samples (or 20%) were analyzed.

➤ Punctuate after, not before, parenthetical expressions.

INCORRECT

> Compound **1,** (7 mg) obtained by typical workup methods, was used without further purification.

CORRECT

> Compound **1** (7 mg), obtained by typical workup methods, was used without further purification.

➤ If a parenthetical sentence is within another sentence, do not use a final period within the closing parenthesis, and do not start the parenthetical sentence with a capital letter.

> Our results (the spectra are shown in Figure 5) justified our conclusions.
>
> Our results justified our conclusions (the spectra are shown in Figure 5).

➤ If a parenthetical sentence is not within another sentence, use a final period inside the closing parenthesis, and start the parenthetical sentence with a capital letter.

> A mechanism involving loss of a CH radical followed by rearrangement was proposed. (The reactions are shown in Scheme 1.)

➤ Use parentheses to enclose numerals in a list. Always use parentheses in pairs, not singly.

> Three applications of this reaction are possible: (1) isomerization of sterically hindered aryl radicals, (2) enol–keto transformation, and (3) sigmatropic hydrogen shift.

➤ Use parentheses to identify the manufacturer of reagents and equipment.

cobalt chloride (Mallinckrodt)
a pH meter with a glass electrode (Corning)

➤ Do not use parentheses when citing a reference number in narrative text. In such a case, the reference number is the point of the sentence, not subsidiary information, and thus not parenthetical.

INCORRECT

in ref (12), in (12)

CORRECT

in ref 12

➤ Use parentheses in mathematical expressions as discussed in Chapter 11 and in chemical nomenclature and notation as discussed in Chapters 12 and 13.

### Square Brackets

➤ Use square brackets within quotation marks to indicate material that is not part of a direct quote.

> In the words of Sir William Lawrence Bragg, "The important thing in science is not so much to obtain new facts as to *discover new ways* [italics added] of thinking about them."

➤ Use square brackets to indicate concentration: $[Ca^+]$.

➤ Use square brackets in mathematical expressions as discussed in Chapter 11 and in chemical nomenclature and notation as discussed in Chapters 12 and 13.

### Dashes

The shortest dash is the hyphen (-); the en dash (–) is longer; and the em dash (—) is the longest. Hyphens are discussed in the section on hyphenation in Chapter 10, starting on p 135.

### En Dash

➤ Use an en dash to mean the equivalent of "and", "to", or "versus" in multiword concepts where the words are of equal weight.

acid–base titration
bromine–olefin complex
carbon–oxygen bond
cis–trans isomerization
cost–benefit analysis
dose–response relationship
ethanol–ether mixture
freeze–pump–thaw degassed
helix–coil transition
host–guest complexation

log–normal function
metal–ligand complex
metal–metal bonding
nickel–cadmium battery
oxidation–reduction potential
producer–user communication
pump–probe technique
red–black dichroic crystals
structure–activity relationship
structure–property relationship
temperature–time curve
vapor–liquid equilibrium
winter–fall maxima
0–*t*–∞ sequence

EXCEPTION Use a hyphen for color combinations such as blue-green. See Chapter 10, page 140.

➤ Use an en dash to mean "to" or "through" with a span of three or more numerals or other types of ranges.

| | | |
|---|---|---|
| 12–20 months | Figures 1–4 | 5–50 kg |
| sections 1b–1f | parts C–E | compounds **A–I** |
| Lyon and co-workers (*23–26*) | Lyon and co-workers[23a–d] | |

EXCEPTION 1 When either one or both numbers are negative or include a symbol that modifies the number, use the word "to" or "through", not the en dash.

| | | |
|---|---|---|
| –20 to +120 K | –145 to –30 °C | ≈50 to 60 |
| 10 to >600 mL | <5 to 15 mg | |

EXCEPTION 2 Do not use an en dash when the word "from" or "between" is used.

from 500 to 600 mL (*not* from 500–600 mL)
between 7 and 10 days (*not* between 7–10 days)

➤ Use an en dash to link the names of two or more persons of equal importance used as a modifier.

Bednorz–Müller theory
Beer–Lambert law
Bose–Einstein statistics
Debye–Hückel theory
Diels–Alder reaction
Fermi–Dirac statistics
Fischer–Tropsch effect
Fisher–Johns hypothesis
Flory–Huggins interaction
Franck–Condon factor
Friedel–Crafts reaction
Geiger–Müller effect
Henderson–Hasselbalch equation
Jahn–Teller effect
Lee–Yang–Parr method
Lineweaver–Burk method
Mark–Houwink plot
Meerwein–Ponndorf theory
Michaelis–Menten kinetics
Stern–Volmer plot
van't Hoff–Le Bel theory
Wolff–Kishner theory
Young–Laplace equation
Ziegler–Natta-type catalyst

Treatment of double surnames is covered in Chapter 10 (p 139).

➤ Use an en dash between components of a mixed solvent. (A slash can also be used.)

> The melting point was unchanged after four crystallizations from hexane–benzene.

## Em Dash

➤ Use em dashes to set off words that would be misunderstood without them.

INCORRECT

> All three experimental parameters, temperature, time, and concentration, were strictly followed.

CORRECT

> All three experimental parameters—temperature, time, and concentration—were strictly followed.

➤ Do not use em dashes to separate phrases or nonrestrictive clauses if another form of punctuation can be used.

INCORRECT

> Knauth—not Stevens—obtained good correlation of results and calculations.

CORRECT

> Knauth, not Stevens, obtained good correlation of results and calculations.

INCORRECT

> The singly charged complexes—which constituted bands 1 and 3—liberated maleate anion upon decomposition.

CORRECT

> The singly charged complexes, which constituted bands 1 and 3, liberated maleate anion upon decomposition.

## Ellipsis Points

➤ Within a quotation, use three periods (points of ellipsis) to indicate deleted words or phrases. These three periods are in addition to other needed punctuation. Thus, if a period is already there, the result will be four periods.

> No science is immune to the infection of politics and the corruption of power…. The time has come to consider how we might bring about a separation, as complete as possible, between Science and Government in all countries.
>
> —Jacob Bronowski

➤ Do not begin or end a quotation with ellipsis points.

➤ Use ellipsis points where part of a series is omitted, when the pattern of the series is unambiguous.

$a = 1, 2, 3, \ldots$
$n = 2, 4, 6, \ldots$
$x = 1, 3, 5, \ldots, 15$

## Spelling

Consult a dictionary to resolve spelling questions. *Merriam-Webster's Collegiate Dictionary* and *Webster's New World College Dictionary* are the desk dictionaries used by the ACS technical editing staff. ACS staff members also use the unabridged *Webster's Third New International Dictionary*. However, whatever your dictionary, choose the first spelling of a word. Use American spellings, except in proper names and direct quotations (including titles). Appendix 9-1 contains a list of the recommended spellings for words that have two or more acceptable spellings.

➤ For the correct spelling and styling of company names, search the Internet for the company and look for the company name on its "contact" page. Do not rely on either a company's home page or the presence of a particular spelling in a search engine, because the former may be informal and the latter common but also incorrect.

### *Tricky Possessives*

➤ Form the possessive of a joint owner by adding an apostrophe and an "s" after the last name only.

Celapino and Marshall's results
Bausch and Lomb's equipment

➤ Form the possessive of plural nouns that do not end in "s" by adding an apostrophe and an "s". Form the possessive of plural nouns that end in "s" by adding an apostrophe only.

people's rights
children's books
compounds' structures

➤ Form the possessive of a proper name ending in "s" by adding an apostrophe and an "s".

Jacobs's laboratory
Mathers's reception

## *Tricky Plurals*

Sometimes, the plural form is so familiar that it is used erroneously instead of the singular, usually with Latin and other non-English words. The following list shows the correct singulars and plurals. The preferred forms are given first.

| SINGULAR | PLURAL |
|---|---|
| alga | algae |
| apparatus | apparatus, apparatuses |
| appendix | appendixes, appendices |
| bacterium | bacteria |
| basis | bases |
| criterion | criteria, criterions |
| erratum | errata |
| focus | focuses, foci |
| formula | formulas, formulae |
| fungus | fungi, funguses |
| genus | genera, genuses |
| helix | helixes, helices |
| hypothesis | hypotheses |
| index | indexes |
| index | indices (mathematical) |
| latex | latices, latexes |
| locus | loci |
| matrix | matrices (mathematical) |
| matrix | matrixes (media) |
| maximum | maximums, maxima |
| medium | media, mediums |
| minimum | minimums, minima |
| phenomenon | phenomena, phenomenons |
| polyhedron | polyhedrons, polyhedra |
| sequela | sequelae |
| spectrum | spectra, spectrums |
| stratum | strata |
| symposium | symposia, symposiums |
| vertex | vertexes, vertices |
| vortex | vortexes, vortices |

➤➤➤➤➤

APPENDIX 9-1

# Recommended Spelling List

Many words in regular usage, as well as many technical terms, have two or more acceptable spellings. The following list gives recommended spellings and capitalizations, where appropriate, for some terms not found in easily accessible dictionaries, words often misspelled, common expressions, and words for which the ACS preference may not match your dictionary's.

absorbance
absorbency
absorbent
accommodate
acknowledgment
adsorbent
aerobic
aging
aglycon
air-dry (verb)
ambiguous
amine ($RNH_2$)
ammine ($NH_3$ complex)
amphiphile
ampule
analog (computer)
analogue (structural derivative)
analyte
analyze
annelation
annulation
antioxidant
appendixes
aqua regia
Arrhenius
artifact
asymmetry
audio frequency
autoxidation
auxiliary
Avogadro
bacitracin
back-bonding
back-donation
back-titrate (verb)
backscatter
backscattering
backward
band gap
bandwidth
baseline
Beckmann (thermometer, rearrangement)
Beer's law
Beilstein
bit
black box
blackbody
blender
Boltzmann
borderline
Bragg scattering
break-seal
break up (verb)
breakup (noun)
bremsstrahlung
bridgehead
broad band (noun)
broad-band (adjective)
Brønsted
Büchner
build up (verb)
buildup (noun)
buret
butanol, 1-butanol (*not* *n*-butanol)

*n*-butyl alcohol
*tert*-butylation
byline
bypass
byproduct
byte
canceled
canister
cannot
Cartesian
catalog
chloramine
chloro amine
chlorophyll *a*
clean up (verb)
cleanup (noun)
clear-cut
close up (verb)
close-up (noun)
co-ion
co-occurrence
co-worker
coauthor
collinear
colorimetric
complexometric
concomitant
condensable
conductometric
conrotatory
constantan
coordination
Coulombic
counter electrode
counteranion
counterion
coverslip
cross-coupling
cross-link
cross over (verb)
cross-react
cross-reaction
cross section (noun)
cross-sectional (adjective)
crossover (noun, adjective)
cuboctahedron
cut off (verb)
cutoff (noun)
cuvette
cytochrome *c*
Darzens
database
deamino (*not* desamino)
deoxy (*not* desoxy)
dependent
desamine (amino acid names only)
desiccator
deuterioxide
deuteroporphyrin
Dewar benzene
Dewar flask
dialogue
diffractometer
disc (electrophoresis, compact disc)
discernible
disk (anatomy, computers)
disrotatory
dissymmetric
distill
dry ice
drybox
dyad
e-mail
ebullioscopic
eigenfunction
eigenvalue
electroless
electron microscope
electronvolt
electrooptic
electropositive
eluant
eluate
eluent
Elvehjem
end point
enzymatic
enzymic
Erlenmeyer (flask)
exchangeable
fall off (verb)
falloff (noun, adjective)

far-infrared
faradic (referring to current, not the person)
fax (noun, verb, adjective)
feedback
fiber-optic (adjective)
fiber optics (noun)
filterable
firebrick
flavin
flow sheet
fluoborate
fluoramine
fluoro amine
focused
follow up (verb)
follow-up (noun, adjective)
forbear (verb, to refrain)
forebear (noun, ancestor)
forego (verb, to go before)
foreword (part of a book)
forgo (verb, to do without)
formulas
forward (direction)
freeze-dry (verb)
fulfill
γ ray
gauge
Gaussian
gegenion
glovebag
glovebox
Gouy
graduated cylinder
gram
Gram-negative
Gram-positive
gray
Grignard
groundwater
half-ester
half-life
half-width
halfway
Hamiltonian
Hantzsch
hazmat
heat-treat (verb)
hemoglobin
hemolysate
heterogeneous
Hoffmann degradation
homogeneous
homologue
Hunsdiecker
hydrindan
hydriodic
hydriodide
hydrolysate
hydrolyzed
ice-cold
ice–water bath (use en dash, see p 124)
inasmuch
indan
indexes (book parts, catalog)
indices (mathematical)
indispensable
inflection
infrared
innocuous
inoculate
insofar
inter-ring
intra-ring
iodometric
iodometry
isooctane
isopiestic
isopropyl alcohol (*not* isopropanol)
isosbestic
judgment
Karl Fischer
kayser
Kekulé
Kjeldahl
Kramers
Kugelrohr
labeled
laser
leukocyte
leveling
levorotatory

lifetime
ligancy
ligate
ligated
line shape
line width
liquefy
liter
lumiflavin
luster
lysate
lysed
make up (verb)
makeup (noun)
Markovnikov
matrices (mathematical)
matrixes (media)
megohm
Mendeleev
mesoporphyrin
metalate
metalation
metallization
metallize
metalloenzyme
meter
micro-Kjeldahl
mid-infrared
midpoint
minuscule
mixture melting point
monochromator
Mössbauer
naphthyl
near-ultraviolet
neopentyl
Nernstian
nuclide
occurred
occurrence
occurring
ortho ester
orthoformate
orthohydrogen
orthopositronium
outgas
outgassing
overall
parametrization
path length
percent
Petri
pharmacopeia
phenolphthalein
phlorin
phosphomonoester
phosphorous (as in phosphorous acid)
phosphorus (element)
phthalic
pipet
pipetted
plaster of Paris
point source
porphine
porphyrin
portland cement
programmed
2-propanol (*not* isopropanol)
pseudo-first-order
pyrolysate
quantitation
radio frequency
radioelement
radioiodine
radionuclide
re-form (to form again)
reform (to amend)
repellent
riboflavin
ring-expand (verb)
rotamer
scale up (verb)
scale-up (noun)
scavengeable
Schwarzkopf
seawater
self-consistent
selfsame
Sephadex
set up (verb)
setup (noun)
side arm (noun)

side chain
sideband
siphon
Soxhlet
spin-label (noun)
spin–lattice (use en dash, see p 124)
spin–orbit (use en dash, see p 124)
steam bath
steam-distill (verb)
stepwise
stereocenter
stereopair
stereoptically
Student's *t* test or the Student *t* test
sulfolane
sulfur
superacid
superhigh frequency
supernatant (adjective)
supernate (noun)
syndet
synthase
synthetase
test tube
θ solvent
thiamin
thioacid
thioester
thioether
thioketone
toward
transmetalation
tropin
Ubbelohde
ultrahigh vacuum
un-ionized
uni-univalent
upfield
urethane
van der Waals
VandenHeuvel
van't Hoff–Le Bel (use en dash, see p 124)
Vigreux
vis-à-vis
voltameter (measures voltaic electricity)
voltammeter (measures ranges of volts and amperes)
voltmeter (measures cell potential)
wastewater
wave function
waveform
wavelength
wavenumber
well-known
work up (verb)
workup (noun)
X-irradiation
X-ray
ylide
zerovalent
zigzag
zinc blende

➤➤➤➤➤ CHAPTER 10

# Editorial Style

This chapter presents recommended stylistic and editorial conventions, mainly but not solely for ACS publications. The style recommended by ACS is, for the most part, taken from established authoritative sources, such as *The Chicago Manual of Style*, *Words into Type*, and the *United States Government Printing Office Style Manual*.

Other points of style are discussed in Chapter 11, "Numbers, Mathematics, and Units of Measure"; Chapter 12, "Names and Numbers for Chemical Compounds"; and Chapter 13, "Conventions in Chemistry".

## Hyphenation

Consult a dictionary to resolve hyphenation questions. *Merriam-Webster's Collegiate Dictionary* and *Webster's New World College Dictionary* are the desk dictionaries used by the ACS technical editing staff. ACS staff also use the unabridged *Webster's Third New International Dictionary.*

### *Prefixes*

➤ Most prefixes are not hyphenated. Do not hyphenate the following prefixes when added to words that are not proper nouns.

| | | | |
|---|---|---|---|
| after | bio | de | hetero |
| ante | by | di | homo |
| anti | co | down | hyper |
| auto | counter | electro | hypo |
| bi | cyber | extra | in |

| | | | |
|---|---|---|---|
| infra | mid | poly | techno |
| inter | mini | post | tele |
| intra | mis | pre | thermo |
| intro | mono | pro | trans |
| iso | multi | pseudo | tri |
| macro | nano | re | ultra |
| mega | neo | retro | un |
| meso | non | semi | under |
| meta | over | stereo | uni |
| metalla | peri | sub | up |
| metallo | photo | super | video |
| micro | physico | supra | visco |

EXAMPLES

| | | |
|---|---|---|
| antibacterial | isospin | precooled |
| cooperation | microorganism | pseudomorph |
| cyberspace | multicolored | superacid |
| extranuclear | nonpolar | transactinide |
| interelectrode | photoredox | viscoelastic |

EXCEPTIONS Hyphens are sometimes used (1) when letters are doubled, (2) when more than one prefix is present, or (3) when the unhyphenated form does not convey the intended meaning.

| | | |
|---|---|---|
| anti-infective | inter-ring | post-reorganization |
| anti-inflammatory | intra-ring | post-translational |
| bi-univalent | mid-infrared | pre-equilibrium |
| co-ion | non-native | sub-bandwidth |
| co-worker | non-nuclear | un-ionize |

➤ Some prefixes may be hyphenated or not, depending on meaning.

recollect *or* re-collect
recover *or* re-cover
reform *or* re-form
retreat *or* re-treat

RARE EXCEPTIONS

autoxidation
counter electrode
hetero group
homo nucleoside

➤ Do not hyphenate multiplying prefixes.

hemi, mono, di, tri, tetra, penta, hexa, hepta, octa, ennea, nona, deca, deka, undeca, dodeca, etc.

semi, uni, sesqui, bi, ter, quadri, quater, quinque, sexi, septi, octi, novi, deci, etc.

bis, tris, tetrakis, pentakis, hexakis, heptakis, octakis, nonakis, decakis, etc.

EXAMPLES

2,2′-bipyridine
1,4-bis(3-bromo-1-oxopropyl)piperazine
divalent
hemihydrate
heptacoordinate tetrahedron
hexachlorobenzene
1,1′:3′,1″:3″,1‴-quaterphenyl
tetrakis(hydroxymethyl)methane
triatomic
triethyl phosphate
tris(ethylenediamine)cadmium dihydroxide

➤ Hyphenate a prefix to a two-word compound.

multi-million-dollar lawsuit
non-diffusion-controlled system
non-English-speaking colleagues
non-radiation-caused effects
non-tumor-bearing organ
pre-steady-state condition
pseudo-first-order reaction

➤ Hyphenate prefixes to chemical terms.

non-alkane
non-phenyl atoms

➤ Hyphenate a prefix to a numeral.

pre-1900s

➤ Hyphenate prefixes to proper nouns and adjectives, and retain the capital letter.

anti-Markovnikov
non-Coulombic
non-Gaussian
non-Newtonian
oxy-Cope

## Suffixes

➤ Most suffixes are not hyphenated. Do not hyphenate the following suffixes when added to words that are not proper nouns.

| | | |
|---|---|---|
| able | less | ship |
| fold | like | wide |
| ful | ment | wise |

EXAMPLES

clockwise
fellowship
lifelike
multifold
rodlike
spoonful
statewide
worldwide

EXCEPTIONS

bell-like
gel-like
shell-like

➤ Hyphenate the suffixes "like" and "wide" when they are added to words of three or more syllables.

bacteria-like
computer-like
radical-like
resonance-like
university-wide

➤ Hyphenate the suffix "like" in two-word compounds used as unit modifiers.

first-order-like
ion-exchange-like
rare-earth-like
transition-metal-like

➤ Hyphenate the suffix "like" to chemical names.

adamantane-like
cycloalkane-like
morphine-like
olefin-like

➤ Hyphenate a numeral and a suffix.

10-fold
25-fold

➤ Hyphenate suffixes to proper nouns, and retain the capital letter.

Asia-wide
Claisen-like
Kennedy-like
Michaelis–Menten-like

### *Compound Words*

Compound words are two or more terms used to express a single idea. Compound words in common usage are listed in most dictionaries. Many are hyphenated, but many are not.

back-reaction
cross hairs
cross-link
crosshatch
half-life
self-consistent

➤ Hyphenate spelled-out fractions.

one-half
one-ninth
three-fourths
two-thirds

➤ Hyphenate two-word verbs.

air-dry
flame-seal
freeze-dry
jump-start
ring-expand
vacuum-dry

➤ Do not hyphenate phrasal verbs. As unit modifiers or nouns, these words are often hyphenated or closed up; check a dictionary.

break down
build up
grow up
hand out
line up
mix up
scale up
set off
set up
slow down
stand by
take off
warm up
wear out

➤ Do not hyphenate foreign phrases used as unit modifiers.

ab initio calculation
ad hoc committee
in situ evaluation
in vivo reactions

EXCEPTION Some foreign phrases are hyphenated in the original language, for example, laissez-faire.

➤ People who have double surnames may choose to hyphenate them or use a space between them. When they are hyphenated, use a hyphen, not an en dash, between the two surnames in a person's name. Some combinations of two given names are also hyphenated.

Robert Baden-Powell
David Ben-Gurion
Cecil Day-Lewis
Chen-Chou Fu
Joseph-Louis Gay-Lussac
Irene Joliot-Curie
Jackie Joyner-Kersee
John Edward Lennard-Jones

### *Unit Modifiers*

Unit modifiers are two words that together describe a noun; they are almost always hyphenated. Most unit modifiers consist of

- a noun and an adjective (e.g., time-dependent reaction, radiation-sensitive compound, water-soluble polymer, halogen-free oscillator);
- an adjective and a noun (e.g., high-frequency transition, small-volume method, first-order reaction, outer-sphere redox couple);

- an adjective and a participle (e.g., slow-growing tree, broad-based support, far-reaching influence);
- a noun and a participle (e.g., time-consuming project, earth-shaking news, silver-coated electrode);
- an adverb and an adjective (e.g., above-average results, still-unproven technique); or
- two nouns (e.g., ion-exchange resin, liquid-crystal polymers, transition-state modeling, charge-transfer reaction, gas-phase hydrolysis).

The following is a short list (by no means complete) of unit modifiers commonly seen in ACS publications. These should be hyphenated when modifying a noun.

air-dried
air-equilibrated
back-bonding
$^{14}$C-labeled
charge-transfer
cost-effective
diffusion-controlled
double-bond
electron-diffraction
electron-transfer
energy-transfer
excited-state
first-order
flame-ionization
fluorescence-quenching
free-energy
free-radical
gas-phase
gel-filtration
Gram-negative
Gram-positive
halogen-free
high-energy
high-frequency
high-performance
high-pressure
high-resolution
high-temperature
inner-sphere
ion-exchange
ion-promoted
ion-selective
large-volume
laser-induced
least-squares
light-catalyzed
long-chain
long-lived
low-energy
low-frequency
low-pressure
low-resolution
low-temperature
moisture-sensitive
nearest-neighbor
oil-soluble
outer-sphere
radiation-caused
radiation-produced
radiation-sensitive
rate-limiting
reversed-phase
room-temperature
round-bottom
rubber-lined
second-harmonic
second-order
short-chain
side-chain
size-dependent
small-volume
solid-phase
solid-state
species-specific
steady-state
structure-specific
temperature-dependent
thin-layer
three-dimensional
three-phase
time-dependent
transition-metal
transition-state
two-dimensional
vapor-phase
water-soluble
weak-field
wild-type

EXCEPTIONS

particle size distribution
water gas shift

➤ Hyphenate combinations of color terms used as unit modifiers.

blue-green solution
bluish-purple solid
red-black precipitate
silver-gray body

➤ Do not hyphenate unit modifiers if the first word is an adverb ending in "ly".

accurately measured values
carefully planned experiment
poorly written report
recently developed procedure

➤ Hyphenate unit modifiers containing the adverbs "well", "still", "ever", "ill", and "little".

ever-present danger
ever-rising costs
ill-fitting stopper
little-known hypothesis
still-new equipment
well-known scientist
well-trained assistants

EXCEPTION Do not hyphenate unit modifiers containing the adverbs "well", "still", "ever", "ill", and "little" if they are modified by another adverb.

most ill advised investment
very high density lipoprotein
very well studied hypothesis

➤ Hyphenate unit modifiers containing a comparative or superlative if the meaning could be different without the hyphen.

best-known processes
best-loved advisor
higher-temperature values
least-squares analysis
lowest-frequency wavelengths
nearest-neighbor interaction

➤ Do not hyphenate a number and a unit of time or measure used as a unit modifier.

$1.2 \times 10^{-4}$ $cm^{-1}$ peak
25 K increments
10 mg sample
a 0.1 mol $dm^{-3}$ solution
20 mL aliquot
12° angle

➤ When two or more unit modifiers with the same ending base modify one noun, use a hyphen after each element, and do not repeat the ending base.

first- and second-order reactions
high-, medium-, and low-frequency measurements

➤ Do not hyphenate unit modifiers that are chemical names.

acetic anhydride concentration
amino acid level
barium sulfate precipitate
sodium hydroxide solution

➤ Hyphenate unit modifiers made up of a single letter or number and a noun or adjective.

| | |
|---|---|
| α-helix structure | π-electron system |
| $^{13}$C-enriched proteins | 3-position substitution |
| $^{14}$C-labeling patterns | s-orbital diagrams |
| D-configuration settings | *t*-test analysis |
| γ-ray spectrometer | U-band transmitter |
| 1-isomer profile | *x*-axis labels |
| O-ring suppliers | X-band radar |

➤ Do not hyphenate unit modifiers if one of the words is a proper name.

Fourier transform technique
Lewis acid catalysis
Schiff base measurement

➤ Hyphenate unit modifiers that contain spelled-out numbers.

| | |
|---|---|
| five-coordinate complex | three-neck flask |
| one-electron transfer | three-stage sampler |
| seven-membered ring | two-compartment model |
| three-dimensional model | two-phase system |

➤ Hyphenate unit modifiers that contain a present or past participle.

| | |
|---|---|
| air-equilibrated samples | laser-induced species |
| English-speaking colleagues | methyl-substituted intermediate |
| fluorescence-quenching solution | photon-induced conversion |
| hydrogen-bonding group | rate-limiting step |
| immobilized-phase method | research-related discussion |
| ion-promoted reaction | steam-distilled sample |

CAUTION Watch for cases where the participle forms a unit with the noun that follows: for example, "ligand binding site" should not be hyphenated.

➤ Hyphenate unit modifiers of three or more words.

| | |
|---|---|
| head-to-head placement | root-mean-square analysis |
| high-molecular-weight compound | signal-to-noise ratio |
| nine-membered-ring species | tried-and-true approach |
| out-of-plane distance | voltage-to-frequency converter |

➤ Hyphenate unit modifiers containing three words when similar two-word modifiers are hyphenated.

acid-catalyzed reaction
general-acid-catalyzed reaction
metal-promoted reaction
transition-metal-promoted reaction

EXCEPTION Do not hyphenate unit modifiers containing three or more words, even if similar two-word modifiers are hyphenated, when doing so would break other rules. For example, do not hyphenate unit modifiers if one of the words is a proper name. Do not hyphenate unit modifiers that are two-word chemical names.

acid-catalyzed reactions (*but* Lewis acid catalyzed reactions)

copper-to-iron ratio (*but* sodium chloride to iron ratio)

➤ Hyphenate unit modifiers used as predicate adjectives. (*Predicate adjectives* are usually used with the verb "to be"; they are adjectives that modify the subject but come after the verb.) Usually, only unit modifiers that consist of nouns and adjectives or nouns and participles can be used as predicate adjectives.

All compounds were light-sensitive and were stored in the dark.

In these cluster reactions, dehydrogenation is size-dependent.

The antibody is species-specific.

The complex is square-planar.

The movie was thought-provoking.

The reaction is first-order.

➤ Hyphenate phrases also containing en dashes (see pp 124–126) when they are used as unit modifiers.

alkyl–heavy-metal complexes
high-spin–low-spin transition
metal–metal-bonded complexes
Michaelis–Menten-like kinetics
retro-Diels–Alder reaction
transition-metal–chalcogen complexes

➤ Hyphenate phrases containing parenthetical expressions when they are used as unit modifiers.

element (silicon or tin)-centered radicals

# Capitalization

## *In Text*

Generally, in text keep all words lowercase, including chemical names and terms, except proper nouns and adjectives. However, there are many exceptions.

➤ Capitalize the words "figure", "table", "chart", and "scheme" only when they refer to a specific numbered item.

Chart 4
Figure 1
Schemes 4–7
Table II

➤ Do not capitalize the "r" in "X-ray" at the beginning of a sentence or in a title.

➤ Capitalize parts of a book when they refer to a specific titled and numbered part.

Appendix I
Chapter 3
Section 4.2

But

the appendix
the chapter
the contents
the preface

➤ Capitalize only the name of an eponym, not the accompanying noun.

Avogadro's number
Boltzmann constant
Einstein's theory
Graham's law
Hodgkin's disease
Lewis acid
nuclear Overhauser effects
Raman spectroscopy
Schiff base

EXCEPTIONS

Nobel Peace Prize
Nobel Prize

➤ Capitalize adjectives formed from proper names.

Boolean
Cartesian
Copernican
Coulombic
Darwinian
Einsteinian
Freudian
Gaussian
Hamiltonian
Laplacian
Lorentzian
Mendelian
Newtonian

➤ Capitalize the first word after a colon if the colon introduces more than one complete sentence, a quotation, or a formal statement.

> Chemists find enzymes attractive as potentially useful synthetic tools for many reasons: Enzymes catalyze reactions with high regio- and stereoselectivity. They cause tremendous rate accelerations under mild reaction conditions. They reduce the need for protecting groups and give enantiomerically pure products.

> An emulsion is a thermodynamically unstable system: it has a tendency to separate into two phases.

> Two types of asymmetric reactions were conducted: synthesis of styrene oxide and reduction of olefinic ketones.

> The editor wishes to make the following point: No papers will receive preferential treatment on the basis of artwork.

➤ Do not capitalize lowercase chemical descriptors hyphenated to chemical names when they are at the beginning of a sentence.

> *cis*-4-Chloro-3-buten-2-one was obtained in 74% yield.
>
> *o*-Dichlorobenzene was the solvent.

➤ When the first word of a sentence is a roman chemical descriptor that is not part of a chemical name, capitalize it.

> Cis and trans isomers are used in pharmaceuticals and agrochemicals.
>
> Erythro diols were obtained in good yield.
>
> Syn hydroxylation of cycloalkenes was attempted.
>
> Trans hydroxyl groups are oxidized biochemically.

➤ Do not capitalize chemical names or nonproprietary drug names unless they are at the beginning of a sentence or are in a title or heading. In such cases, capitalize the first letter of the English word, not the locant, stereoisomer descriptor, or positional prefix. (See Chapter 12, "Names and Numbers for Chemical Compounds".)

➤ Some reaction names are preceded by element symbols; they may be used as nouns or adjectives. When they are the first word of a sentence or appear in titles and headings, the first letter of the word is capitalized.

> N-Oxidation of the starting compounds yielded compounds **3–10**.
>
> N-Benzoylated amines undergo hydroxylation when incubated with yeast.
>
> Preparation of S-Methylated Derivatives
>
> O-Substituted Structural and Functional Analogues

➤ When a sentence begins with a symbol that is not hyphenated to the following word, the word is not capitalized.

> $\pi$-Electron contributions are evident.
>
> $\pi$ electrons make significant contributions in this system.
>
> $\sigma$ values were calculated from eq 3.

➤ Always capitalize kingdom, phylum, class, order, family, and genus taxonomic names, as well as names of cultivars. Subclassifications follow the same presentation as the main category.

> Animalia, Planta (kingdom)
> Chordata (phylum)
> Vertebrata (subphylum)
> Mammalia, Reptilia (class)
> Primates, Testudines (order)

Hominidae, Apiaceae (family)
*Homo, Drosophila* (genus)
*Lycopersicon esculentum* Mill. cv. Jennita (cultivar)

➤ Use lowercase for species, subspecies, and varieties, even in titles.

*Escherichia coli*
*Achromobacter haemolyticus* subsp. *alcaligenes*
*Zea mays* var. *rugosa* (variety)

Three New Dihydroisocoumarins from the Greek Endemic Species *Scorzonera cretica*

➤ Do not capitalize the abbreviation for species, singular or plural (sp. or spp., respectively), subspecies (subsp.), variety (var.), or cultivar (cv.).

*Salmonella* sp.
*Polygonum* spp.
*Petroselinum crispum* Mill. subsp. *tuberosum*
*Zea mays* var. *rugosa*
*Lycopersicon esculentum* Mill. cv. Jennita

➤ Do not capitalize genus names used as common nouns except at the beginning of a sentence or in a title or heading.

bacillus
gorilla, a member of the genus *Gorilla*
hippopotamus, a member of the genus *Hippopotamus*
klebsiella
pseudomonad, a member of the genus *Pseudomonas*
streptococcus

➤ Do not capitalize the adjectival or plural form of a genus name unless it is at the beginning of a sentence or in a title or heading.

bacilli
pneumococcal
streptococcal

➤ In text, do not capitalize polymer names that contain the names of the polymerizing species in parentheses following the prefix "poly". At the beginning of a sentence, capitalize only the "P" in "poly".

Poly(vinyl chloride) is a less useful polymer than poly(ethylene glycol).

➤ Capitalize trademarks; use them as adjectives with the appropriate nouns.

| | | |
|---|---|---|
| Ficoll | Pyrex | Teflon |
| Novocain (*but* novocaine) | Sephadex | Triton |
| Plexiglas (*but* plexiglass) | Styrofoam | Tween |

➤ Do not capitalize the word "model" with a number or code.

$\gamma$ counter (Beckman model 5500B)
mass spectrometer (PerkinElmer model 240C)
multichannel spectrometer (Otsuka model MCPD-1000)
spectrometer (Varian model XL-200)
Waters model 660 gradient controller

➤ Do not capitalize the common names of equipment.

dynamic mechanical analyzer
electron-diffraction chamber
flame-ionization detector
gas chromatograph
mass spectrometer
mercury lamp
spectrophotometer
temperature controller unit

➤ Use only an initial capital letter, not all capitals, for company names that are not acronyms, because company names are not trademarks and are not protected by law.

Valspar Corporation
Xerox Corporation

But

EMD Chemicals Inc.
IBM Corporation

➤ Capitalize the names of specific organizations or entities, including ACS local sections, committees, and governing bodies, but not the general terms for them.

| | |
|---|---|
| ACS Board of Directors | the board |
| ACS Committee on Analytical Reagents | an ACS committee |
| ACS Division of Fuel Chemistry | the division |
| American Chemical Society | the society |
| Clean Water Act | the act |
| Environmental Protection Agency | the agency |
| the Milwaukee Section | a local section |
| University of Michigan | the university |

➤ Capitalize the names of specific titles when they appear with a person's name, but not the general terms for them.

the professor
the general
the mayor

Professor Perry Key
General James Shore
Mayor Ralph Estes
Walter Baldwin, Professor of Chemistry

The well-known professor Perry Key will give a tutorial.

James Shore, a general in the U.S. Army, will teach a graduate course.

Our speaker will be the retired general James Shore.

Isaac Bickford is an assistant professor.

Ralph Estes is the mayor of a small town in upstate New York.

➤ Capitalize the names of special events but not the general terms for them.

229th ACS National Meeting
39th ACS Western Regional Meeting
79th ACS Colloid and Surface Science Symposium
the regional meeting
the spring national meeting
the symposium

➤ Capitalize sections of the country but not the corresponding adjectives.

the Midwest, *but* midwestern
the Northeast, *but* northeastern

➤ Do not capitalize the names of the four seasons: summer, fall, autumn, winter, spring.

➤ Capitalize Earth, Sun, and Moon only when used in an astronomical sense.

Venus and Mars are the closest planets to Earth.

The Earth rotates on its axis and revolves around the Sun.

The Moon is the only body that orbits the Earth.

But

Water bodies on the earth's surface contain a variety of chromophoric substances.

Pollution occurs to some extent everywhere on earth.

The sun is the primary source of radiation that can cause chemical transformations.

The next full moon will be on Thursday.

### *In Titles and Headings*

These guidelines apply to titles and headings at all levels; that is, they apply to subtitles and subheadings. They also apply to table, scheme, and chart titles.

➤ In titles and headings that are typeset in capital and lowercase letters, capitalize the main words, which are nouns, pronouns, verbs, adjectives, adverbs, and subordinating conjunctions, regardless of the number of letters. Do not capitalize coordinating conjunctions ("and", "but", "or", "nor", "yet", "so"), articles ("a", "an", "the"), or prepositions. Do capitalize the "to" in infinitives. Do capitalize the first and last words of a title or heading, regardless of part of speech, unless the word is mandated to be lowercase (e.g., pH, d Orbital).

Changes in the Electronic Properties of a Molecule When It Is Wired into a Circuit

Derivatives from a Chiral Borane–Amine Adduct

In Situ Nutrient Analyzer

In Vitro and in Vivo Antiestrogenic Effects of Polycyclic Musks in Zebrafish

Nickel-Catalyzed Addition of Grignard Reagents: Ring-Opening Reactions with Nucleophiles

Phosphonolipids with and without Purified Hydrophobic Lung Surfactant Proteins

Properties of Organometallic Fragments in the Gas Phase

Reactions of Catalyst Precursors with Hydrogen and Deuterium

Scope of the Investigations: The First Phase

The Computer as a Tool To Improve Chemistry Teaching

Vibrations in Situ

EXCEPTION 1 In titles and headings, capitalize small words that are parts of phrasal verbs.

Break Down
Build Up
Set Off
Set Up
Slow Down
Wear Out

EXCEPTION 2 In titles and headings, capitalize small words that are parts of phrasal adjectives.

End-On Bonding
In-Plane Atoms
Side-On Bonding

(*but* Out-of-Plane Vibrations [only the first preposition is capitalized])

➤ In titles and headings, capitalize "as" when it is used as a subordinating conjunction but not when it is used as a preposition.

Alumina as a Catalyst Support

Kinetics of Cyanocobalamin As Determined by Binding Capacity

➤ Do not capitalize the "r" in "X-ray" in titles and headings. Do capitalize the "r" in "$\gamma$ ray" and the "p" in "$\alpha$ particle" and "$\beta$ particle" in titles and headings.

➤ Do not capitalize lowercase chemical descriptors in titles and headings, but do capitalize the first letter of the English word.

Reaction of *trans*-4-(Phenylsulfonyl)-3-buten-2-one

➤ When abbreviated units are acceptable in titles and headings, do not capitalize those that are ordinarily lowercase.

Analysis of Milligram Amounts

Determination of *N*-Nitrosodimethylamine at Concentrations <7 ng/L

➤ Always capitalize genus names, but never capitalize species names, in titles and headings.

Active-Site Nucleophile of *Bacillus circulans* Xylanase

Novel Metabolites of *Siphonaria pectinata*

➤ In titles and headings, capitalize all main words in a unit modifier.

Base-Catalyzed Cyclization
Cross-Linked Polymer
Deuterium-Labeling Experiment
High-Temperature System
Non-Hydrogen-Bonding Molecules
Thyrotropin-Releasing Hormone

➤ In titles and headings, capitalize each component of compound words if the component would be capitalized when standing alone.

Cross-Link
Half-Life
Quasi-Elastic

➤ Do not capitalize hyphenated suffixes.

Synthesis of Cubane-like Clusters

➤ In titles and headings, capitalize only the first letter ("P") of polymer names that contain the names of the polymerizing species in parentheses following the prefix "poly".

IR Spectroscopic Analysis of Poly(1*H*,1*H*-fluoroalkyl α-fluoroacrylate)

Light-Scattering Studies of Poly(ethylene-*co*-butylene)

New Uses for Poly(ethylene terephthalate)

Polystyrene-*block*-poly(2-cinnamoylethyl methacrylate) Adsorption

Reactions of Poly(methyl methacrylate)

Synthesis and Characterization of Poly(isobutylene-*b*-methyl vinyl ether)

➤ Capitalize only the first letter in a chemical name containing complex substituents in parentheses or brackets.

2-[3-[2-[(2*S*)-2-Cyano-1-pyrrolidinyl]-2-oxoethylamino]-3-methyl-1-oxobutyl]-1,2,3,4-tetrahydroisoquinoline: A Potent, Selective, and Orally Bioavailable Dipeptide-Derived Inhibitor of Dipeptidyl Peptidase IV

Preparations of (Methyl isocyanide)iron Compounds

Structures of Tetrakis(methyl isocyanide)iron Complexes

➤ Capitalize parenthetical phrases in titles and headings as if they were not parenthetical.

Versatile Organic (Fullerene)–Inorganic (CdTe Nanoparticle) Nanoensembles

# Surnames

## *Capitalization*

Although a current trend is to lowercase the surnames of persons when these names are used as modifiers and have become very familiar, many are still capitalized. The following is a list (by no means complete) of names that should be capitalized.

Avogadro
Beckmann
Beilstein
Boltzmann
Bragg
Brønsted
Büchner
Bunsen
Claisen
Dewar benzene, flask
Dreiding
Erlenmeyer
Gram
Kekulé
Kjeldahl
Mahalanobis
Markovnikov
Mössbauer
Petri
Poiseuille
Poisson
Priestley
Scatchard
VandenHeuvel

EXCEPTIONS

de Broglie
van der Waals
van't Hoff

➤ Surnames that are used as units of measure are lowercase.

ampere
angstrom
coulomb
curie
dalton
darcy
debye
einstein
faraday
gauss
gilbert
gray
hartree
henry
hertz
joule
kelvin
langmuir
newton
ohm
pascal
poise
siemens
sievert
stokes
tesla
watt
weber

In the temperature–current curves, temperature is given in kelvins and current is shown in amperes.

NMR coupling constants are reported in hertz.

## Hyphenation

➤ Hyphenation of double surnames is discussed on p 139.

➤ Hyphenate prefixes and suffixes to proper names as nouns and adjectives, and retain the capital letter.

anti-Markovnikov
hetero-Diels–Alder
Kennedy-like
Michaelis–Menten-like
non-Coulombic
non-Gaussian
non-Newtonian
oxy-Cope
retro-Diels–Alder

## Foreign Surnames

Some foreign surnames follow a format different from the American system. The Chinese use their surnames first, followed by their given names. For example, Sun Yat-sen's surname is Sun. However, the problem of identifying surnames extends to many other cultures. This multiplicity of usage can create problems in bibliographic indexes and in reference citations. A reference citation in a bibliography should always list the surname first, followed by given name or initials. In a byline, the author names should be presented in standard American format (given names first and surnames last) to ensure consistency of citation practice. If a footnote would clarify the situation or eliminate any perceived confusion, use a footnote.

In most cultures, the surname is the family name, but it may not be the formal name, that is, the name or shortest string of names that are properly used following a title (Mr., Dr., Professor, etc.). Presented here are some cases in which different customs are used for the order of surnames, given names, and formal names. This list is by no means complete, but at least it will help you to be aware of these differences.

ARABIC Often many names; the position of the surname is highly variable. The formal name often consists of two or three names, including articles that can be joined. Examples: Ibn Saud, Abd al-Qadir.

CHINESE Two or three names; the surname is first. Examples: Chiang Kai-shek is Dr. Chiang; Chou En-lai is Dr. Chou.

HEBREW Two or three names; the surname is the last one or two and is the formal name. Examples: David Ben-Gurion is Dr. Ben-Gurion; Moshe Bar-Even is Dr. Bar-Even.

HUNGARIAN Two names; the surname is first, and it is the formal name. However, the second name is accepted as formal internationally.

JAPANESE Two names; the surname is the formal name. The surname is first in Japanese. However, when the names are translated into non-Asian languages, surnames appear last. Example: Taro Yamada is Dr. Yamada.

KOREAN Usually three names; the surname is first and is the formal name. In North Korean names, all three parts start with a capital letter. Examples: Kim Il Sung is Dr. Kim. In South Korean names, the two parts of the given name are hyphenated, and the second part is lowercase. Example: Kim Young-sum is Dr. Kim.

SPANISH Frequently three or more names; the last two are surnames, sometimes connected by "y". The second surname is often dropped or abbreviated. The formal name begins with the first surname and includes the second surname only in very formal usage. Example: Juan Perez Avelar is Dr. Perez or Dr. Perez Avelar, but never Dr. Avelar. The two surnames may also be hyphenated. Example: José Gregorio Angulo Vivas is Dr. Angulo or Dr. Angulo-Vivas, but never Dr. Vivas.

THAI Two names; the surname is last, but the formal name is first.

VIETNAMESE Two or three names; the first is the surname and formal name.

INITIALS Some foreign names are abbreviated with two-letter initials that reflect transliteration from a non-Latin alphabet: Ch., Kh., Ph., Sh., Th., Ts., Ya., Ye., Yu., and Zh.

Some foreign names are abbreviated with hyphenated initials: C.-C. Yu.

## Special Typefaces

Special typefaces help the reader quickly distinguish certain letters, words, or phrases from the rest of the text.

### *Italic Type*

Chapter 11 describes the use of italic type in mathematical material, and Chapters 12 and 13 give guidelines for the use of italic type in chemical names and conventions in chemistry.

➤ Use italic type sparingly to emphasize a word or phrase. Do not use italics for long passages.

➤ Use italic type for a word being defined or for a newly introduced term the first time it appears in text.

> In an *outer-sphere transfer*, an electron moves from reductant to oxidant with no chemical alteration of the primary coordination spheres.
>
> We call these materials *microcapsules.*

➤ Use italic type for the titles and abbreviations of periodicals, books, and newspapers. If "the" is the first word of the title, italicize and capitalize it.

> An article on a promising cholesterol biosynthesis inhibitor appeared in *The Journal of Organic Chemistry* this month.
>
> *Enough for One Lifetime* is the biography of Wallace Carothers.
>
> I read three articles on that new chiral compound in the *Journal of the American Chemical Society.*
>
> *The Washington Post* did a feature story on the president's daughter.

➤ Do not use italic type for common Latin terms and abbreviations.

| | | |
|---|---|---|
| a priori | e.g. | in vitro |
| ab initio | et al. | in vivo |
| ad hoc | etc. | status quo |
| ca. | i.e. | vs |
| de novo | in situ | |

➤ Use italic type for genus, species, subspecies, and variety names of all animals, plants, and microorganisms, but not when these names are used as singular or plural common nouns or when they are adjectival.

> a bacterium of the genus *Salmonella*
>
> *Bacillus coagulans* and *Bacillus dysenteriae* are two species of bacilli.
>
> *Staphylococcus aureus* is the bacterium that causes staphylococcal infection.
>
> *Streptococcus pneumoniae* (genus and species)
>
> The red rhododendron, *Rhododendron arboreum*, needs bright sun.
>
> *Achromobacter haemolyticus* subsp. *alcaligenes* (subspecies)
>
> *Zea mays* var. *rugosa* (variety)

➤ Use italic type for genotypes (representations of genes) and roman type for phenotypes (representations of proteins).

> The *vanA* gene that encodes one such inducible resistance protein is designated VanA.
>
> *E. coli* DH5α Δ(*lacZYA–argF*)U169 *deoR recA1 endA1 hsdR17*($r_K^+$, $m_K^+$) *supE44* $\lambda^-$ *thi-1 gyrA96 relA1* F′ *proAB*$^+$ *lac1*$^q$*Z* Δ M15 ssf::Tn5 [$Km^r$]

➤ Names of restriction endonucleases should follow the typeface conventions of the names from which they were derived: use italic type for the three-letter portion derived from the genus and species name; use roman type for additional strain designators (letters and/or arabic numerals) and for the roman numeral identifiers.

*Bam*HI
*Eco*RI
*Hae*II
*Hae*III
*Hin*dIII
*Sau*3AI
*Sma*I
*Xho*I

EXCEPTION Use roman type for abbreviations of general enzyme types.

Exo III *for* exonuclease III

Pol α *for* DNA polymerase α

Topo II *for* topoisomerase II

➤ Do not use italic type for "pH"; "p" is always lowercase, and "H" is always capitalized.

➤ Do not use italic type for M (molar) or N (normal). Do use italic type for *m* (molal).

### *Greek Letters*

➤ Use Greek letters, not the spelled-out words, for chemical and physical terms. Do not italicize Greek letters.

α helix (*not* alpha helix)

β particle (*not* beta particle)

β sheet (*not* beta sheet)

γ radiation (*not* gamma radiation)

NFκB (nuclear factor κB) (*not* NF kappa B)

EXCEPTIONS

delta opioid receptor
mu opioid receptor

## Computer-Related Usage

➤ Capitalize the first letter of the names of computer languages.

AP
Basic
Cobol
Eiffel
Fortran
Java
Logo
Pascal
Perl
Python
Smalltalk

➤ Capitalize the first letter of the names of programs, and follow the manufacturer's or creator's usage within the name.

| | |
|---|---|
| Acrobat | Microsoft Excel |
| Alchemy | Microsoft PowerPoint |
| ChemDraw | Microsoft Word |
| ChemIntosh | Molecular Presentation Graphics |
| ChemPlus | MULTAN78 |
| EasyPlot | Oracle |
| EndNote | Photoshop |
| FileMaker Pro | ProCite |
| Freehand | SigmaPlot |
| HyperChem | SIMI4A |
| ISIS/Draw | Symphony |
| KaleidaGraph | TK Solver |
| LaTeX | Un-Plot-It |
| Mathematica | UniVersions |
| MathType | WordPerfect |

➤ Use lowercase letters for the spelled-out forms of protocols, except as the first word of a sentence and in titles and headings.

network news-transfer protocol (NNTP)

Appendix 10-1 contains a list of some common computer and Internet terms.

## *Uniform Resource Locators*

A typical uniform resource locator (URL), which is an address on the World Wide Web, takes the following forms:

http://www.domain.zone

http://www.domain.zone/name1/~name2/

http://domain.zone/name1/~name2/name3.html

The number of names varies. For example,

http://www.chemistry.org

http://nobelprize.org/chemistry/index.html

These examples are short, but URLs can be quite long, and in narrative text they often will need to be broken at the end of a line. If the URL does not fit on one line, it can be broken according to the following guidelines:

➤ Break after an ampersand, a slash, or a period, but keep two slashes together.

➤ Do not add a hyphen to the end of the line.

➤ Do not break after a hyphen to avoid confusion as to the hyphen's purpose.

### *E-Mail Addresses*

A typical e-mail address usually takes one of these forms:

personname@companyname.zone
initial_surname@companyname.zone
surnameinitial@companyname.zone

All kinds of variations on the person's name and initials are possible, and besides the underscore, other types of punctuation are used. Long names are often truncated.

➤ Break e-mail addresses in text after the @ or a period. Do not insert a hyphen or any other character.

Chapter 14, "References", presents the editorial style for electronic sources listed in reference lists and bibliographics.

## Trademarks

A trademark is an adjective that describes a material or product (e.g., Teflon resin, Kleenex tissue). The term "brand name" is a synonym for trademark. In ACS publications, do not use the trademark symbols ™ and ® or the service mark symbol ℠. They are not necessary to ensure legal protection for the trademark.

➤ Capitalize trademarks; use them as adjectives with the appropriate nouns. Do not use them in titles.

Ficoll
Novocain (*but* novocaine)
Plexiglas (*but* plexiglass)
Pyrex
Sephadex
Styrofoam
Teflon
Triton
Tween

➤ In general, however, use generic names rather than trade names.

cross-linked dextran polymer beads (*not* Sephadex)
diatomaceous earth (*not* Celite)
4,4′-isopropylidenediphenol (*not* Bisphenol A)
2-methoxyethanol (*not* Methyl Cellosolve)
mineral oil (*not* Nujol)
paclitaxel (*not* Taxol)
petroleum jelly (*not* Vaseline)
photocopy (*not* Xerox)
poly(ethylene glycol) (*not* Carbowax)
tensile testing machine (*not* Instron tester)

➤ Use trademarks as adjectives only, never as nouns or verbs. Because trademarks are adjectives, they do not have plural forms.

# Abbreviations and Acronyms

An abbreviation is a short form of a word; often the individual letters are pronounced. In an acronym, the letters always form a pronounceable word. ACS is an abbreviation; *CASSI* is an acronym.

A list of ACS-recommended abbreviations, acronyms, and symbols is given in Appendix 10-2. Check the list to find an abbreviation. If no abbreviation is listed for the term you are using, you may devise an abbreviation provided that (1) it is not identical to an abbreviation of a unit of measure, (2) it will not be confused with the symbol of an element or a group, (3) it does not hamper the reader's understanding, and (4) you do not use the same abbreviation for more than one spelled-out form.

➤ If a very long name or term is repeated many times throughout a paper, an abbreviation is warranted. Place the abbreviation in parentheses following the spelled-out form the first time it appears in the text. If it is used in the abstract, define it in the abstract and again in the text. After defining the abbreviation in the text, you may use it throughout the paper.

EXCEPTIONS The following list shows abbreviations that never need to be defined. Refer to Appendix 10-2 for all other abbreviations.

| | |
|---|---|
| a.m. | before noon (Latin ante meridiem) |
| anal. | analysis |
| at. wt | atomic weight |
| bp | boiling point |
| ca. | about (Latin circa) |
| cf. | compare (Latin confer) |
| CP | chemically pure |
| DNA | deoxyribonucleic acid |
| e.g. | for example (Latin exempli gratia) |
| ed., eds. | edition, editions |
| Ed., Eds. | Editor, Editors |
| eq(s) | equation(s) [with number(s)] |
| equiv | equivalent(s) [with number(s)] |
| equiv wt | equivalent weight |
| et al. | and others (Latin et alii) |
| etc. | and so forth (Latin et cetera) |
| fp | freezing point |
| GLC | gas–liquid chromatography |
| i.d. | inside diameter |
| i.e. | that is (Latin id est) |
| in. | inch, inches |
| IR | infrared |
| *m* | molal |
| M | molar |

| | |
|---|---|
| mmp | mixture melting point |
| mp | melting point |
| $M_r$ | relative molecular mass (molecular weight) |
| N | normal |
| NMR | nuclear magnetic resonance |
| no., nos. | number, numbers |
| o.d. | outside diameter |
| p, pp | page, pages |
| p.m. | after noon (Latin post meridiem) |
| P.O. | Post Office (with Box and number) |
| ref(s) | reference(s) [with number(s)] |
| RNA | ribonucleic acid |
| sp., spp. | species, singular and plural |
| sp gr | specific gravity |
| sp ht | specific heat |
| sp vol | specific volume |
| U.K. | United Kingdom |
| U.S. | United States |
| USP | United States Pharmacopeia |
| UV | ultraviolet |
| v/v | volume per volume |
| vol | volume |
| vs | versus |
| w/v | weight per volume |
| w/w | weight per weight |
| wt | weight |

➤ Avoid abbreviations in the title of a paper.

➤ For some, but not all, abbreviations, case is important; that is, if they are capitalized, they must never be made lowercase; if they are lowercase, they must never be capitalized. This guideline applies to abbreviations that would lose their meanings or change meanings if their forms were changed, such as units of measure (e.g., mg cannot be changed to Mg, min cannot be changed to Min), mathematical symbols (e.g., pH cannot be changed to PH or ph), and chemical symbols (e.g., *o* for ortho cannot be changed to *O*).

However, if the meaning would not be affected, some abbreviations can be capitalized at the beginning of a sentence and in titles and headings, especially if they are so common that they are more like words than abbreviations. For example, you could use “e-mail” in text and “E-mail” at the beginning of a sentence.

➤ Symbols for the chemical elements are not treated as abbreviations. They need not be defined, and they are typeset in roman type. (*See* Table 13-1 on p 270 f.)

➤ Abbreviate units of measure and do not define them when they follow a number. Without a number, spell them out.

9 V/s or 9 V·s$^{-1}$ (*but* measured in volts per second)

For exceptions, see p 225.

➤ Abbreviations that are common to a specific field may be permitted without identification in books and journals in that field only, at the discretion of the editor.

➤ For genus and species names, spell out the full genus name in the title, in the abstract, and the first time it appears in text. Abbreviate it thereafter with the same species name, but spell it out again with each different species name. Form the abbreviation with the initial of the genus name. If the paper contains more than one genus name that starts with the same initial letter, devise abbreviations that distinguish them. Use italic type for all names and abbreviations.

| FIRST TIME | SUBSEQUENTLY |
|---|---|
| *Bacillus stearothermophilus* | *B. stearothermophilus* |
| *Bacillus subtilis* | *B. subtilis* |
| *Escherichia coli* | *E. coli* |
| *Salmonella typhimurium* | *S. typhimurium* |
| *Staphylococcus aureus* | *Staph. aureus* |

➤ Use "e.g.", "i.e.", "vs", and "etc." only in figure captions, in tables, and in parentheses in text. Elsewhere, spell out "for example", "that is", "versus", and "and so forth".

➤ Do not confuse abbreviations and mathematical symbols. An abbreviation is usually two or more letters; a mathematical symbol should generally be only one letter, possibly with a subscript or superscript. An abbreviation may be used in narrative text but seldom appears in equations; a mathematical symbol is preferred in equations and may also be used in text. For example, in text with no equations, PE may be used for potential energy, but in mathematical text and equations, $E_{\mathrm{p}}$ is preferred. Abbreviations are typeset in roman type; most mathematical symbols are typeset in italic type.

➤ In text, spell out all months with or without a specific day.

On August 3, 1996, we completed the second phase of the experiment.

The final results will be available in January 1997.

➤ Use the following abbreviations (with no periods) or spelled-out forms for months with a day or with a day and year in footnotes, tables, figure captions, bibliographies, and lists of literature cited.

| | | |
|---|---|---|
| Jan | May | Sept |
| Feb | June | Oct |
| March | July | Nov |
| April | Aug | Dec |

➤ Use the abbreviations U.S. and U.K. as adjectives only; spell out United States and United Kingdom as the noun forms in text. Either United Kingdom or U.K. may be used in addresses.

U.K. educational system
educational system in the United Kingdom

U.S. science policy
chemical industry in the United States

➤ Form the plurals of multiletter, all-capital abbreviations and abbreviations ending in a capital letter by adding a lowercase "s" only, with no apostrophe.

HOMOs
JPEGs
PAHs
PCBs
PCs
pHs

➤ To avoid ambiguity or poor appearance, add an apostrophe and a lowercase "s" to form the plurals of lowercase abbreviations, single-capital-letter abbreviations, abbreviations ending in a subscript or superscript, and abbreviations ending in an italic letter.

cmc's
O's (or oxygens; Os is the symbol for osmium)
p$K$'s
p$K_a$'s
$T_g$'s

➤ Use two-letter abbreviations for U.S. state and territory names and the District of Columbia and for Canadian province and territory names on all letters going through the U.S. Postal Service and most express delivery services. Use them after the name of a city in text, footnotes, and references.

U.S. STATES, TERRITORIES, AND THE DISTRICT OF COLUMBIA

| | | | |
|---|---|---|---|
| Alabama | AL | Colorado | CO |
| Alaska | AK | Connecticut | CT |
| American Samoa | AS | Delaware | DE |
| Arizona | AZ | District of Columbia | DC |
| Arkansas | AR | Federated States of Micronesia | FM |
| California | CA | Florida | FL |

| | | | |
|---|---|---|---|
| Georgia | GA | New York | NY |
| Guam | GU | North Carolina | NC |
| Hawaii | HI | North Dakota | ND |
| Idaho | ID | Northern Mariana Islands | MP |
| Illinois | IL | Ohio | OH |
| Indiana | IN | Oklahoma | OK |
| Iowa | IA | Oregon | OR |
| Kansas | KS | Palau | PW |
| Kentucky | KY | Pennsylvania | PA |
| Louisiana | LA | Puerto Rico | PR |
| Maine | ME | Rhode Island | RI |
| Marshall Islands | MH | South Carolina | SC |
| Maryland | MD | South Dakota | SD |
| Massachusetts | MA | Tennessee | TN |
| Michigan | MI | Texas | TX |
| Minnesota | MN | Utah | UT |
| Mississippi | MS | Vermont | VT |
| Missouri | MO | Virgin Islands | VI |
| Montana | MT | Virginia | VA |
| Nebraska | NE | Washington | WA |
| Nevada | NV | West Virginia | WV |
| New Hampshire | NH | Wisconsin | WI |
| New Jersey | NJ | Wyoming | WY |
| New Mexico | NM | | |

CANADIAN PROVINCES AND TERRITORIES

| | | | |
|---|---|---|---|
| Alberta | AB | Nunavut | NU |
| British Columbia | BC | Ontario | ON |
| Manitoba | MB | Prince Edward Island | PE |
| New Brunswick | NB | Quebec | QC |
| Newfoundland and Labrador | NL | Saskatchewan | SK |
| Northwest Territories | NT | Yukon Territory | YT |
| Nova Scotia | NS | | |

➤ Spell out and capitalize “company” and “corporation” as part of company names when they appear in an author’s affiliation. Abbreviate them elsewhere in text. After the first mention, drop Co. and Corp. and use only the company name.

➤➤➤➤➤

APPENDIX 10-1

# Computer and Internet Terms

This appendix lists the spelling, capitalization, and abbreviations of some common computer and Internet terms. This list is not intended to be exclusive. Alternative choices, in many cases, are acceptable. Proscribed usages are specifically indicated.

An excellent source for definitions of these terms may be found at http://www.google.com/. In the search statement, use the syntax "define:xyz" where xyz is the term for which a definition is sought.

active matrix
ADSL (asymmetric digital subscriber line)
AI (artificial intelligence)
anonymous FTP
applet
application
archive
ASCII (American Standard Code for Information Interchange)
ASP (application service provider)
asynchronous
AVI (audio video interleaved)

back up (verb)
backup (noun, adjective)
bandwidth
Base 64
batch processing
baud
baud rate
BBS (bulletin board system)
BinHex (binary hexadecimal)
BIOS (basic input/output system)
bit
bitmap
Bitnet (Because It's There NETwork)
bitstream
blog, blogger
bookmark
boot
bounce (e-mail)
bps (bits per second)
broadband
browser
bulletin board
byte

C (programming language)
C++ (programming language)
cable modem
cache
CAD (computer-assisted design)
CAD/CAM (computer-assisted design and manufacturing)
CCD (charge-coupled device)
CD (compact disc)
CD key
CD-R (compact disc read-only)
CD-ROM (compact disc with read-only memory)
CD-RW (compact disc read–write)
CDMA (code division multiple access)
CGI (common gateway interface)
CGM (computer graphics metafile)
CIF (crystallographic information file)
codec
compact disc (CD)
compiler
CPU (central processing unit)
CRT (cathode ray tube)

CSS (cascading style sheet)
cursor
CVC (color video controller)
cyberspace

daemon
data domain
data log
data parse, data parsing
data processing
data set
database
DBMS (database management system)
debug (verb)
default
defragment
desktop
DHCP (dynamic host configuration protocol)
DHTML (dynamic hypertext markup language)
dialog box
digital signal
directory
disc (compact disc only)
disk (floppy or hard disk)
disk drive
disk space
diskette
DNS (domain name system or server)
domain name
DOI (digital object identifier)
DOS (disk operating system)
double-click (as verb)
download
dpi (dots per inch)
drag and drop
DSL (digital subscriber line)
DTD (document-type definition)
duplex
DVD (digital video disc)
DVD-R (digital video disc read-only)
DVD-RW (digital video disc read–write)

e-book
e-journal
e-mail (electronic mail)
e-money
e-publish
e-zine (electronic magazine)
EBCDIC (Extended Binary-Coded Decimal Interchange Code)
emoticons
encryption
end user (noun)
EPS (encapsulated PostScript)
Ethernet (*but* an ethernet)
extranet

FAQ (frequently asked question)
FDDI (fiber distributed data interface)
fiber optics
file compression
file name
filter
finger
Firefox
firewall
FireWire
flash memory
floppy disk
flowchart
format, formatted, formatting
Fortran
FreeNet (*but* a freenet)
freeware
front end
FTP (file transfer protocol)

gateway
GB (gigabyte, equal to 1024 megabytes; always a space between number and GB)
GDDM (graphical data display manager)
GDI (graphics device interface)
GIF (graphics interchange format)
4GL (fourth-generation language)
glyph
Google
GPIB (general purpose interface bus)
graphic (noun)
graphical interface

graphics (adjective)
graphics conversion
graphics files
graphics terminal emulation
GUI (graphical user interface)

hard disk
hard disk drive
hardware
hardwired
high-level-language compiler
home directory
home page (lowercase, but capitalized when part of a specific name, e.g., ACS Home Page)
hot key
hotline
HTML (hypertext markup language)
HTTP (hypertext transfer protocol)
HTTPS (hypertext transfer protocol secure)
hyperlink
hypermedia
hypertext

IBM-compatible
icon
iconization
iconize
IM (instant messaging)
IMAP (Internet message access protocol)
information superhighway
input
integrated circuit
interdomain conversion
Internet
intranet
I/O (input/output)
IP (Internet protocol)
IP address
IRC (Internet relay chat)
ISDN (integrated services digital network)
ISP (Internet service provider)

Java
Javascript
JDK (Java Development Kit)
joystick
JPEG (Joint Photographic Experts Group)

K (kilobyte, equal to 1024 bytes; always closed up to number; as in 8K or 16K disk drive; kB is preferred)
kB (kilobyte, equal to 1024 bytes; always a space between number and kB)
KB (kilobyte; kB is preferred)
kbps (kilobits per second)
kBps (kilobytes per second)
keyboard
keypad
keystroke
kilobit
kilobyte

LAN (local area network)
laptop
LaTeX (pronounced "lahtek")
LCD (liquid-crystal display)
Lexis
LexisNexis
list-administration software
list-management software
list owner
list server
Listserv (software)
local area network (LAN)
log in, logging in (verb)
log off, logging off (verb)
log on, logging on (verb)
log out, logging out (verb)
login name
logon name

Macintosh, Macintoshes
macro, macros
mainframe
math coprocessor
MB (megabyte, equal to 1024 kilobytes; always a space between number and MB)
megapixel
meta-list

metadata
microchip
microcomputer, microcomputing
microprocessor
Microsoft Excel
Microsoft PowerPoint
Microsoft Windows
Microsoft Word
MIDI (musical instrument digital interface)
MIME (multipurpose Internet mail extension)
minicomputer
minifloppy disk
modem
monitor (noun)
motherboard
motif
mouse (plural: mouse devices)
Mozilla
MP3 (MPEG audio layer 3)
MPEG (Motion Picture Experts Group)
MS-DOS (Microsoft disk operating system, always hyphenated)
MTA (mail-transfer agent)
MUA (mail-user agent)
multitasking

NCP (network control program)
NCSA (National Center for Supercomputing Applications)
Net (when referring to the Internet; lowercase when referring to any network)
netiquette
netizen
Netscape
netware
network
newsgroup
NIC (Network Information Center)
NNTP (network news transfer protocol)
node, nodes
NREN (National Research and Education Network)

OCR (optical character recognition)
ODBC (open database connectivity)
off-site (always hyphenated)
offline (one word in computer context)
on-site (always hyphenated)
online (one word in computer context)
open source
OS (operating system)
OSX (Macintosh Operating System X)
output

PageMaker
PAM (pulse amplitude modulation)
parallel port
parser
password
path
PC (personal computer)
PCMCIA (Personal Computer Memory Card International Association)
pdb (Protein Data Bank) format
PDF (portable document format)
Perl (programming language)
PHP (personal home page)
PIF (picture interchange format)
pixel
plaintext
plug and play
plug-in
PNG (portable network graphics)
POP (post office protocol)
popup
PostScript
PPP (point-to-point protocol)
primary domain
print queue
programmer, programming
PROM (programmable read-only memory)
protocol
proxy server
PSTN (public switched telephone network)
pull-down (adjective)

QuarkXPress
queue
QuickTime

RAM (random-access memory)
raster, rasterize

RDBMS (relational database management system)
read/write permission
real time (noun)
real-time (unit modifier)
reboot
RFC (request for comments)
rich text
RJE (remote job entry)
ROM (read-only memory)
router
RPG (report program generator)
RSS (rich site summary)
RTF (rich text format)
run time (noun)
run-time (adjective)

Safari
scale up (verb)
scanner
screen dump
script
SCSI (small computer system interface, pronounced "skuzzy")
search engine
security certificate
serial communication
serial port
server
servlet
set up (verb)
setup (noun)
SGML (standard generalized markup language)
shared user
shareware
shortcut
shut down (verb)
shutdown (noun, adjective)
sign off (verb)
sign-off (noun, adjective)
simplex
SLIP (serial-line Internet protocol)
SMB (server message block)
SMDS (switched multimegabit data service)
smiley (the :) or ☺ symbol)
SMTP (simple mail transfer protocol)
SNMP (simple network management protocol)
SOAP (simple object access protocol)
software
source code
spam
spelling checker
spreadsheet
spyware
SQL (structured query language)
SSL (secure sockets layer)
stand-alone (always hyphenated)
start up (verb)
startup (noun)
strikethrough
submenu
SVGA (super video graphics adapter)
systems programs

T-1, T-3
TB (terabyte)
Tcl (programming language, pronounced "tickle")
TCP (transmission control protocol)
TCP/IP (transmission control protocol/ Internet protocol)
telecommute
Telnet
terminal emulation program
terminal server
TeX (pronounced "tek")
TFT (thin-film transistor)
throughput
TIFF (tagged image file format)
time-sharing (always hyphenated)
TLD (top-level domain)
toolbar
toolbox
trackball
Trojan horse
TTL (transistor–transistor logic)
TTY (teletype)

UDP (user datagram protocol)
UGA (ultra graphics accelerator)
Unicode

Unify
Unix
upload
UPS (uninterruptible power source)
URC (uniform resource characteristic)
URI (uniform resource identifier)
URL (uniform resource locator)
URN (uniform resource name)
USB (universal serial bus)
Usenet
user id, user ids
utility program
uuencode (Unix-to-Unix encoding)

vector, vectorize
VGA (video graphics adapter)
video adapter
Visual Basic
VoIP (voice over Internet protocol)
VPN (virtual private network)
VRML (virtual reality modeling language)

W3C (World Wide Web Consortium)
WAIS (wide-area information service or server)
WAN (wide-area network)
the Web
Web browser
Web page
Web server
Web site
Webmaster
Webzine
Wi-Fi (wireless fidelity)
window (general term, not specific program)
Windows (Microsoft program)
WinZIP
word-processing software
word processor
WordPerfect
wordwrap
workstation
World Wide Web (three words, no hyphens)
World Wide Web Consortium (W3C)
worm
WORM (write once, read many times)
WWW (World Wide Web)
WYSIWYG (what you see is what you get)

Xbase
Xenix
XML (extensible markup language)

ZIP archive

➤➤➤➤➤

APPENDIX 10-2

# Abbreviations, Acronyms, and Symbols

This list is not intended to be exclusive. Alternative choices, in many cases, are acceptable. Proscribed usages are specifically indicated. If no abbreviation is listed for the term you are using, you may devise an abbreviation provided that (1) it is not identical to an abbreviation of a unit of measure, (2) it will not be confused with the symbol of an element or a group, (3) it does not hamper the reader's understanding, and (4) you do not use the same abbreviation for more than one spelled-out form.

| | |
|---|---|
| $\alpha$ | fine structure constant |
| | optical rotation |
| | stereochemical descriptor |
| $[\alpha]^t_{\mathrm{D}}$ | specific rotation at temperature *t* and wavelength of sodium D line |
| $[\alpha]^t_\lambda$ | specific rotation at temperature *t* and wavelength $\lambda$ |
| a | antisymmetric |
| | are (unit of area, 100 $m^2$) |
| | atto ($10^{-18}$) |
| | axial [*use* 2(a)-methyl in names] |
| *a* | *a* axis |
| | absorptivity |
| | axial chirality [as in (*aR*)-6,6′-dinitrodiphenic acid] |
| $a_0$ | Bohr radius (0.52917 Å) |
| A | adenosine |
| | alanine |
| | ampere |
| | ring (italic in steroid names) |
| Å | angstrom |
| *A* | absorbance [as in $A = \log(1/T)$] |
| | anticlockwise (chirality symbol) |
| | Helmholtz energy |
| | mass number |
| A.D. | anno Domini |
| a.m. | ante meridiem |
| AAS | atomic absorption spectroscopy |
| abs | absolute |
| ac | alternating current |
| *ac* | anticlinal |

| | |
|---|---|
| Ac | acetyl |
| | actinium |
| acac | acetylacetonato (ligand) |
| acam | acetamide (ligand) |
| AcCh | acetylcholine |
| AcChE | acetylcholinesterase |
| AcO | acetate |
| ACTH | adrenocorticotropin; adrenocorticotropic hormone |
| Ade | adenine |
| Ado | adenosine |
| ADP | adenosine 5′-diphosphate |
| AEM | analytical electron microscopy |
| AES | atomic emission spectroscopy |
| | Auger electron spectroscopy |
| af | audio frequency |
| AFM | atomic force microscopy |
| AFS | atomic fluorescence spectroscopy |
| AGU | anhydroglucose unit |
| *ala* | alanyl in genetics |
| Ala | alanyl, alanine |
| *alt* | alternating, as in poly(A-*alt*-B) |
| AM | amplitude modulation |
| AMP | adenosine 5′-monophosphate, adenosine 5′-phosphate |
| amu | atomic mass unit [amu, reference to oxygen, is deprecated; u (reference to mass of $^{12}$C) should be used] |
| anal. | analysis (Anal. in combustion analysis presentations) |
| anhyd | anhydrous |
| ANN | artificial neural network |
| ANOVA | analysis of variance |
| Ans | ansyl |
| ansyl | 8-anilino-1-naphthalenesulfonyl |
| antilog | antilogarithm |
| AO | atomic orbital |
| *ap* | antiperiplanar |
| AP | appearance potential |
| APIMS | atmospheric pressure ionization mass spectrometry |
| APS | appearance potential spectroscopy |
| aq | aqueous |
| $A_r$ | relative atomic mass (atomic weight) |
| Ar | aryl |
| AR | analytical reagent (e.g., AR grade) |
| Ara | arabinose |
| *ara*-A | adenosine, with arabinose rather than ribose (arabinoadenosine, also ara-A, araA) |
| *ara*-C | cytidine, with arabinose rather than ribose (arabinocytidine, also ara-C, araC) |
| arb unit | arbitrary unit (clinical) |

| | |
|---|---|
| Arg | arginyl, arginine |
| ARPES | angle-resolved photoelectron spectroscopy |
| ARPS | angle-resolved photoelectron spectroscopy |
| *as* | asymmetrical |
| AS | absorption spectroscopy |
| Asa | β-carboxyaspartic acid |
| ASIS | aromatic solvent-induced shift |
| Asn | asparaginyl, asparagine |
| Asp | aspartyl, aspartic acid |
| Asx | "Asn or Asp" |
| *asym* | asymmetrical |
| at. wt | atomic weight |
| ATCC | American Type Culture Collection |
| atm | atmosphere |
| atom % | atom percent |
| ATP | adenosine 5′-triphosphate |
| ATPase | adenosinetriphosphatase |
| ATR | attenuated total reflection |
| au | atomic unit |
| AU | absorbance unit |
| | astronomical unit (length) |
| AUFS | absorbance units at full scale |
| av | average |
| | |
| β | stereochemical descriptor |
| b | barn (neutron capture area, $10^{-24}$ $cm^2$) |
| | bohr (unit of length) |
| | broad or broadened (spectra) |
| *b* | *b* axis |
| | block, as in poly(A-*b*-B) |
| B | "aspartic acid or asparagine" |
| | bel |
| | buckingham ($10^{-26}$ esu $cm^2$) |
| | ring (italic in steroid names) |
| *B* | boat (conformation) |
| B.C. | before Christ |
| B.C.E. | before the common era |
| b.i.d. | twice a day |
| bar | unit of pressure; unit and abbreviation are the same |
| bbl | barrel |
| bcc | body-centered cubic (crystal structure) |
| bccub | body-centered cubic (crystal structure) |
| BCD | binary coded decimal |
| Bd | baud |
| BEHP | bis(2-ethylhexyl) phthalate |
| BET | Brunauer–Emmett–Teller (adsorption isotherm) |
| BeV | billion electronvolts |

| | |
|---|---|
| bGH | bovine growth hormone |
| Bi | biot |
| binap | 2,2′-bis(diphenylphosphino)-1,1′-binaphthyl (ligand) |
| binol | 1,1′-bi-2-naphthol (ligand) |
| biol | biological(ly) |
| bipy | 2,2′-bipyridine, 2,2′-bipyridyl (bpy preferred) |
| | 4,4′-bipyridine, 4,4′-bipyridyl (bpy preferred) |
| bis-Tris | [bis(2-hydroxyethyl)amino]tris(hydroxymethyl)methane (also bistris, Bis-Tris, bis-tris) |
| bit | binary digit |
| BL | bioluminescence |
| BM | Bohr magneton (*use* $\mu_B$) |
| Bn | benzyl (also Bzl) |
| BN | bond number |
| BO | Born–Oppenheimer |
| BOD | biological oxygen demand |
| bp | base pair |
| | boiling point |
| bps | bits per second |
| Bps | bytes per second |
| bpy | 2,2′-bipyridine, 2,2′-bipyridyl |
| | 4,4′-bipyridine, 4,4′-bipyridyl |
| *BPY* | bipyramidal (coordination compounds) |
| Bq | becquerel |
| br | broad or broadened (spectra) |
| BSA | bovine serum albumin |
| BSSE | basis set superposition error |
| Btu | British thermal unit |
| bu | bushel |
| Bu | butyl |
| BWR | Benedict–Webb–Rubin (equation) |
| Bz | benzoyl |
| Bzac | benzoylacetone |
| Bzl | benzyl (also Bn) |

| | |
|---|---|
| $\chi$ | magnetic susceptibility |
| c | candle |
| | centered (crystal structure) |
| | centi ($10^{-2}$) |
| | cyclo [as in c-$C_6H_{11}$, c-Hx (cyclohexyl)] |
| *c* | *c* axis |
| | concentration, for rotation, e.g., $[\alpha]^{20}_{489}$ +25 (*c* 0.13, $CHCl_3$) |
| | *cyclo* [as in *c*-$S_6$ (*cyclo*-hexasulfur)] |
| | specific cytochrome (i.e., cytochrome *c*) |
| C | Celsius (*use* °C as unit abbreviation) |
| | coulomb |
| | cysteine |

| | |
|---|---|
| C | cytidine |
| | ring (italic in steroid names) |
| *C* | chair (conformation) |
| | clockwise (chirality symbol) |
| C.E. | common era |
| $c/m^2$ | candles per square meter |
| $^{13}C$ NMR | carbon nuclear magnetic resonance |
| ca. | circa, about [used before an approximate date or figure (ca. 1960)] |
| CAD | computer-assisted design |
| cal | calorie |
| $cal_{IT}$ | International Table calorie |
| calcd | calculated |
| CAM | computer-assisted manufacturing |
| cAMP | adenosine cyclic 3′,5′-phosphate |
| | adenosine 3′,5′-cyclic phosphate |
| CAN | ceric ammonium nitrate |
| CARS | coherent anti-Stokes Raman spectroscopy |
| CAT | computed axial tomography |
| | computer-averaged transients |
| cB | conjugate base, counterbase (also CB) |
| Cbz | carbobenzoxy, carbobenzyloxy, (benzyloxy)carbonyl, benzyloxycarbonyl (also Z) |
| cc | cubic centimeter (*do not use*; *use* $cm^3$ or mL) |
| CCD | charge-coupled device |
| CCGC | capillary column gas chromatography |
| ccp | cubic close-packed (crystal structure) |
| cd | candela |
| | current density |
| CD | circular dichroism |
| CDH | ceramide dihexoside [$Cer(Hex)_2$] |
| cDNA | complementary DNA |
| CDP | cytidine 5′-diphosphate |
| CE | Cotton effect |
| CE–MS | capillary electrophoresis–mass spectrometry |
| cf. | compare |
| CFC | chlorofluorocarbon |
| cfm | cubic feet per minute |
| CFSE | crystal field stabilization energy (also cfse) |
| cfu | colony-forming units (bacterial inocula) |
| cgs | centimeter–gram–second (as in cgs system) |
| cgsu | centimeter–gram–second unit(s) |
| ChE | cholinesterase |
| CHF | coupled Hartree–Fock |
| Ci | curie |
| CI | chemical ionization |
| | configuration interaction |
| CIDEP | chemically induced dynamic electron polarization |

| | |
|---|---|
| CIDNP | chemically induced dynamic nuclear polarization |
| CIMS | chemical ionization mass spectrometry |
| CL | cathodoluminescence |
| | chemiluminescence |
| CM | carboxymethyl (as in CM-cellulose) |
| cmc | critical micelle concentration |
| CMH | ceramide monohexoside [Cer(Hex)] |
| CMO | canonical molecular orbital |
| CMP | cytidine 5′-monophosphate, cytidine 5′-phosphate |
| cmr | carbon magnetic resonance (*do not use*; *use* $^{13}$C NMR) |
| CMR | carbon magnetic resonance (*do not use*; *use* $^{13}$C NMR) |
| CN | coordination number |
| CNDO | complete neglect of differential overlap |
| CNS | central nervous system |
| *co* | copoly (as in A-*co*-B) |
| CoA | coenzyme A |
| cod | 1,5-cyclooctadiene (ligand) |
| COD | chemical oxygen demand |
| coeff | coefficient |
| colog | cologarithm |
| compd | compound |
| con | conrotatory (may be italic) |
| concd | concentrated |
| concn | concentration |
| const | constant |
| cor | corrected |
| cos | cosine |
| cosh | hyperbolic cosine |
| COSY | correlation spectroscopy |
| cot | cotangent |
| | 1,3,5,7-cyclooctatetraene (ligand) |
| coth | hyperbolic cotangent |
| counts/s | counts per second |
| $C_p$ | heat capacity at constant pressure |
| cp | candlepower |
| cP | centipoise |
| Cp | cyclopentadienyl |
| CP | chemically pure |
| | cross-polarization |
| CP/MAS | cross-polarization/magic-angle spinning (also CP-MAS, CP–MAS, CPMAS, CP MAS) |
| cpd | contact potential difference |
| CPE | controlled-potential electrolysis |
| CPK | Corey–Pauling–Koltun (molecular models) |
| | creatine phosphokinase |
| CPL | circular polarization of luminescence |
| cpm | counts per minute |

| | |
|---|---|
| cps | counts per second (*use* counts/s) |
| | cycles per second (*use* Hz or $s^{-1}$) |
| CRAMPS | combined rotation and multiple-pulse spectroscopy |
| CRIMS | chemical reaction interface mass spectrometry |
| crit | critical |
| cRNA | complementary RNA |
| CRT | cathode ray tube |
| CRU | constitutional repeating unit |
| cryst | crystalline |
| csc | cosecant |
| csch | hyperbolic cosecant |
| CT | charge transfer |
| CTEM | conventional transmission electron microscopy |
| CTH | ceramide trihexoside [$Cer(Hex)_3$] |
| CTP | cytidine 5′-triphosphate |
| *CU*-8 | cubic, coordination number 8 |
| cub | cubic (crystal structure) |
| $C_v$ | heat capacity at constant volume |
| CV | coefficient of variation |
| | cyclic voltammetry |
| CVD | chemical vapor deposition |
| CW | constant width |
| | continuous wave (as in CW ESR) |
| cwt | hundredweight |
| Cy | cyclohexyl |
| cyclam | 1,4,8,11-tetraazacyclotetradecane |
| Cyd | cytidine |
| Cys | cysteinyl, cysteine |
| cyt | cytochrome |
| Cyt | cytosine |
| cytRNA | cytoplasmic RNA |
| CZE | capillary zone electrophoresis |
| | |
| δ | NMR chemical shift in parts per million downfield from a standard |
| ∂ | partial differential |
| d | day (spelled-out form is preferred) |
| | deci ($10^{-1}$) |
| | deoxy |
| | deuteron |
| | differential (mathematical) |
| | diffuse |
| | doublet (spectra) |
| d. | diameter, with i. and o. (inside and outside) |
| *d* | density |
| | dextrorotatory |
| | distance |
| | spacing (X-ray) |

| | |
|---|---|
| D | absolute configuration |
| D | aspartic acid |
| | debye |
| | deuterium |
| | ring (italic in steroid names) |
| *D* | diffusion coefficient |
| | symmetry group [e.g., $D_3$; also used in names, such as (+)-$D_3$-trishomocubane] |
| 2-D | two-dimensional (also 2D) |
| 3-D | three-dimensional (also 3D) |
| da | deca or deka (10) |
| Da | dalton |
| daf | dry ash free |
| dAMP | 2′-deoxyadenosine 5′-monophosphate or phosphate (the A can be replaced with C, G, U, etc.) |
| dansyl | 5-(dimethylamino)-1-naphthalenesulfonyl |
| dB | decibel |
| dc | direct current |
| *DD*-8 | dodecahedral, coordination number 8 |
| DD NMR | dipolar decoupling NMR |
| DDT | 1,1,1-trichloro-2,2-bis(*p*-chlorophenyl)ethane |
| de | diastereomeric excess |
| DEAE | (diethylamino)ethyl (as in DEAE-cellulose) |
| dec | decomposition |
| decomp | decompose |
| DEFT | driven equilibrium Fourier transform |
| deg | degree (*use* °B, degrees Baumé; °C, °F, *but* K) |
| DEG | diethylene glycol |
| DEHP | bis(2-ethylhexyl) phthalate (BEHP is preferred) |
| DES | diethylstilbestrol |
| det | determinant |
| df | degrees of freedom |
| DF | degrees of freedom |
| DFT | density functional theory |
| diam | diameter |
| dil | dilute |
| dis | disrotatory (may be italic) |
| distd | distilled |
| DLVO | Derjaguin–Landau–Verwey–Overbeek |
| DMA | dynamic mechanical analyzer |
| DMBA | 9,10-dimethylbenz[*a*]anthracene |
| DME | 1,2-dimethoxyethane |
| | dropping mercury electrode |
| DMEM | Dulbecco's modified Eagle's medium |
| DMF | dimethylformamide |
| DMN | diaminomaleonitrile |
| dmr | deuterium magnetic resonance (*do not use; use* $^2$H NMR) |

| | |
|---|---|
| DMR | deuterium magnetic resonance (*do not use; use* $^2$H NMR) |
| DMSO | dimethyl sulfoxide (also $Me_2SO$) |
| DMTA | dynamic mechanical thermal analyzer |
| DNA | deoxyribonucleic acid |
| DNase | deoxyribonuclease |
| DNMR | dynamic nuclear magnetic resonance |
| DNP | deoxynucleoprotein |
| | dynamic nuclear polarization (NMR) |
| DNPH | (2,4-dinitrophenyl)hydrazine |
| Dns | dansyl |
| Dopa | 3-(3,4-dihydroxyphenyl)alanine (also DOPA) |
| DP | degree of polymerization (also dp) |
| dpm | disintegrations per minute |
| DPPH | 2,2-diphenyl-1-picrylhydrazyl |
| dps | disintegrations per second |
| *Dq* | crystal field splittings |
| DQF | double quantum filtered |
| DRIFT | diffuse reflectance Fourier transform |
| *Ds* | crystal field splittings |
| DSC | differential scanning calorimetry |
| dT | thymidine |
| *Dt* | crystal field splittings |
| DTA | differential thermal analysis |
| DTC | depolarization thermocurrent |
| | differential thermal calorimetry |
| dTDP | thymidine 5′-diphosphate |
| DTE | dithioerythritol |
| dThd | thymidine |
| dTMP | thymidine 5′-monophosphate, thymidine 5′-phosphate |
| DTT | dithiothreitol |
| dTTP | thymidine 5′-triphosphate |
| dyn | dyne |
| | |
| $\varepsilon$ | dielectric constant |
| | molar absorptivity |
| $\varepsilon^*$ | complex permittivity |
| e | base of natural logarithm |
| | electron |
| | equatorial [in names, e.g., 2(e)-methyl] |
| $e_{aq}^-$ | hydrated electron |
| $e^-(aq)$ | hydrated electron |
| $e_s^-$ | solvated electron |
| $e^-(s)$ | solvated electron |
| *e* | electronic charge |
| E | exa ($10^{18}$) |
| | glutamic acid |

| | |
|---|---|
| *E* | electromotive force |
| | energy |
| | entgegen (configuration) |
| | envelope (conformation) |
| | potential energy |
| | specific extinction coefficient ($E_{280nm}^{1\%,1cm}$) |
| | Young's modulus |
| $E°$ | standard electrode potential |
| | standard electromotive force |
| $E_{1/2}$ | half-wave potential |
| E1 | first-order elimination |
| E2 | second-order elimination |
| e.g. | for example |
| $E_a$, $E_A$ | Arrhenius or activation energy |
| $ea_0$ | electronic charge in electrostatic units × Bohr radius or atomic units for dipole moment |
| EC | exclusion chromatography |
| ECD | electron-capture detector, detection |
| ECE | electrochemical, chemical, electrochemical (mechanisms) |
| ECG | electrocardiogram |
| ecl | electrochemical luminescence |
| ECL | electrochemical luminescence |
| ECP | effective core potential |
| ed. | edition, edited |
| Ed. | editor |
| ED | effective dose |
| $ED_{50}$ | dose that is effective in 50% of test subjects (also ED50) |
| edda | ethylenediaminediacetato (ligand) |
| eds. | editions |
| Eds. | editors |
| EDS | energy-dispersive system (or spectrometry) |
| edta | ethylenediaminetetraacetato (ligand) |
| EDTA | ethylenediaminetetraacetic acid, ethylenediaminetetraacetate |
| EDXS | energy-dispersive X-ray spectrometry |
| ee | enantiomeric excess |
| EEG | electroencephalogram |
| EELS | electron energy loss spectroscopy |
| EFG | electric field gradient |
| EGA | evolved gas analysis |
| EGD | evolved gas detection |
| EGR | exhaust gas recirculation |
| $E_h$ | hartree (unit); Hartree energy |
| EH | extended Hückel |
| EI | electron impact |
| | electron ionization |
| EIA | enzyme immunoassay |
| $E_k$ | kinetic energy |

| | |
|---|---|
| EKC | electrokinetic chromatography |
| EKG | electrocardiogram |
| EL | electroluminescence |
| ELISA | enzyme-linked immunosorbent (immunoadsorbent) assay |
| *e/m* | ratio of electron charge to mass |
| EM | electron microscopy |
| e-mail | electronic mail |
| EMC | equilibrium moisture content |
| emf | electromotive force |
| EMIS | electromagnetic isotope separation |
| emu | electromagnetic unit |
| en | ethylenediamine (ligand) |
| ENDOR | electron–nuclear double resonance |
| *ent* | reversal of stereocenters |
| $E_p$ | potential energy |
| EPA | ether–isopentane–ethanol (solvent system) |
| epi | inversion of normal configuration (italic with a number, as in 15-*epi*-prostaglandin A) |
| EPMA | electron probe microanalysis |
| EPR | electron paramagnetic resonance |
| EPXMA | electron probe X-ray microanalysis |
| eq | equation |
| equiv | equivalent |
| equiv wt | equivalent weight |
| erf | error function |
| erfc | error function complement |
| $erfc^{-1}$ | inverse error function complement |
| ESCA | electron spectroscopy for chemical analysis |
| esd | estimated standard deviation |
| ESE | electron spin echo |
| ESEEM | electron spin echo envelope modulation |
| ESI | electrospray ionization |
| ESIMS | electrospray ionization mass spectrometry |
| ESP | elimination of solvation procedure |
| ESR | electron spin resonance |
| esu | electrostatic unit |
| Et | ethyl |
| et al. | and others |
| etc. | and so forth |
| eu | entropy unit |
| EU | enzyme unit |
| eV | electronvolt |
| EXAFS | extended X-ray absorption fine structure |
| exch | exchangeable (spectra) |
| exp | exponential |
| expt | experiment |
| exptl | experimental |

| | |
|---|---|
| f | and page following (as in p 457 f) |
| | femto ($10^{-15}$) |
| | fermi (unit of length, also fm) |
| | fine (spectral) |
| *f* | focal length |
| | frequency (in statistics) |
| | function [as in $f(x)$] |
| | furanose form |
| F | Fahrenheit (*use* °F as unit abbreviation) |
| | farad |
| | formal (*use* judiciously; M is preferred) |
| | phenylalanine |
| *F* | Faraday constant |
| | variance ratio (in statistics) |
| FAAS | flame atomic absorption spectroscopy |
| FABMS | fast atom bombardment mass spectrometry |
| *fac* | facial |
| FAD | flavin adenine dinucleotide |
| FAES | flame atomic emission spectroscopy |
| FAFS | flame atomic fluorescence spectroscopy |
| FAS | flame absorption spectroscopy |
| fcc | face-centered cubic (crystal structure) |
| FCC | fluid catalytic cracking |
| Fd | ferredoxin |
| FEM | field emission microscopy or spectroscopy |
| FES | field emission spectroscopy |
| | flame emission spectrometry |
| ff | and pages following (as in p 457 ff) |
| FFEM | freeze–fracture electron microscopy |
| FFF | field flow fractionation |
| FFS | flame fluorescence spectroscopy |
| FFT | fast Fourier transform |
| FHT | Fisher–Hirschfelder–Taylor (space-filling models) |
| FI | field ionization |
| FIA | flow-injection analysis |
| | fluorescence immunoassay |
| fid | free induction decay (in Fourier transform work) |
| FID | flame ionization detector, detection |
| | free induction decay (in Fourier transform work) |
| FIK | field ionization kinetics |
| FIR | far-infrared |
| FLC | ferroelectric liquid crystal |
| fm | femtometer |
| | fermi (unit of length, also f) |
| FM | frequency modulation |
| FMN | flavin mononucleotide |
| FMO | frontier molecular orbital |

| | |
|---|---|
| FOPPA | first-order polarization propagator approach |
| fp | freezing point |
| FPC | fixed partial charge |
| FPT | finite perturbation theory |
| Fr | franklin |
| *Fr* | Froude number |
| Fru | fructose |
| FSGO | floating spherical Gaussian orbital |
| FSH | follicle-stimulating hormone |
| ft | foot |
| FT | Fourier transform |
| ft-c | foot-candle |
| ft-lb | foot-pound |
| ft-lbf | foot-pound-force |
| FTICR | Fourier transform ion cyclotron resonance |
| FTIR | Fourier transform infrared (also FT/IR, FT-IR, and FT IR) |
| FTIRS | Fourier transform infrared spectroscopy |
| FTP | file transfer protocol |
| FTS | Fourier transform spectroscopy |
| Fuc | fucose |
| fw | formula weight |
| fwhh | full width at half-height |
| fwhm | full width at half-maximum |
| | |
| $\gamma$ | microgram (*use* μg) |
| | photon |
| | surface tension |
| $\Gamma$ | surface concentration |
| g | gas [as in $H_2O(g)$] |
| | gram |
| *g* | acceleration due to gravity (closed up to number preceding) |
| | splitting factor (ESR and NMR spectroscopy) |
| G | gauss |
| | generally labeled |
| | giga ($10^9$) |
| | glycine |
| | guanosine |
| *G* | Gibbs energy |
| | gravitational constant |
| g-atom | gram-atom (*use* mol) |
| *Ga* | Galileo number |
| gal | gallon |
| Gal | galactose |
| | galileo (unit of acceleration) |
| GalNAc | *N*-acetylgalactosamine |
| GC | gas chromatography |
| GDC | gas displacement chromatography |

| | |
|---|---|
| GDMS | glow discharge mass spectrometry |
| GDP | guanosine 5′-diphosphate |
| *gem* | geminal |
| GFAAS | graphite furnace atomic absorption spectroscopy |
| GFC | gas frontal chromatography |
| gfw | gram formula weight |
| GH | growth hormone (somatotropin) |
| GHz | gigahertz |
| Gi | gilbert |
| GIAO | gauge-invariant atomic orbital |
| Glc | glucose |
| GLC | gas–liquid chromatography |
| GlcNAc | *N*-acetylglucosamine |
| Gln | glutaminyl, glutamine |
| GLPC | gas–liquid partition chromatography |
| Glu | glutamyl, glutamic acid |
| Glx | "Gln or Glu" |
| gly | glycine (ligand) |
| Gly | glycyl, glycine |
| GMP | guanosine 5′-monophosphate, guanosine 5′-phosphate |
| GPC | gel permeation chromatography |
| gr | grain (unit of weight) |
| GSC | gas–solid chromatography |
| GSH | reduced glutathione |
| GSL | glycosphingolipid |
| GSSG | oxidized glutathione |
| GTP | guanosine 5′-triphosphate |
| Gua | guanine |
| Guo | guanosine |
| Gy | gray (international unit of absorbed dose) |
| | |
| $\eta$ | hapto |
| | viscosity |
| h | hecto ($10^2$) |
| | helion |
| | hour |
| *h* | crystallographic index (*hkl*) |
| | Planck's constant |
| $\hbar$ | Planck's constant divided by $2\pi$ |
| H | henry |
| | histidine |
| *H* | enthalpy |
| | half-chair (conformation) |
| | Hamiltonian |
| $\mathcal{H}$ | Hamiltonian |
| $^1$H NMR | proton nuclear magnetic resonance |

| | |
|---|---|
| $^2$H NMR | deuterium nuclear magnetic resonance |
| $H_0$ | magnetic field (ESR and NMR spectroscopy) |
| ha | hectare |
| Hb | hemoglobin |
| Hbg | biguanide |
| HCG | human chorionic gonadotropin |
| hcp | hexagonal close-packed (crystal structure) |
| HCP | hexachlorophene |
| HCS | hazard communication standard |
| HDPE | high-density polyethylene |
| Hedta | ethylenediaminetetraacetate(3–) (-ato as ligand in full name) |
| $H_2$edta | ethylenediaminetetraacetate(2–) (-ato as ligand in full name) |
| $H_3$edta | ethylenediaminetetraacetate(1–) (-ato as ligand in full name) |
| $H_4$edta | ethylenediaminetetraacetic acid |
| HEEDTA | *N*-(2-hydroxyethyl)ethylenediaminetriacetate |
| Hepes | *N*-(2-hydroxyethyl)piperazine-*N*′-ethanesulfonic acid (also HEPES, hepes) |
| Hepps | *N*-(2-hydroxyethyl)piperazine-*N*′-propanesulfonic acid (also HEPPS, hepps) |
| hex | hexagonal (crystal structure) |
| HF | Hartree–Fock |
| hfs | hyperfine splitting |
| hfsc | hyperfine splitting constant |
| hGH | human growth hormone |
| HIPS | high-impact polystyrene |
| His | histidyl, histidine |
| HIV | human immunodeficiency virus |
| *hkl* | crystallographic index |
| HMDS | hexamethyldisilane |
| | hexamethyldisiloxane |
| HMO | Hückel molecular orbital |
| HMPA | hexamethylphosphoramide |
| HMPT | hexamethylphosphoric triamide |
| hnRNA | heterogeneous nuclear RNA |
| $h\nu$ | indicates light; $h$ is Planck's constant, and $\nu$ is the photon frequency |
| HOHAHA | homonuclear Hartmann–Hahn |
| HOMO | highest occupied molecular orbital |
| $H_2$ox | oxalic acid |
| hp | horsepower |
| HPCE | high-performance capillary electrophoresis |
| HPLC | high-performance liquid chromatography |
| | high-pressure liquid chromatography |
| HREELS | high-resolution electron energy loss spectroscopy |
| HREM | high-resolution electron microscopy |
| HRMS | high-resolution mass spectrometry |
| HSP | heat shock protein |

| | |
|---|---|
| Hyl | hydroxylysyl, hydroxylysine |
| Hyp | hydroxyprolyl, hydroxyproline |
| | hypoxanthine |
| Hz | hertz |
| | |
| i | $(-1)^{1/2}$ |
| *i* | iso (as in *i*-Pr; *do not use* *i*-propyl) |
| I | inosine |
| | isoleucine |
| *I* | electric current (also *i*) |
| | ionic strength |
| | moment of inertia |
| | spin quantum number (ESR and NMR spectroscopy) |
| i.d. | inside diameter |
| i.e. | that is |
| I/O | input–output |
| ibid. | in the same place (in the reference cited; *use is discouraged*) |
| ic | intracerebrally |
| IC | integrated circuit |
| | ion chromatography |
| ICP | inductively coupled plasma |
| ICR | ion cyclotron resonance |
| ics | internal chemical shift |
| ICSH | interstitial-cell-stimulating hormone |
| ICT | International Critical Tables |
| $i_d$ | diffusion current |
| ID | infective dose |
| $ID_{50}$ | dose that is infective in 50% of test subjects (also ID50) |
| IDAS | isotope dilution α spectrometry |
| IDMS | isotope dilution mass spectrometry |
| IDP | inosine 5′-diphosphate |
| IE | ionization energy |
| IEC | ion-exchange chromatography |
| IEF | isoelectric focusing |
| IEP | isoelectric point |
| IETS | inelastic electron-tunneling spectroscopy |
| IFQ | interfacial fluorescence quenching |
| IKES | ion kinetic energy spectroscopy |
| Ile | isoleucyl, isoleucine |
| ILS | increased life span |
| im | intramuscularly |
| IMMA | ion microprobe mass analysis |
| IMP | inosine 5′-monophosphate, inosine 5′-phosphate |
| in. | inch |
| INDO | intermediate neglect of differential overlap |
| INDOR | internal nuclear double resonance |
| | internucleus (nucleus–nucleus) double resonance |

| | |
|---|---|
| INH | inhibitor |
| | isonicotinic acid hydrazide |
| Ino | inosine |
| INO | iterative natural orbital |
| insol | insoluble |
| ip | intraperitoneally |
| IP | ionization potential |
| ips | iron pipe size |
| IR | infrared |
| IRDO | intermediate retention of differential overlap |
| IRMA | immunoradiometric assay |
| IRMS | isotopic ratio mass spectrometry |
| IRP | internal reflection photolysis |
| IRRAS | infrared reflection–absorption spectroscopy |
| IRS | internal reflection spectroscopy |
| isc | intersystem crossing |
| ISCA | ionization spectroscopy for chemical analysis |
| ISE | ion-selective electrode |
| iso | inversion of normal chirality (not as in isopropyl, but in uses such as 8-*iso*-prostaglandin $E_1$; generally italic with a number) |
| ISS | ion-scattering spectroscopy |
| ITP | inosine 5′-triphosphate |
| | isotachophoresis |
| IU | international unit |
| iv | intravenous, intravenously |
| | |
| J | joule |
| *J* | coupling constant (NMR and ESR spectroscopy) |
| JT | Jahn–Teller |
| | |
| k | kilo ($10^3$) |
| *k* | Boltzmann constant (also $k_B$) |
| | crystallographic index (*hkl*) |
| | rate constant |
| K | 1000 (as in 60K protein) |
| | kayser (*use* $cm^{-1}$) |
| | kelvin (*do not use* °K) |
| | kilobyte (kB is preferred) |
| | lysine |
| *K* | equilibrium constant |
| Kα | spectral line |
| kat | katal (unit of enzyme catalytic activity) |
| Kβ | spectral line |
| $k_B$ | Boltzmann constant |
| kb | kilobase |
| | kilobit |

| | |
|---|---|
| kB | kilobel |
| | kilobyte |
| kbar | kilobar |
| kbp | kilobase pair |
| kD | kilodebye |
| kDa | kilodalton |
| KE | kinetic energy |
| kg | kilogram |
| kgf | kilogram-force |
| kHz | kilohertz |
| $K_m$ | Michaelis constant |
| $K_{oc}$ | carbon-referenced sediment partition coefficient |
| | organic chemicals partition coefficient |
| $K_{ow}$ | octanol–water partition coefficient |
| $K_{SP}$ | solubility product constant |
| $K_w$ | autoionization constant |
| | |
| λ | absolute activity |
| | microliter (*use* μL) |
| | wavelength |
| $\lambda_{ex}$ | excitation wavelength |
| $\lambda_{max}$ | wavelength of maximum absorption |
| l | liquid [as in $NH_3(l)$] |
| *l* | crystallographic index (*hkl*) |
| | levorotatory |
| L (small capital) | absolute configuration |
| L | leucine |
| | ligand |
| | liter |
| $L_I$ | spectral line |
| $L_{II}$ | spectral line |
| $L_{III}$ | spectral line |
| Lac | lactose |
| LAMMA | laser microprobe mass spectrometry |
| lat | latitude |
| lb | pound |
| lbf | pound-force |
| LC | liquid chromatography |
| LCAO | linear combination of atomic orbitals |
| LCD | liquid-crystal display |
| LCICD | liquid-crystal-induced circular dichroism |
| LCVAO | linear combination of virtual atomic orbitals |
| LD | lethal dose |
| $LD_{50}$ | dose that is lethal to 50% of test subjects (also LD50) |
| LDH | lactic dehydrogenase |
| LDMS | laser desorption mass spectrometry |
| LE | locally excited |

| | |
|---|---|
| LED | light-emitting diode |
| LEED | low-energy electron diffraction |
| LEEDS | low-energy electron diffraction spectroscopy |
| LEISS | low-energy ion-scattering spectroscopy |
| LEMF | local effective mole fraction |
| Leu | leucyl, leucine |
| LFER | linear free-energy relationship |
| LH | luteinizing hormone |
| LIF | laser-induced fluorescence |
| lim | limit |
| LIMS | laboratory information management system |
| LIS | lanthanide-induced shift |
| lit. | literature |
| LJ, L-J | Lennard-Jones |
| LLC | liquid–liquid chromatography |
| lm | lumen |
| LMCT | ligand-to-metal charge transfer |
| ln | natural logarithm |
| LNDO | local neglect of differential overlap |
| log | logarithm to the base 10 |
| Log | principal logarithm |
| long. | longitude |
| Lp | Lorentz–polarization (effect) |
| *Lp* | Lorentz factor × polarization factor |
| LSC | liquid–solid chromatography |
| LSD | lysergic acid diethylamide |
| LSR | lanthanide shift reagent |
| LUMO | lowest unoccupied molecular orbital |
| lut | lutidine (ligand) |
| Lut | lutidine |
| lx | lux |
| LYP | Lee–Yang–Parr |
| Lys | lysyl, lysine |
| | |
| μ | chemical potential |
| | dipole moment |
| | electrophoretic mobility |
| | micro ($10^{-6}$) |
| | micron (*do not use*; *use* μm or micrometer) |
| $\mu^{\pm}$ | muon |
| $\mu_B$ | Bohr magneton |
| $\mu_N$ | nuclear magneton |
| $\mu_W$ | Weiss magneton |
| m | medium (spectra) |
| | meter |
| | mile (in mpg and mph; otherwise mi) |
| | milli ($10^{-3}$) |

| | |
|---|---|
| m | multiplet (spectra) |
| *m* | isotopic mass |
| | magnetic quantum number (ESR and NMR spectroscopy) |
| | meta |
| | molal (mol $kg^{-1}$) |
| M | mega ($10^6$) |
| | mesomeric |
| | metal (*do not use* Me) |
| | methionine |
| | molar (mol $dm^{-3}$, mol $L^{-1}$) |
| *M* | minus (left-handed helix) |
| [M] | molecular rotation |
| *m/e* | mass-to-charge ratio (*m/z* is preferred) |
| mAb | monoclonal antibody (also Mab, MAb) |
| Mal | maltose |
| MALDI | matrix-assisted laser desorption ionization |
| MALDI-TOFMS | matrix-assisted laser desorption ionization time-of-flight mass spectrometry (also MALDI-TOF MS) |
| Man | mannose |
| MAO | monoamine oxidase |
| MAS | magic-angle spinning |
| MASS | magic-angle sample spinning |
| max | maximum |
| Mb | myoglobin |
| MBE | molecular beam epitaxy |
| MCD | magnetic circular dichroism |
| mCi | millicurie |
| MCT | mercury cadmium telluride |
| MD | molecular dynamics |
| $m_e$ | electron rest mass |
| Me | methyl (*not* metal) |
| MED | mean effective dose |
| MEKC | micellar electrokinetic capillary chromatography |
| MEM | minimum Eagle's essential medium |
| mequiv | milliequivalent |
| mer | polymer notation (as in 16-mer) |
| *mer* | meridional |
| Mes | mesityl (2,4,6-trimethylphenyl), 2-morpholinoethanesulfonic acid, 2-morpholinoethanesulfonate (also MES) |
| Met | methionyl, methionine |
| MetHb | methemoglobin |
| MetMb | metmyoglobin |
| MeV | million electronvolts |
| mho | reciprocal ohm ($\Omega^{-1}$ is preferred) |
| MHz | megahertz |
| mi | mile |

| | |
|---|---|
| min | minimum |
| | minute |
| MINDO | modified intermediate neglect of differential overlap |
| MIR | mid-infrared |
| MIRS | multiple internal reflection spectroscopy |
| ML | monolayer |
| MLCT | metal-to-ligand charge transfer |
| MLR | multiple linear regression |
| mmHg | millimeters of mercury (measure of pressure) |
| mmp | mixture melting point |
| mmu | millimass unit |
| $m_\mathrm{n}$ | neutron rest mass |
| $M_\mathrm{n}$ | number-average molecular weight |
| MO | molecular orbital |
| mol | mole |
| mol % | mole percent |
| molar equiv | molar equivalent |
| mol wt | molecular weight ($M_\mathrm{r}$ is preferred) |
| MOM | methoxymethyl |
| mon | monoclinic (crystal structure) |
| $m_\mathrm{p}$ | proton rest mass |
| mp | melting point |
| MP | Møller–Plesset |
| MP2 | second-order Møller–Plesset perturbation theory |
| mpg | miles per gallon |
| mph | miles per hour |
| MPI | multiphoton ionization |
| MPV | Meerwein–Ponndorf–Verley |
| MQ ENDOR | multiple-quantum electron nuclear double resonance |
| $M_\mathrm{r}$ | relative molecular mass (molecular weight) |
| MR | molecular refraction |
| MRI | magnetic resonance imaging |
| mRNA | messenger RNA |
| Ms | mesyl (methylsulfonyl) |
| MS | mass spectrometry |
| | mass spectrum |
| | microwave spectroscopy |
| MSDS | manufacturer's safety data sheet |
| | material safety data sheet |
| MSG | monosodium glutamate |
| MSH | melanocyte-stimulating hormone, melanotropin |
| Mt | megaton |
| MTD | mean therapeutic dose |
| mtDNA | mitochondrial DNA |
| mtRNA | mitochondrial RNA |
| mu | mass unit |

| | |
|---|---|
| MVA | mevalonic acid |
| MVS | multiple-variable storage |
| $M_w$ | weight-average molecular weight |
| MW | molecular weight ($M_r$ is preferred) |
| MWD | molecular weight distribution |
| Mx | maxwell |
| $M_z$ | *z*-average molecular weight |
| *m/z* | mass-to-charge ratio |
| | |
| ν | frequency |
| $\tilde{\nu}$ | wavenumber |
| $\nu_{1/2}$ | full width at half-maximum height (NMR spectra) |
| $\nu_e$ | neutrino |
| $\nu_{max}$ | frequency of maximum absorption |
| n | nano ($10^{-9}$) |
| | neutron |
| *n* | normal (as in *n*-butyl, *n*-Bu) |
| | refractive index ($n_D^{20}$, at 20 °C, Na D line) |
| | total number of individuals |
| N | asparagine |
| | newton |
| | normal (concentration) |
| | unspecified nucleoside |
| N.B. | nota bene (note well) |
| $N_A$ | Avogadro's number |
| NAA | neutron activation analysis |
| [Na]ATPase | sodium ion activated ATPase (also Na-ATPase, NaATPase) |
| NAD | nicotinamide adenine dinucleotide |
| NADH | reduced nicotinamide adenine dinucleotide |
| NADP | nicotinamide adenine dinucleotide phosphate |
| NADPH | reduced nicotinamide adenine dinucleotide phosphate |
| [Na,K]ATPase | sodium and potassium ion activated ATPase (also Na,K-ATPase) |
| NBS | *N*-bromosuccinimide |
| NDA | New Drug Application |
| NDDO | neglect of diatomic differential overlap |
| nDNA | nuclear DNA |
| NEMO | nonempirical molecular orbital |
| neut equiv | neutralization equivalent |
| NHE | normal hydrogen electrode |
| NIR | near-infrared |
| Nle | norleucyl, norleucine |
| NLO | nonlinear optical (optics) |
| nm | nanometer |
| NM | nuclear magneton (*use* $\mu_N$) |
| NMN | nicotinamide mononucleotide |
| NMR | nuclear magnetic resonance (*do not use* nmr) |
| no. | number |

| | |
|---|---|
| NO | natural orbital (as in CNDO/2-NO) |
| NOCOR | neglect of core orbitals |
| NOE | nuclear Overhauser effect |
| NOESY | nuclear Overhauser enhancement spectroscopy |
| $NO_x$ | nitrogen oxides |
| Np | neper |
| NPR | net protein retention |
| NQR | nuclear quadrupole resonance |
| nRNA | nuclear RNA |
| NRTL | nonrandom two-liquid |
| NSOM | near-field scanning optical microscopy |
| NTP | normal temperature and pressure |
| | unspecified nucleoside 5′-triphosphate |
| Nuc | nucleoside (unspecified) |
| Nva | norvalyl, norvaline |
| | |
| ω | angular frequency |
| Ω | ohm |
| *o* | ortho |
| O | orotidine |
| o.d. | outside diameter |
| o-rh | orthorhombic (crystal structure) |
| o/w | oil in water (emulsion) |
| O/W | oil in water (emulsion) |
| OAc | acetate |
| obsd | observed |
| *OC*-6 | octahedral, coordination number 6 |
| OCR | optical character recognition |
| OD | optical density |
| ODMR | optically detected magnetic resonance |
| ODU | optical density unit |
| Oe | oersted |
| OES | optical emission spectroscopy |
| OFDR | off-frequency decoupling resonance |
| OMVPE | organometallic vapor-phase epitaxy |
| Ord | orotidine |
| ORD | optical rotary dispersion |
| Orn | ornithyl, ornithine |
| Oro | orotic acid |
| ORTEP | Oak Ridge thermal ellipsoid plot |
| osm | osmolar (also osM, Osm) |
| OTTLE | optically transparent thin-layer electrode |
| ox | oxalato (ligand) |
| | oxidized or oxidation (in subscripts and superscripts) |
| oxidn | oxidation |
| oz | ounce |

| | |
|---|---|
| % | percent |
| ‰ | per thousand (parts per thousand) |
| π | pros (near) in NMR measurements (as in $N^{\pi}$ of histidine) |
| | type of orbital, electron |
| $\pi^{\pm}$ | pion |
| $\pi^{0}$ | pion |
| ψ | pseudouridine |
| ψrd | pseudouridine |
| p | negative logarithm (as in pH) |
| | page |
| | pico ($10^{-12}$) |
| | proton |
| *p* | angular momentum (ESR and NMR spectroscopy) |
| | para |
| | probability (in statistics) |
| | pyranose form |
| P | peta ($10^{15}$) |
| | poise |
| | proline |
| *P* | plus (right-handed helix) |
| | probability (in statistics) |
| p.m. | post meridiem |
| P450 | specific cytochrome designation (i.e., cytochrome P450) |
| P-450 | specific cytochrome designation (i.e., cytochrome P-450) |
| $P_{450}$ | specific cytochrome designation (i.e., cytochrome $P_{450}$) |
| $^{31}$P NMR | phosphorus-31 nuclear magnetic resonance |
| Pa | pascal |
| PAC | perturbed angular correlation |
| PAD | perturbed angular distribution |
| PAGE | polyacrylamide gel electrophoresis |
| $pa_H$ | negative logarithm of hydrogen ion activity |
| PAH | polycyclic aromatic hydrocarbon |
| PAN | polyacrylonitrile |
| PBS | phosphate-buffered saline |
| pc | parsec (unit of length) |
| PC | paper chromatography |
| | personal computer |
| | planar chromatography |
| PCB | polychlorobiphenyl, polychlorinated biphenyl |
| PCDD | polychlorinated dibenzo-*p*-dioxin |
| | polychlorodibenzo-*p*-dioxin |
| PCDF | polychlorodibenzofuran |
| PCILO | perturbed configuration interaction with localized orbitals |
| PCP | pentachlorophenol |
| PCR | polymerase chain reaction |
| PCTFE | poly(chlorotrifluoroethylene) |

| | |
|---|---|
| PDL | pumped dye laser |
| PDMS | plasma desorption mass spectrometry |
| PE | polyethylene |
| | potential energy |
| PEG | poly(ethylene glycol) |
| PEL | permissible exposure limit |
| PEO | poly(ethylene oxide) |
| PES | photoelectron spectroscopy |
| PET | positron emission tomography |
| PETP | poly(ethylene terephthalate) |
| PFU | plaque-forming unit |
| PG | prostaglandin |
| pH | negative logarithm of hydrogen ion concentration |
| Ph | phenyl (for $C_6H_5$ only) |
| Phe | phenylalanyl, phenylalanine |
| phen | 1,10-phenanthroline, *o*-phenanthroline |
| phr | parts per hundred parts of resin (or rubber) |
| $P_i$ | inorganic phosphate |
| PIB | polyisobutylene |
| PIXE | proton-induced X-ray emission |
| p*K* | negative logarithm of equilibrium constant |
| $pK_a$ | p*K* for association |
| PL | photoluminescence |
| PLOT | porous-layer open-tubular |
| PMMA | poly(methyl methacrylate) |
| PMO | perturbational molecular orbital |
| PMR | phosphorus magnetic resonance (*do not use; use* $^{31}$P NMR) |
| | polymerization of monomeric reactants |
| | proton magnetic resonance (*do not use; use* $^{1}$H NMR) |
| PNA | polynuclear aromatic hydrocarbon |
| PNDO | partial neglect of differential overlap |
| po | per os (orally) |
| POM | poly(oxymethylene), polyformaldehyde |
| POPOP | 1,4-bis(5-phenyl-2-oxazolyl)benzene |
| pp | pages |
| PP | polypropene |
| ppb | parts per billion |
| ppbv | parts per billion by volume |
| $PP_i$ | inorganic pyrophosphate, phosphoric acid |
| ppm | parts per million |
| ppmv | parts per million by volume |
| PPO | 2,5-diphenyloxazole |
| PPP | Pariser–Parr–Pople |
| PPS | photophoretic spectroscopy |
| ppt | parts per trillion |
| | precipitate |

| | |
|---|---|
| pptv | parts per trillion by volume |
| Pr | propyl |
| PRDDO | partial retention of diatomic differential overlap |
| prepn | preparation |
| PRFT | partially relaxed Fourier transform |
| Pro | prolyl, proline |
| *pro-R* | stereochemical descriptor (also pro-*R*) |
| *pro-S* | stereochemical descriptor (also pro-*S*) |
| PRT | platinum resistance thermometer |
| Ps | positronium |
| PS | polystyrene |
| psi | pounds per square inch |
| psia | pounds per square inch absolute |
| psig | pounds per square inch gauge |
| pt | pint |
| | point |
| PTC | phase-transfer catalysis |
| PTFE | poly(tetrafluoroethylene) |
| PTH | parathyroid hormone |
| | phenylthiohydantoin |
| PTV | programmed-temperature vaporizer |
| PU | polyurethane |
| PVA | poly(vinyl alcohol) |
| PVAC | poly(vinyl acetate) |
| PVAL | poly(vinyl alcohol) |
| PVC | poly(vinyl chloride) |
| PVDC | poly(vinylidene dichloride) |
| PVDF | poly(vinylidene difluoride) |
| PVE | poly(vinyl ether) |
| PVF | poly(vinyl fluoride) |
| PXRD | powder X-ray diffraction |
| py | pyridine (ligand) |
| Py | pyridine |
| *PY* | pyramidal (coordination compounds) |
| Py–GC–MS | pyrolysis–gas chromatography–mass spectrometry |
| pyr | pyrazine (ligand) |
| pyrr | pyrrolidine (ligand) |
| pz | pyrazole (ligand) |
| | |
| q | quartet (spectra) |
| *q* | heat, electric charge (also *Q*) |
| Q | glutamine |
| *Q* | heat, electric charge (also *q*) |
| QCPE | Quantum Chemistry Program Exchange |
| QELS | quasi-elastic light scattering |
| QSAR | quantitative structure–activity relationship |
| qt | quart |

| | |
|---|---|
| $\rho$ | density |
| $r$ | correlation coefficient |
| R | arginine |
| | Rankine (temperature scale, *use* °R as unit abbreviation) |
| | roentgen |
| $R$ | gas constant |
| | rectus (configurational) |
| | regression coefficient |
| | resistance |
| *rac* | racemic |
| rad | radian |
| | unit of radiation |
| RBS | Rutherford backscattering spectrometry |
| rd | rad |
| RDE | rotating disk electrode |
| $r_e$ | electron radius |
| *re* | stereochemical descriptor (as in the *re* face) |
| recryst | recrystallized |
| red | reduced or reduction (in subscripts and superscripts) |
| redn | reduction |
| redox | reduction–oxidation |
| ref | reference |
| rel | relative |
| *rel* | relative (stereochemical descriptor) |
| REL | recommended exposure limit |
| rem | roentgen equivalent man |
| REM | rapid eye movement |
| rep | roentgen equivalent physical |
| rf | radio frequency |
| $R_f$ | retention factor (ratio of distance traveled by the center of a zone to the distance simultaneously traveled by the mobile phase) |
| RFC | request for comments |
| RH | relative humidity |
| Rha | rhamnose |
| RI | refractive index |
| RIA | radioimmunoassay |
| Rib | ribose |
| RIMS | resonance ionization mass spectrometry |
| RIS | resonance ionization spectrometry |
| rms | root mean square |
| RNA | ribonucleic acid |
| RNase | ribonuclease |
| ROA | Raman optical activity |
| RPLC | reversed-phase liquid chromatography |
| rpm | revolutions per minute |
| RQ | respiratory quotient |
| RRDE | rotating ring-disk electrode |

| | |
|---|---|
| RRKM | Rice–Ramsperger–Kassel–Marcus |
| rRNA | ribosomal RNA |
| RRS | resonance Raman spectroscopy |
| RRT | relative retention time |
| RS | Raman spectroscopy |
| RSD | relative standard deviation |
| | risk-specific dose |
| Ry | rydberg |
| | |
| σ | standard deviation |
| | surface charge density |
| | surface tension |
| | tensile strength |
| | type of orbital, electron |
| Σ | summation |
| s | second |
| | single bond [as in s-cis (italic in compound names)] |
| | singlet (spectra) |
| | solid [as in NaCl(s)] |
| | strong (spectra) |
| *s* | secondary (as in *s*-Bu; *but sec*-butyl) |
| | sedimentation coefficient |
| | standard deviation (analytical) |
| | symmetrical |
| $s^0_{20,w}$ | sedimentation coefficient measured at 20 °C in water and extrapolated to 0 °C |
| $s^2$ | sample variance |
| S | serine |
| | siemens |
| *S* | entropy |
| | sinister (configurational) |
| | skew (conformation) |
| S/N | signal-to-noise ratio |
| SAM | self-assembled monolayer |
| SANS | small-angle neutron scattering |
| *SAPR*-8 | square antiprismatic, coordination number 8 |
| sar | sarcosine (*N*-methylglycine) (ligand) |
| Sar | sarcosyl, sarcosine (*N*-methylglycine) |
| SAR | structure–activity relationship |
| SARISA | surface analysis by resonance ionization of sputtered atoms |
| SAXS | small-angle X-ray scattering (or spectroscopy) |
| sc | subcutaneously |
| *sc* | synclinal |
| sccm | standard cubic centimeters per minute |
| SCE | saturated calomel electrode |
| SCF | self-consistent field |
| SCF–HF | self-consistent field, Hartree–Fock |

| | |
|---|---|
| scfh | standard cubic feet per hour |
| SCOT | support-coated open-tubular |
| SD | standard deviation |
| SDS | sodium dodecyl sulfate |
| SE | standard error |
| $S_E2$ | second-order electrophilic substitution |
| sec | secant |
| *sec* | secondary (as in *sec*-butyl; *but s*-Bu) |
| SEC | size exclusion chromatography |
| sech | hyperbolic secant |
| SECM | scanning electrochemical microscopy |
| SECS | simulation and evaluation of chemical synthesis |
| SEM | scanning electron microscopy |
| | standard error of the mean |
| Ser | seryl, serine |
| SERS | surface-enhanced Raman spectroscopy (or scattering) |
| SEW | surface electromagnetic wave |
| $S_{ex}$ | exciplex substitution |
| SFC | supercritical-fluid chromatography |
| sh | sharp (spectra) |
| | shoulder (spectra) |
| *Sh* | Sherwood number |
| SHC | shape and Hamiltonian constant |
| SHE | standard hydrogen electrode |
| *si* | stereochemical descriptor (as in the *si* face) |
| SI | International System of Units (Système International) |
| | secondary ion (as in SIMS) |
| SIM | selected-ion monitoring |
| SIMS | secondary-ion mass spectrometry |
| sin | sine |
| sinh | hyperbolic sine |
| SLR | spin–lattice relaxation |
| SMOSS | surface Mössbauer |
| SMSI | strong metal support interaction |
| *sn* | stereospecific numbering |
| SN | separation number |
| $S_N1$ | first-order nucleophilic substitution |
| $S_N2$ | second-order nucleophilic substitution |
| $S_Ni$ | internal nucleophilic substitution |
| SNO | semiempirical natural orbital |
| sol | solid |
| soln | solution |
| sp | specific |
| sp. | species (singular) |
| *sp* | synperiplanar |
| *SP*-4 | square planar, coordination number 4 |
| sp gr | specific gravity |

| | |
|---|---|
| sp ht | specific heat |
| sp vol | specific volume |
| SPECT | single-photon-emission computed tomography |
| spp. | species (plural) |
| SPR | stroboscopic pulse radiolysis |
| *SPY*-5 | square pyramidal, coordination number 5 |
| sq | square |
| SQF | single quantum filtered |
| SQUID | superconducting quantum interference device |
| sr | steradian |
| $S_{RN}1$ | first-order nucleophilic substitution triggered by electron transfer |
| SRS | stimulated Raman scattering |
| SSC | standard saline citrate (NaCl–citrate) |
| St | stokes |
| std | standard |
| STEM | scanning transmission electron microscopy |
| STM | scanning tunneling microscopy |
| STO | Slater-type orbital |
| STO-3G | Slater-type orbital, three Gaussian |
| STP | standard temperature and pressure |
| subsp. | subspecies |
| Suc | sucrose |
| Sv | sievert |
| | svedberg |
| SVL | single vibrational level |
| swg | standard wire gauge |
| *sym* | symmetrical |
| | |
| τ | tele (far) in NMR measurements (as in $N^{\tau}$ of histidine) |
| θ | angle |
| [θ] | ORD measurement, deg $cm^2$/dmol |
| Θ | temperature (e.g., in Curie–Weiss expressions) |
| t | metric ton |
| | triplet (spectra) |
| | triton |
| *t* | Student distribution (the Student *t* test in statistics) |
| | temperature (in degrees Celsius) |
| | tertiary (as in *t*-Bu; but *tert*-butyl) |
| | time |
| $t_{1/2}$ | half-life |
| T | ribosylthymine |
| | tautomeric |
| | tera ($10^{12}$) |
| | tesla |
| | threonine |
| | tritium |

| | |
|---|---|
| *T* | temperature (in kelvins) |
| | twist (conformation) |
| *T*-4 | tetrahedral, coordination number 4 |
| T/C | treated vs cured |
| tan | tangent |
| tan δ | mechanical loss factor |
| tanh | hyperbolic tangent |
| TBP | tri-*n*-butyl phosphate |
| *TBPY*-5 | trigonal bipyramidal, coordination number 5 |
| TCA | tricarboxylic acid cycle (citric acid cycle, Krebs cycle) |
| | trichloroacetic acid |
| TCD | thermal conductivity detector |
| TCP/IP | transmission control protocol/Internet protocol |
| TDS | total dissolved solids |
| TEA | tetraethylammonium |
| | transversely excited atmospheric |
| TEAE | triethylaminoethyl (as in TEAE-cellulose) |
| TEM | transmission electron microscopy |
| temp | temperature |
| *tert* | tertiary (as in *tert*-butyl; but *t*-Bu) |
| tetr | tetragonal (crystal structure) |
| TFA | trifluoroacetyl |
| $T_g$ | glass-transition temperature |
| TGA | thermogravimetric analysis |
| Tham | tris(hydroxymethyl)aminomethane (also Tris) |
| THC | tetrahydrocannabinol |
| Thd | ribosylthymine |
| theor | theoretical |
| THF | tetrahydrofuran |
| Thr | threonyl, threonine |
| Thy | thymine |
| TIMS | thermal ionization mass spectrometry |
| TIP | temperature-independent paramagnetism |
| TL | triboluminescence |
| TLC | thin-layer chromatography |
| TMA | thermomechanical analysis |
| TMS | tetramethylsilane |
| | trimethylsilyl |
| TMV | tobacco mosaic virus |
| TnL | tunnel luminescence |
| TOC | total organic carbon |
| TOD | total oxygen demand |
| TOFMS | time-of-flight mass spectrometry (also TOF MS) |
| tol | tolyl (also Tol) |
| TOM | transmitted optical microscopy |
| Torr | torr |

| | |
|---|---|
| tosyl | 4-toluenesulfonyl (also Ts) |
| TPD | temperature-programmed desorption |
| TPDE | temperature-programmed decomposition |
| TPR | temperature-programmed reduction |
| *TPR*-6 | trigonal prismatic, coordination number 6 |
| TQMS | triple-quadrupole mass spectrometry |
| $t_R$ | retention time |
| tr | trace |
| Tr | trace |
| tric | triclinic (crystal structure) |
| triflate | trifluoromethanesulfonate |
| trig | trigonal (crystal structure) |
| TRIR | time-resolved infrared |
| Tris | tris(hydroxymethyl)aminomethane (also Tham) |
| tRNA | transfer RNA |
| Trp | tryptophyl, tryptophan |
| Ts | tosyl (4-toluenesulfonyl) |
| TSC | thermal stimulated current |
| TSH | thyroid-stimulating hormone |
| tu | thiourea (ligand) |
| TVA | thermal volatilization analysis |
| Tyr | tyrosyl, tyrosine |
| | |
| u | unified atomic mass unit |
| U | uniformly labeled |
| | uridine |
| *U* | internal energy |
| UCST | upper critical solution temperature |
| UDP | uridine 5′-diphosphate |
| uhf | ultrahigh frequency |
| UHF | ultrahigh frequency |
| | unrestricted Hartree–Fock |
| UHV | ultrahigh vacuum |
| ULSI | ultra-large-scale integration |
| UMP | uridine 5′-monophosphate, uridine 5′-phosphate |
| uncor | uncorrected |
| uns | unsymmetrical |
| UPS | ultraviolet photoelectron spectroscopy |
| ur | urea (ligand) |
| Ura | uracil |
| Urd | uridine |
| USP | United States Pharmacopeial Convention |
| UTP | uridine 5′-triphosphate |
| UV | ultraviolet |
| UV PES | ultraviolet photoelectron spectroscopy |
| UV–vis | ultraviolet–visible |

| | |
|---|---|
| v | vendeko ($10^{-30}$) |
| *v* | scan rate |
| | velocity |
| V | valine |
| | vendeca ($10^{30}$) |
| | volt |
| v/v | volume per volume |
| Val | valyl, valine |
| VASS | variable-angle sample spinning |
| VB | valence bond |
| VCD | vibrational circular dichroism |
| VDT | video display terminal |
| VEELS | vibrational electron energy loss spectroscopy |
| VESCF | variable electronegativity self-consistent field |
| vhf | very high frequency |
| VHF | very high frequency |
| *vic* | vicinal |
| vis | visible |
| viz. | namely |
| VLE | vapor–liquid equilibrium |
| VLSI | very large scale integration |
| VOA | vibrational optical activity |
| VOC | volatile organic compound |
| vol | volume |
| vol % | volume percent |
| vp | vapor pressure |
| VPC | vapor-phase chromatography |
| VPO | vapor pressure osmometry |
| VRML | virtual reality modeling language |
| vs | versus (v in legal expressions) |
| | very strong (spectra) |
| VSIP | valence-state ionization potential |
| VUV | vacuum ultraviolet |
| VVk | Van Vleck |
| vw | very weak (spectra) |

| | |
|---|---|
| w | weak (spectra) |
| *w* | weighting factor |
| | work |
| W | tryptophan |
| | watt |
| *W* | work |
| w/v | weight per volume |
| w/w | weight per weight |
| WAN | wide-area network |
| WAXS | wide-angle X-ray scattering |

| | |
|---|---|
| Wb | weber |
| WCOT | wall-coated open-tubular |
| WDS | wavelength-dispersive spectroscopy |
| WHSV | weight-hourly space velocity |
| WLF | Williams–Landel–Ferry (molecular models) |
| wt | weight |
| wt % | weight percent |
| | |
| x | xenno ($10^{-27}$) |
| *x* | *x* axis |
| X | xanthosine (*use* N for unknown nucleoside) |
| | xenna ($10^{27}$) |
| Xan | xanthine |
| XANES | X-ray absorption near-edge spectroscopy |
| | X-ray absorption near-edge structure |
| Xao | xanthosine |
| XEDS | X-ray energy-dispersive spectrometry |
| XES | X-ray emission spectroscopy |
| XMP | xanthosine 5′-monophosphate, xanthosine 5′-phosphate |
| XPS | X-ray photoelectron spectroscopy |
| XRD | X-ray diffraction |
| XRDF | X-ray radial distance function |
| XRF | X-ray fluorescence |
| Xyl | xylose |
| | |
| y | yocto ($10^{-24}$) |
| *y* | *y* axis |
| Y | tyrosine |
| | yotta ($10^{24}$) |
| | |
| z | zepto ($10^{-21}$) |
| *z* | charge number of an ion |
| | *z* axis |
| Z | benzyloxycarbonyl (also Cbz) |
| | "glutamic acid or glutamine" |
| | zetta ($10^{21}$) |
| *Z* | atomic number |
| | zusammen (configurational) |
| zfs | zero-field splitting |
| zfsc | zero-field-splitting constant |

➤➤➤➤➤ CHAPTER 11

# Numbers, Mathematics, and Units of Measure

## Numbers

Both numerals and words can be used to express numbers. The usage and style conventions for numerals and words are different for technical and nontechnical material.

### *Numeral and Word Usage*

➤ Use numerals with units of time or measure, and use a space between the numeral and the unit, except %, $, ° (angular degrees), ′ (angular minutes), and ″ (angular seconds).

| | | |
|---|---|---|
| 6 min | 25 mL | 125 V/s |
| 0.30 g | 50% | $250 |
| 273 K | 47°8′23″ | 180° (*but* 180 °C) |
| 90 °F | 50 μg of compound/dL of water | |

EXCEPTION Spell out numbers with units of measure used in a nontechnical sense.

If you take five minutes to read this article, you'll be surprised.

➤ With items other than units of time or measure, use words for cardinal numbers less than 10; use numerals for 10 and above. Spell out ordinals "first" through "ninth"; use numerals for 10th or greater.

three flasks 30 flasks

| | |
|---|---|
| third flask | 12th flask |
| seven trees | 10 trees |
| eighth example | 33rd example |
| first century | 21st century |
| sixfold | 20-fold |

EXCEPTION 1 Use all numerals in a series or range containing numbers 10 or greater, even in nontechnical text.

5, 8, and 12 experiments
2nd and 20th samples
5–15 repetitions

EXCEPTION 2 Use all numerals for numbers modifying nouns in parallel construction in the same sentence if one of the numbers is 10 or greater.

Activity was reduced in 2 pairs, not significantly changed in 11 pairs, and increased in 6 pairs.

We present new results pertaining to 12 phenanthrolines and 3 porphyrins.

EXCEPTION 3 For very large numbers used in a nontechnical sense, use a combination of numerals and words.

1 billion tons
180 million people
2 million pounds (*not* lb)
4.5 billion years
$15 million (*not* 15 million dollars)

➤ When a sentence starts with a specific quantity, spell out the number as well as the unit of measure.

Twelve species were evaluated in this study.

Twenty slides of each blood sample were prepared.

Fifteen milliliters of supernate was added to the reaction vessel.

Twenty-five milliliters of acetone was added, and the mixture was centrifuged.

However, if possible, recast the sentence.

Acetone (25 mL) was added, and the mixture was centrifuged.

A 25 mL portion of acetone was added, and the mixture was centrifuged.

➤ Even when a sentence starts with a spelled-out quantity, use numerals when appropriate in the rest of the sentence.

Twenty-five milliliters of acetone and 5 mL of HCl were added.

Three micrograms of sample was dissolved in 20 mL of acid.

Fifty samples were collected, but only 22 were tested.

➤ Use numerals for expressions used in a mathematical sense.

The incidence of disease increased by a factor of 4.

The yield of product was decreased by 6 orders of magnitude.

The efficiency of the reaction was increased 2-fold.

After 2 half-lives, the daughter product could be measured.

The control group had 3 times the risk for colon cancer.

The values are determined with 5 degrees of freedom.

➤ When the suffix "fold" is used in a nonmathematical sense, spell out the accompanying number if it is less than 10.

The purpose of this discussion is twofold.

➤ When the word "times" is used in a nonmathematical sense, spell out the accompanying number if it is less than 10.

The beaker was rinsed four times.

➤ Use numerals in ratios.

a ratio of 1:10
a ratio of 1/10

a 1:1 (v/v) mixture
a 1/1 (v/v) mixture

➤ In dates, use numerals without ordinal endings.

January 3, Jan 3 (*not* January 3rd, Jan 3rd)
September 5, Sept 5 (*not* September 5th, Sept 5th)

➤ Use numerals for decades, and form their plurals by adding an "s". Do not use apostrophes in any position.

the 1960s (*not* the 1960's, *not* the '60s)
values in the 90s (*not* the 90's)
She is in her 20s. (*not* her 20's)

➤ Use numerals with a.m. and p.m.

12:15 a.m. 4:00 p.m.

➤ Spell out and hyphenate fractions whose terms are both less than 10. If one of the terms is 10 or greater, use a piece fraction.

one-quarter of the experiments ⅟20 of the subjects
two-thirds of the results ⅟12 of the volume

➤ Use numerals to label figures, tables, schemes, structures, charts, equations, and references. Number sequentially; do not skip numbers or number out of sequence. Use arabic numerals for references, but for the other items, the use of arabic and roman numerals varies among ACS publications. Consult a recent issue or author instructions to determine what system is preferred.

➤ In journal articles and book chapters, instead of repeating chemical names over and over, use numerals in boldface (not italic) type to identify chemical species. Use these identifiers only in text, not in article or chapter titles, and number consecutively.

> This paper describes the syntheses, structures, and stereodynamic behavior of the novel hexacoordinate silicon complexes **1–4.**
>
> The cyclization of 1,3,5-hexatriene (**6**) to 1,3-cyclohexadiene (**7**) is predicted to proceed more rapidly in an electrostatic field.
>
> Complexes **8–12,** in the presence of monoamine oxidase, produce active catalysts for propylene polymerization.
>
> Primary amines **2–5, 7,** and **9** gave the same Cotton effect signs, depending on the configuration.
>
> Monomer **III** reacts with the initiator (**I,** Ar = 2,6-diisopropylphenyl) via a ring-opening metathesis polymerization mechanism.

➤ Numerals may be used to name members of a series.

> Sample 1 contained a high level of contamination, but samples 2 and 3 were relatively pure.
>
> Methods 1 and 2 were used for water-soluble compounds, and methods 3 and 4 were used for oil-soluble compounds.

➤ When numerals are used as names and not enumerators, form their plurals by adding an apostrophe and “s” to avoid confusion with mathematical expressions and to make it clear that the “s” is not part of the name.

> The athlete received five 9’s from the judges.
>
> Boeing 747’s are among the largest airplanes.

➤ Arabic numerals in parentheses may be used to enumerate a list of phrases or sentences in text. Always use an opening and a closing parenthesis, not one alone.

> Some advantages of these materials are (1) their electrical properties after pyrolysis, (2) their ability to be modified chemically before pyrolysis, and (3) their abundance and low cost.
>
> The major conclusions are the following: (1) We have further validated the utility of molecular mechanical methods in simulating the kinetics of these reactions.

(2) A comparison of the calculated structures with available X-ray structures revealed satisfactory agreement. (3) The combined use of different theoretical approaches permitted characterization of the properties of a new isomer.

➤ Arabic numerals followed by periods or enclosed in parentheses may be used to enumerate a displayed list of sentences or to number paragraphs. Here are two acceptable ways to format a list.

These results suggest the following:
1. Ketones are more acidic than esters.
2. Cyclic carboxylic acids are more acidic than their acyclic analogues.
3. Alkylation of the active methylene carbon reduces the acidity.

These results suggest the following:
(1) Ketones are more acidic than esters.
(2) Cyclic carboxylic acids are more acidic than their acyclic analogues.
(3) Alkylation of the active methylene carbon reduces the acidity.

## *Style for Numbers*

➤ For very large numbers with units of measure, use scientific notation or choose an appropriate multiplying prefix for the unit to avoid numbers of more than four digits.

$1.2 \times 10^6$ s
$3.0 \times 10^4$ kg
$5.8 \times 10^{-5}$ M *or* 58 μM
42.3 L (*not* 42,300 mL *or* 42 300 mL)

EXCEPTION 1 In tables, use the same unit and multiplying prefix for all entries in a column, even if some entries therefore require four or more digits.

EXCEPTION 2 Use the preferred unit of a discipline, even when the numbers require four or more digits:

| | |
|---|---|
| g/L | for mass density of fluids |
| kg/m$^3$ | for mass density of solids |
| GPa | for modulus of elasticity |
| kPa | for fluid pressure |
| MPa | for stress |

➤ In four-digit numbers, use no commas or spaces.

EXCEPTION Spaces or commas are inserted in four-digit numbers when alignment is needed in a column containing numbers of five or more digits.

➤ When a long number cannot be written in scientific notation, the digits must be grouped. For grouping of digits in long numbers (five digits or greater), check the publication in which the manuscript will appear. Two styles are possible.

STYLE 1 In some publications, for numbers with five or more digits, the digits are grouped with commas placed between groups of three counting to the left of the decimal point.

4837
10,000
930,582
6,398,210
85,798.62578

STYLE 2 In some publications (including ACS journals), for numbers with five or more digits, the digits are grouped with a thin space between groups of three, counting both to the left and to the right of the decimal point.

9319.4
74 183.0629
0.508 27
501 736.293 810 4

EXCEPTIONS

- U.S. monetary values are always written with commas: $5,000.
- U.S. patent numbers are always written with commas: U.S. Patent 6,555,655. The patent numbers of other countries should be presented as on the original patent document.
- Page numbers in reference citations are always printed solid: p 11597.

➤ Use the period as the decimal point, never a comma.

➤ Use numerals before and after a decimal point.

0.25 (*not* .25)
78.0 *or* 78 (*not* 78.)

➤ Use a decimal and a zero following a numeral only when such usage truly represents the precision of the measurement: 27.0 °C and 27 °C are not interchangeable.

➤ Use decimals rather than fractions with units of time or measure, except when doing so would imply an unwarranted accuracy.

3.5 h (*not* 3½ h)
5.25 g (*not* 5¼ g)

➤ Standard deviation, standard error, or degree of accuracy can be given in two ways:

- with only the deviation in the least significant digit(s) placed in parentheses following the main number and closed up to it or
- with all digits preceded by a ± and following the main number. Spaces are left on each side of the ±.

2.0089(1) means 2.0089 ± 0.0001
1.4793(23) means 1.4793 ± 0.0023

The shorter version is better in tables. Always specify which measure (e.g., standard deviation or standard error) of uncertainty is being used.

➤ When two numbered items are cited in narrative, use "and".

Figures 1 and 2
refs 23 and 24
compounds **I** and **II**

➤ Use a comma between two reference callouts in parentheses or as superscripts.

Lewis (*12, 13*) found
Lewis[12,13] found

When the reference numbers are on the line, the comma is followed by a space; when the numbers are superscripts, the comma is not followed by a space.

➤ Use an en dash in ranges or series of three or more numbered items, whether on the line or in a superscript.

43–49
325–372
2005–2008
Tables 1–4
temperatures of 100–125 °C
refs 3–5

aliquots of 50–100 mL
eqs 6–9
samples 5–10
past results (*27–31*)
past results[27–31]
pp 165–172

EXCEPTION 1 Do not use an en dash in expressions with the words "from ... to" or "between ... and".

from 20 to 80 (*not* from 20–80)
between 50 and 100 mL (*not* between 50–100 mL)

EXCEPTION 2 When either one or both numbers are negative or include a symbol that modifies the number, use the word "to" or "through", not the en dash.

–20 to +120 K
–145 to –30 °C
≈50 to 60
10 to >600 mL
<5 to 15 mg

➤ For ranges in scientific notation, retain all parts of all numbers or avoid ambiguity by use of parentheses or other enclosing marks.

$9.2 \times 10^{-3}$ to $12.6 \times 10^{-3}$ *or* $(9.2–12.6) \times 10^{-3}$ (*not* 9.2 to $12.6 \times 10^{-3}$)

➤ For very large numbers in ranges, retain all parts of all numbers.

26 million to 35 million

➤ Do not use e or E to mean "multiplied by the power of 10".

$3.7 \times 10^5$ (*not* 3.7e5, 3.7E+5)

# Mathematics

## *Mathematical Concepts*

VARIABLE A variable is a quantity that changes in value, substance, or amount, such as $V$ for volume, $m$ for mass, and $t$ for time.

CONSTANT A constant is a quantity that has a fixed value, such as $h$ for the Planck constant and $F$ for the Faraday constant.

FUNCTION The function $f(x) = y$ represents a rule that assigns a unique value of $y$ to every $x$. The *argument* of the function is $x$.

OPERATOR An operator is a symbol, such as a function (d, derivative; ln, natural logarithm; and $\mathcal{H}$, the Hamiltonian operator) or an arithmetic sign (+, −, =, ÷, and ×), denoting an operation to be performed.

PHYSICAL QUANTITY A physical quantity is a product of a numerical value (a pure number) and a unit. Physical quantities may be scalars or vectors, variables or constants.

SCALAR A scalar is an ordinary number without direction, such as length, temperature, or mass. Any quantity that is not a vector quantity is a scalar quantity.

VECTOR A vector is a quantity with both magnitude and direction, such as force or velocity. For the vector $\mathbf{V} = [a, b]$ (also denoted as $\vec{V} = [a, b]$), $a$ and $b$ are the *components* of the vector.

TENSOR A tensor represents a generalized vector with more than two components.

MATRIX A matrix is represented by a rectangular array of *elements*; an *array* consists of rows and columns. The elements of matrix **U** are $u_{11}$, $u_{12}$, etc.

$$\mathbf{U} = \begin{bmatrix} u_{11} & \cdots & u_{1n} \\ \vdots & \ddots & \vdots \\ u_{n1} & \cdots & u_{nn} \end{bmatrix}$$

DETERMINANT The determinant of a matrix is a function that assigns a number to a matrix. For example, the determinant of the $n \times n$ matrix **B** is represented by

$$\det \mathbf{B} = \begin{vmatrix} b_{11} & \cdots & b_{1n} \\ \vdots & & \vdots \\ b_{n1} & \cdots & b_{nn} \end{vmatrix}$$

INDEX An index is a subscript or superscript character in an element of a matrix, vector, or tensor; indices usually represent numbers. For example, $i$ and $j$ are indices in $b_{ij}$.

Do not confuse abbreviations and mathematical symbols. An abbreviation is usually two or more letters; a mathematical symbol is generally only one letter, possibly with a subscript or superscript. An abbreviation is used in narrative text but seldom appears in equations; a mathematical symbol is preferred in equations and may also be used in text. For example, in text with no equations, PE for potential energy is acceptable, but in mathematical text and equations, $E_\mathrm{p}$ is preferred.

## Usage and Style for Symbols

➤ Define all symbols for mathematical constants, variables, and unknown quantities the first time you use them in the text. If you use them in the abstract, define them there and then again at their first appearance in text. Do not define standard mathematical constants such as π, i, and e.

➤ Form the plurals of mathematical symbols by adding an apostrophe and "s" if you cannot use a word such as "values" or "levels".

large *r* values *is better than* large *r*'s

➤ Do not use an equal sign as an abbreviation for the word "is" or the word "equals" in narrative text.

$PV = nRT$, where $P$ is pressure (*not* where $P$ = pressure)

when the temperature is 50 °C (*not* when the temperature = 50 °C)

➤ Do not use a plus sign as an abbreviation for the word "and" in narrative text.

a mixture of A and B (*not* a mixture of A + B)

➤ Do not use an asterisk to indicate multiplication except in computer language expressions.

## *Italic Type*

➤ Use italic type for

- variables: *T* for temperature, *x* for mole fraction, *r* for rate
- axes: the *y* axis
- planes: plane *P*
- components of vectors and tensors: $a_1 + b_1$
- elements of determinants and matrices: $g_n$
- constants: $k_B$, the Boltzmann constant; *g*, the acceleration due to gravity
- functions that describe variables: $f(x)$

➤ Even when you use mathematical constants, variables, and unknown quantities in adjective combinations, retain the italic type.

> In this equation, $V_i$ is the frequency of the *i*th mode.
>
> In eq 4, *n* is the number of extractions and *M* is the mass remaining after the *n*th extraction.

➤ Use italic type for two-letter variables defining transport properties.

| | | | |
|---|---|---|---|
| *Al* | Alfvén number | *Ma* | Mach number |
| *Bi* | Biot number | *Nu* | Nusselt number |
| *Co* | Cowling number | *Pe* | Péclet number |
| *Da* | Damkohler number | *Pr* | Prandtl number |
| *Eu* | Euler number | *Ra* | Rayleigh number |
| *Fo* | Fourier number | *Re* | Reynolds number |
| *Fr* | Froude number | *Sc* | Schmidt number |
| *Ga* | Galileo number | *Sh* | Sherwood number |
| *Gr* | Grashof number | *Sr* | Strouhal number |
| *Ha* | Hartmann number | *St* | Stanton number |
| *Kn* | Knudsen number | *We* | Weber number |
| *Le* | Lewis number | *Wi* | Weissenberg number |

## *Roman Type*

➤ Use roman type for

- numerals;
- punctuation and enclosing marks such as square brackets, parentheses, and braces;
- most operators;
- units of measure and time: mg, milligram; K, kelvin; Pa, pascal; mmHg, millimeters of mercury;
- nonmathematical quantities or symbols: R, radical in chemical nomenclature; $S_1$, molecular state; s, atomic orbital;

- multiple-letter abbreviations for variables: IP, ionization potential; cmc, critical micelle concentration;
- mathematical constants:

  e, the base of the natural logarithm, 2.71828…
  i, the imaginary number, $(-1)^{1/2}$
  π, 3.14159…

- transposes of matrices: $\mathbf{A}^{\mathrm{T}}$ (T is the transpose of matrix **A**);
- points and lines: point A, line $\overline{\mathrm{AB}}$;
- determinants: det **A** is the determinant of matrix **A**; and
- trigonometric and other functions:

| | | | |
|---|---|---|---|
| Ad | adjoint | lim | limit |
| Ai | Airy function | lim inf | limit inferior |
| arg | argument | lim sup | limit superior |
| Bd | bound | ln | natural logarithm (base e) |
| cl | closure | log | logarithm (base 10) |
| Coker | cokernel | Log | principal logarithm |
| cos | cosine | lub | least upper bound |
| cosh | hyperbolic cosine | max | maximum |
| cot | cotangent | min | minimum |
| coth | hyperbolic cotangent | mod | modulus |
| csc | cosecant | P | property |
| csch | hyperbolic cosecant | Re | real |
| det | determinant | sec | secant |
| dim | dimension | sech | hyperbolic secant |
| div | divergence | sign, sgn | sign |
| erf | error function | sin | sine |
| erfc | complement of error function | sinh | hyperbolic sine |
| exp | exponential | SL | special linear |
| GL | general linear | sp | spin |
| glb | greater lower bound | Sp | symplectic |
| grad | gradient | sup | superior |
| hom | homology | Sz(g) | Suzuki group |
| Im | imaginary | tan | tangent |
| inf | inferior | tanh | hyperbolic tangent |
| int | interior | tr | trace |
| ker | kernel | wr | wreath |

## *Boldface Type*

➤ Use boldface type for

- vectors;
- tensors;
- matrices; and
- multidimensional physical quantities: **H**, magnetic field strength.

## Greek Letters

Greek letters (lightface or boldface) can be used for variables, constants, and vectors and anywhere a Latin letter can be used.

| NAME | UPPERCASE | LOWERCASE |
|---|---|---|
| Alpha | Α | α |
| Beta | Β | β |
| Gamma | Γ | γ |
| Delta | Δ | δ, ∂ |
| Epsilon | Ε | ϵ, ε |
| Zeta | Ζ | ζ |
| Eta | Η | η |
| Theta | Θ | θ, ϑ |
| Iota | Ι | ι |
| Kappa | Κ | κ |
| Lambda | Λ | λ |
| Mu | Μ | μ |
| Nu | Ν | ν |
| Xi | Ξ | ξ |
| Omicron | Ο | ο |
| Pi | Π | π |
| Rho | Ρ | ρ |
| Sigma | Σ | σ |
| Tau | Τ | τ |
| Upsilon | Υ | υ |
| Phi | Φ | φ, ϕ |
| Chi | Χ | χ |
| Psi | Ψ | ψ |
| Omega | Ω | ω |

## Script and Open-Faced Letters

➤ Script ($\Re$) and open-faced ($\mathbb{R}$; also known as blackboard boldface) letters are available but should not be used routinely. For open-faced letters, only uppercase is available.

## Spacing

➤ Leave a space before and after functions set in roman type, unless the argument is enclosed in parentheses, brackets, or braces.

| | |
|---|---|
| $\log 2$ | $\exp(-x)$ |
| $-\log x$ | $\cosh(\beta e_0 \phi)$ |
| $4 \sin \theta$ | $4 \tan(2y)$ |
| $\tan^2 y$ | $\mathrm{erfc}(y)$ |

➤ Leave a space before and after mathematical operators that function as verbs or conjunctions; that is, they have numbers on both sides or a symbol for a variable on one side and a number on the other.

| | |
|---|---|
| $20 \pm 2\%$ | $p < 0.01$ |
| $3.24 \pm 0.01$ | $T_{\mathrm{g}} = 176\ °\mathrm{C}$ |
| $4 \times 5$ cm | $n = 25$ |
| $8 \times 10^{-4}$ | 1 in. = 2.54 cm |
| $k \geq 420\ \mathrm{s}^{-1}$ | |

EXCEPTION 1 Leave no space around mathematical operators in subscripts and superscripts.

$\Delta H^{n-1}$

$E_{\lambda>353}$

$\mathrm{M}^{(x+y)+}$

EXCEPTION 2 Leave no space around a slash ($a/b$), a ratio colon (1:10), or a centered dot ($\mathbf{P}_{\mathrm{M}}{\cdot}\mathbf{V}$).

➤ Leave no space between simple variables being multiplied: $xy$. Do not use a centered dot (·) or the times sign (×) with single-letter scalar variables.

➤ In multiplication involving the two-letter symbols for transport properties, use a space, enclose them in parentheses, or use the times sign. When superscripts or subscripts are present, the symbols can be closed up.

$Re\ Nu$
$(Re)(Nu)$
$Re \times Nu$
$Re_x Nu_y$

➤ Use a space for simple multiplication of functions of the type $f(x)$ (one-dimensional) or $g(y, z)$ (multidimensional). Close up multipliers to such functions where applicable. You may also use additional enclosing marks instead of spaces.

$W = 2f(x)\ g(y, z)$

$W = 2[f(x)][g(y, z)]$

➤ When mathematical symbols are used as adjectives, that is, with one number that is not part of a mathematical operation, do not leave a space between the symbol and the number.

| | |
|---|---|
| −12 °C | a conversion of >50% |
| 25 g (±1%) | a probability of <0.01 |
| at 400× magnification | |
| The level can vary from −15 to +25 m. | |

## Enclosing Marks

➤ Use enclosing marks (parentheses, brackets, and braces, also called fences) in accordance with the rules of mathematics. Enclose parentheses within square brackets, and square brackets within braces: {[( )]}.

➤ Use enclosing marks around arguments when necessary for clarity.

$\sin(x + 1)$

$\sin[2\pi(x - y)/n]$

$\log[-V(r)/kT]$

➤ Do not use square brackets, parentheses, or braces around the symbol for a quantity to make it represent any other quantity.

INCORRECT

where $V$ is volume and $(V)$ is volume at equilibrium

CORRECT

where $V$ is volume and $V_\mathrm{e}$ is volume at equilibrium

## Subscripts and Superscripts

➤ Use italic type for subscripts and superscripts that are themselves symbols for physical quantities or numbers. Use roman type for subscripts and superscripts that are abbreviations and not symbols.

$C_p$ for heat capacity at constant pressure

$C_\mathrm{B}$ for heat capacity of substance B

$C_\mathrm{g}$ where g is gas

$E_i$ for energy of the $i$th level, where $i$ is a number

$g_\mathrm{n}$ where n is normal

$\mu_\mathrm{r}$ where r is relative

$E_\mathrm{k}$ where k is kinetic

$\xi_\mathrm{e}$ where e is electric

➤ In most cases, staggered subscripts and superscripts are preferred. Exponents should follow subscripts.

$x_1{}^2$ $\quad T_{2\mathrm{m}}{}^{-1}$ $\quad E_\mathrm{ads}{}^\circ$

$C_x{}^{1/2}$ $\quad \Delta H_1{}^\ddagger$

EXCEPTIONS

$\lambda_+^\infty$ $\quad \sigma_\mathrm{p}^+$ $\quad B_2^\mathrm{exptl}$

➤ Use a slash (/) in all subscript and superscript fractions, with no space on either side.

$t_{1/2}$ $x^{1/2}$ $M^{2/3}$ $f_{a/b}$

➤ Leave no space around operators in subscripts and superscripts.

$\mathrm{M}^{(2-n)+}$ $E_{T+\theta}$

➤ Leave no space around other expressions in subscripts and superscripts, unless confusion or misreading would result.

$Q_{n\text{-Bu}(750°\text{C})}$ $\beta_{\text{zero level}}$ $E^{365\text{nm}}$

➤ The terms $\mathrm{e}^a$ and exp $a$ have the same meaning and can be interchanged. When an exponent to the base e is very long or complicated, replace the e with exp and place the exponent on line and in enclosing marks. Leave no space between exp and the opening enclosing mark.

$\exp(\int y\,\mathrm{d}t)$ (*not* $\mathrm{e}^{\int y \mathrm{d}t}$)

$\exp\{½kT[Y(a+b)-Z]\}$ (*not* $\mathrm{e}^{½kT[Y(a+b)-Z]}$)

➤ In running text, do not use the radical sign ($\sqrt{\ }$) with long terms. Use enclosing marks around the term and a superscript 1/2, 1/3, 1/4 (etc.) for square, cube, fourth root (etc.), respectively.

$(x-y^2)^{1/3}$

$[\sinh^2 u + (\cosh u - 1)^2]^{1/2}$

## *Abbreviations and Symbols*

➤ Certain abbreviations are used only in the context of mathematical equations. Define all of these the first time they are used.

| | |
|---|---|
| lhs | left-hand side (of an equation) |
| rhs | right-hand side (of an equation) |
| ODE | ordinary differential equation |
| rms | root mean square |
| rmsd | root-mean-square deviation |
| s.t. | subject to |
| wrt, WRT | with respect to |

➤ Some standard usages and symbols for mathematical operations and constants need never be defined. They include the following:

| | |
|---|---|
| e | natural base (approximately 2.7183) |
| exp $x$, $\mathrm{e}^x$ | exponential of $x$ |
| i | $\sqrt{-1}$ |
| ln $x$ | natural logarithm of $x$ |

| | |
|---|---|
| $\log x$ | logarithm to the base 10 of $x$ |
| $\log_a x$ | logarithm to the base $a$ of $x$ |
| $\approx$ | approximately equal to |
| $\simeq$ | asymptotically equal to |
| $\propto$, $\sim$ | proportional to |
| $\rightarrow$ | approaches (tends to) |
| $\equiv$ | identically equal to |
| $\infty$ | infinity |
| $\Sigma$ | summation |
| $\Pi$ | product |
| $\cup$ | union |
| $\int$ | integral |
| $\oint$ | line integral around a closed path |
| $\nabla$ | del (or nabla) operator, gradient |
| $\nabla^2$ | Laplacian operator |
| $<$ | less than |
| $\leq$ | less than or equal to |
| $\ll$ | much less than |
| $>$ | greater than |
| $\geq$ | greater than or equal to |
| $\gg$ | much greater than |
| $\neq$ | not equal to |
| $\parallel$ | parallel to |
| $\perp$ | perpendicular to |
| $\lvert a \rvert$ | absolute magnitude of $a$ |
| $a^{1/2}$, $\sqrt{a}$ | square root of $a$ |
| $a^{1/n}$, $\sqrt[n]{a}$ | $n$th root of $a$ |
| $\bar{a}$, $\langle a \rangle$ | mean value of $a$ |
| $\Delta x$ | finite increment of $x$ |
| $\partial x$ | partial differential, infinitesimal increment of $x$ |
| $\mathrm{d}x$ | total differential of $x$ |
| $f(x)$ | function of $x$ |
| $\int y\,\mathrm{d}x$ | integral of $y$ with respect to $x$ |
| $\int_a^b y\,\mathrm{d}x$ | integral of $y$ from $x = a$ to $x = b$ |
| $\mathbf{A}$ | vector of magnitude $A$ |
| $\mathbf{A}\cdot\mathbf{B}$ | scalar product of **A** and **B** |
| $\mathbf{A} \times \mathbf{B}$, $\mathbf{AB}$ | vector product of **A** and **B** |
| $\overline{\mathrm{AB}}$ | length of line from A to B |

## Equations

Mathematical equations can be presented within running text or displayed on lines by themselves. Follow the guidelines for style and usage just described under "Usage and Style for Symbols" (starting on p 211).

➤ Leave a space

- before and after mathematical signs used as operators ($=$, $\neq$, $\equiv$, $\sim$, $\approx$, $\cong$, $>$, $<$, $+$, $-$, $\times$, $\div$, $\cup$, $\supset$, $\subset$, $\in$, etc., but not slash (/), ratio colon, or centered dot), except when they appear in superscripts or subscripts
- before trigonometric and other functions set in roman type
- after trigonometric and other functions set in roman type when their arguments are not in enclosing marks
- before and after derivatives: $\int\int f(x)\ \mathrm{d}x\ f(y)\ \mathrm{d}y$ or $\int x\ \partial x$
- between built-up (display) fractions as components of products:

  $$\frac{a}{b}\ \frac{c}{d}$$

  or write on one line and clarify with enclosing marks and no space: $(a/b)(c/d)$
- between functions as components of products: $W = 2f(x)\ g(y, z)$

➤ Leave no space

- between single-item variables being multiplied
- in any part of a superscript or subscript, unless confusion or misreading would result
- between any character and its own superscript, prime, or subscript
- on either side of a colon used for a ratio
- on either side of a centered dot
- on either side of a slash (/)
- after mathematical operators used as adjectives: $-10$
- after functions when the argument is in parentheses: $\tanh(\lambda/2)$
- between an opening parenthesis, bracket, or brace and the next character: $(2x)y$
- between a closing parenthesis, bracket, or brace and the previous character: $2(xy)$
- between back-to-back parentheses, brackets, and braces, e.g., ](
- between nested parentheses, brackets, and braces, e.g., [(
- in any part of limits to summations, products, and integrals
- in any part of lower limits to min, max, lim, and inf

➤ Use or do not use spaces around ellipses, depending on the treatment of other items in the series.

| | |
|---|---|
| no spaces: | $a_na_{n+1}a_{n+2}{\cdots}a_{n+36}$ |
| spaces: | $a_n + a_{n+1} + a_{n+2} + \ldots + a_{n+36}$ |
| space before: | $a, b, \ldots, x$ |

➤ Use enclosing marks in accordance with the rules of mathematics. If the slash (/) is used in division and if there is any doubt where the numerator ends or where the denominator starts, use enclosing marks for one or the other or both.

$(x + y)/(3x - y)$

$(a/b)/c$, or $a/(b/c)$ but never $a/b/c$

$\frac{x+y}{2} = z$ would be better as $(x + y)/2 = z$

$\frac{x+y}{z} + 2a$ would be better as $[(x + y)/z] + 2a$

➤ If an equation is very short and will not be referred to again, you may run it into the text.

A fluid is said to be Newtonian when it obeys Newton's law of viscosity, given by $\tau = \eta\gamma$, where $\tau$ is the shear stress, $\eta$ is the fluid dynamic constant, and $\gamma$ is the shear rate.

➤ You may use mathematical expressions as part of a sentence when the subject, verb, and object are all part of the mathematical expression.

When $V = 12$, eq 15 is valid.

($V$ is the subject, = is the verb, and 12 is the object.)

➤ When an equation is too long to fit on one line, break it *after* an operator that is not within an enclosing mark (parentheses, brackets, or braces) or break it between sets of enclosing marks. Do not break equations after integral, product, and summation signs; after trigonometric and other functions set in roman type; or before derivatives.

➤ Number displayed equations by using any consistent system of sequencing.

1, 2, 3, ...
1a, 1b, 2, ...
I, II, III, ...
A, B, C, ...
A-1, A-2, A-3, ...
B.1, B.2, B.3, ...
C1, C2, C3, ...

➤ Use equation identifiers in the proper sequence according to appearance in text. Do not skip numbers or letters in the sequence.

➤ Place identifiers in parentheses, flush right on the same line as the equation.

$$V = 64\pi kT\gamma^2 \exp(-\kappa h) \quad (3)$$

➤ Do not use any closing punctuation on the line with displayed equations.

➤ When introducing a displayed equation, do not automatically use a colon; in most cases a colon is incorrect because the equation finishes a phrase or sentence.

> An ideal gas law analogy is
>
> $$\pi A = nRT$$
>
> If the principal radii are $R_1$ and $R_2$, then
>
> $$\Delta = R_1 - R_2$$
>
> The area per adsorbed molecule can be calculated from
>
> $$a = N_\mathrm{A}\Gamma_\mathrm{S}$$
>
> The attractive energy can be approximated by
>
> $$V_\mathrm{A} = Ar(12H)^{-1}$$
>
> The simplest method is to use a mapping potential of the form
>
> $$\varepsilon_\mathrm{m} = (1 - \lambda_\mathrm{m})\varepsilon_\mathrm{A} + \lambda_\mathrm{m}\varepsilon_\mathrm{B}$$
>
> Marshall developed an equation for rapid coagulation:
>
> $$n = 1 + S\pi Drt$$

➤ Following a displayed equation that is part of a sentence, punctuate the text as if it were a continuation of a sentence including the equation but do not punctuate at the end of the equation. Note the absence of a comma at the end of the equation in the example. Punctuation that would normally be present at the end of an equation in text is absent but implicit at the end of a displayed equation.

> The capillary pressure $P$ depends significantly on the wetting contact angle according to
>
> $$P = 2\gamma \cos(\theta / r)$$
>
> where $\gamma$ is the surface tension, $\theta$ is the wetting contact angle, and $r$ is the radius of the capillary.

➤ To cite an equation in text, use the abbreviation "eq" if it is not the first word of the sentence. Spell out "equation" when it is the first word of a sentence or when it is not accompanied by a number. The plural of "eq" is "eqs".

> The number of independent points can be calculated from eq 3.
>
> The number of independent points can be calculated from eqs 3 and 4.
>
> Equation 1 is not accurate for distances greater than 10 μm.
>
> Equations 1 and 2 are not accurate for distances greater than 10 μm.

➤ Some notations differ in text and in display:

| *In display* | *In text* |
|---|---|
| $\sum\limits_{i=1}^{N}$ | $\Sigma_{i=1}^{N}$ *or* $\Sigma_{i=1}{}^{N}$ |
| $\prod\limits_{k=2}^{n}$ | $\Pi_{k=2}^{n}$ *or* $\Pi_{k=2}{}^{n}$ |
| $\lim\limits_{i=1}$ | $\lim_{i=1}$ |
| $\max\limits_{j=2}$ | $\max_{j=2}$ |
| $\min\limits_{j=k,n}$ | $\min_{j=k,n}$ |

## Ratio and Mixture Notation

➤ Use either a colon or a slash (/) to represent a ratio, but not an en dash. Use either a slash or an en dash between components of a mixture, but not a colon.

dissolved in 5:1 glycerin/water
dissolved in 5:1 glycerin–water

the metal/ligand (1:1) reaction mixture
the metal–ligand (1:1) reaction mixture
the metal–ligand (1/1) reaction mixture

the methane/oxygen/argon (1/50/450) matrix
the methane/oxygen/argon (1:50:450) matrix

## Set Notation

The following symbols are used in set notation. Leave a space before and after all operators, but not before and after braces.

| | |
|---|---|
| $A = \{a, b\}$ | set *A*; *A* is italic; braces are used |
| $A \cup B$ | union of sets *A* and *B* |
| $A \cap B$ | intersection of sets *A* and *B* |
| $A \in B$ | *A* is a member (element) of *B* |
| $A \notin B$ | *A* is not a member (element) of *B* |
| $A \subset B$ | *A* is contained in *B* |
| $A \not\subset B$ | *A* is not contained in *B* |
| $A \supset B$ | *A* contains *B* |
| $A \not\supset B$ | *A* does not contain *B* |
| $\forall A$ | for all (every) *A* |
| $\exists$ | there exists |
| $\ni$ | such that |
| $\therefore$ | therefore |

### *Geometric Notation*

➤ Leave no spaces around geometric notation. Use italic type for planes and axes and roman type for points and lines.

| | |
|---|---|
| $X \perp Y$ | *X* is perpendicular to *Y* |
| $X \| Y$ | *X* is parallel to *Y* |
| $\angle \mathrm{AB}$ | the angle between A and B |
| $\overline{\mathrm{AB}}$ | length of line from A to B |

### *Statistics*

Certain statistical symbols are standard.

| | |
|---|---|
| CV | coefficient of variation |
| df, DF | degrees of freedom |
| *f* | frequency |
| *F* | variance ratio |
| *n*, *N* | total number of individuals or random variables |
| *p*, *P* | probability |
| *r* | correlation coefficient |
| *R* | regression coefficient |
| RSD | relative standard deviation |
| σ, SD | standard deviation |
| Σ | summation |
| $s^2$ | sample variance |
| SE | standard error |
| SEM | standard error of the mean |
| *t* | Student distribution (the Student *t* test) |
| $\bar{x}$ | arithmetic mean |

A common statistical measurement is the Student *t* test or Student's *t* test. Student was the pseudonym of W. Gossett, an eminent mathematician.

## Units of Measure

➤ Where possible, use metric and SI units (discussed in Appendix 11-1) in all technical documents. The following conventions apply to all units of measure:

- Abbreviate units of measure when they accompany numbers.
- Leave a space between a number and its unit of measure.
- Do not use a period after an abbreviated unit of measure (exception: in. for inch).

- Do not define units of measure.

| | |
|---|---|
| 500 mL | 200 mV |
| 3 min | $4.14 \times 10^{-9}$ m$^2$/(V s) |
| 4 Å | $2.6 \times 10^4$ J |
| 9 V/s | 3 min interval |
| 9 V s$^{-1}$ | 2 μm droplet |
| 9 V·s$^{-1}$ | 500 mL flask |

EXCEPTION Do not leave a space between a number and the percent, angular degree, angular minute, or angular second symbols.

50%
90°
75′
18″

➤ Use °C with a space after a number, but no space between the degree symbol and the capital C: 15 °C.

➤ Do not add an "s" to make the plural of any abbreviated units of measure. The abbreviations are used as both singular and plural.

50 mg (*not* 50 mgs)

3 mol (*not* 3 mols)

➤ Write abbreviated compound units with a centered dot or a space between the units to indicate multiplication and a slash (/) or negative exponent for division. Enclose compound units following a slash in parentheses. Usage should be consistent within a paper.

| | |
|---|---|
| watt per meter-kelvin is | W·m$^{-1}$·K$^{-1}$ *or*<br>W/(m·K) *or*<br>W m$^{-1}$ K$^{-1}$ *or*<br>W (m K)$^{-1}$ *or*<br>W/(m K) |
| cubic decimeter per mole-second is | dm$^3$·mol$^{-1}$·s$^{-1}$ *or*<br>dm$^3$/(mol·s) *or*<br>dm$^3$ mol$^{-1}$ s$^{-1}$ *or*<br>dm$^3$ (mol s)$^{-1}$ *or*<br>dm$^3$/(mol s) |
| joule per mole-kelvin is | J·mol$^{-1}$·K$^{-1}$ *or*<br>J/(mol·K) *or*<br>J mol$^{-1}$ K$^{-1}$ *or*<br>J (mol K)$^{-1}$ *or*<br>J/(mol K) |

➤ Spell out units of measure that do not follow a number. Do not capitalize them unless they are at the beginning of a sentence or in a title.

several milligrams (*not* several mg)
a few milliliters (*not* a few mL)
degrees Celsius
reciprocal seconds
milligrams per kilogram
volts per square meter

EXCEPTION 1 Abbreviate units of measure in parentheses after the definitions of variables directly following an equation.

$$L = D/P_{\mathrm{O}}$$

where $L$ is the distance between particles (cm), $D$ is the particle density (g/cm$^3$), and $P_{\mathrm{O}}$ is the partial pressure of oxygen (kPa).

EXCEPTION 2 Certain units of measure have no abbreviations: bar, darcy, einstein, erg, faraday, and langmuir. The symbol for the unit torr is Torr. The unit rad is abbreviated rd; the unit radian is abbreviated rad.

EXCEPTION 3 In column headings of tables and in axis labels of figures, abbreviate units of measure, even without numbers.

➤ Add an "s" to form the plural of spelled-out units: milligrams, poises, kelvins, amperes, watts, newtons, and so on.

EXCEPTIONS bar, hertz, lux, stokes, siemens, and torr remain unchanged; darcy becomes darcies; henry becomes henries.

➤ Do not capitalize surnames that are used as units of measure.

| | | |
|---|---|---|
| ampere | franklin | newton |
| angstrom | gauss | ohm |
| coulomb | gilbert | pascal |
| curie | gray | poise |
| dalton | hartree | siemens |
| darcy | henry | sievert |
| debye | hertz | stokes |
| einstein | joule | tesla |
| erg | kelvin | watt |
| faraday | langmuir | weber |

Celsius and Fahrenheit are always capitalized. They are not themselves units; they are the names of temperature scales.

➤ Do not use a slash (/) in spelled-out units of measure. Use the word "per".

Results are reported in meters per second.

The fluid density is given in kilograms per cubic meter.

➤ Do not mix abbreviations and spelled-out units within units of measure.

newtons per meter (*not* N per meter)
100 F/m (*not* 100 farad/m)

EXCEPTION in more complex situations

50 mL of water and 20 mg of NaOH per gram of compound

➤ Use a slash (/), not the word "per", before the abbreviation for a unit in complex expressions.

50 μg of peptide/mL
25 mg of drug/kg of body weight

➤ When the first part of a unit of measure is a word that is not itself a unit of measure, use a slash (/) before the final abbreviated unit.

10 counts/s
12 domains/$cm^3$
$2 \times 10^3$ ions/min
125 conversions/$mm^2$

➤ When the last part of a unit of measure is a word that is not itself a unit of measure, use either a slash (/) or the word "per" before the word that is not a unit.

0.8 keV/channel
0.8 keV per channel
7 $\mu_B$/boron
7 $\mu_B$ per boron

➤ Leave no space between the multiplying prefix and the unit, whether abbreviated or spelled out.

kilojoule *or* kJ
milligram *or* mg
microampere *or* μA

➤ Use only one multiplicative prefix per unit.

nm (*not* mμm)

➤ In ranges and series, retain only the final unit of measure.

10–12 mg
5, 10, and 20 kV
60–90°
between 25 and 50 mL
from 10 to 15 min

➤ Do not use the degree symbol with kelvin: 115 K.

➤ In titles and headings, do not capitalize abbreviated units of measure that are ordinarily lowercase.

> Analysis of 2 mg Samples
>
> A 50 kDa Protein To Modulate Guanine Nucleotide Binding

➤➤➤➤➤

APPENDIX 11-1

# The International System of Units (SI)

Before the 1960s, four systems of units were commonly used in the scientific literature: the English system (centuries old, using yard and pound), the metric system (dating from the 18th century, using meter and kilogram as standard units), the CGS system (based on the metric system, using centimeter, gram, and second as base units), and the MKSA or Giorgi system (using meter, kilogram, second, and ampere as base units).

The International System of Units (SI, Système International d'Unités) is the most recent effort to develop a coherent system of units. It is coherent because there is only one unit for each base physical quantity, and units for all other quantities are derived from these base units by simple equations. It has been adopted as a universal system to simplify communication of numerical data and to restrict proliferation of systems. SI units are used by the National Institute of Standards and Technology (NIST). More information on SI can be found at http://www.physics.nist.gov/cuu/index.html.

The SI is constructed from seven base units for independent quantities (ampere, candela, kelvin, kilogram, meter, mole, and second) plus two supplementary units for plane and solid angles (radian and steradian). Most physicochemical measurements can be expressed in terms of these units.

Certain units not part of the SI are so widely used that it is impractical to abandon them (e.g., liter, minute, and hour) or are so well established that the International Committee on Weights and Measures has authorized their continued use (e.g., bar, curie, and angstrom). In addition, quantities that are expressed in terms of the fundamental constants of nature, such as elementary charge, proton mass, Bohr magneton, speed of light, and Planck constant, are also acceptable. However, broad terms such as "atomic units" are not acceptable, although atomic mass unit, u, is acceptable and relevant to chemistry.

Follow all usage conventions given for units of measure. Use the abbreviations for SI units with capital and lowercase letters exactly as they appear in Tables 11A-1 to 11A-6.

## Table 11A-1. SI Units

| *Name* | *Symbol* | *Physical Quantity* |
|---|---|---|
| ***Base units*** | | |
| ampere | A | electric current |
| candela | cd | luminous intensity |
| kelvin | K | thermodynamic temperature |
| kilogram | kg | mass |
| meter | m | length |
| mole | mol | amount of substance |
| second | s | time |
| ***Supplementary units*** | | |
| radian | rad | plane angle |
| steradian | sr | solid angle |

## Table 11A-2. Multiplying Prefixes

| *Factor* | *Prefix* | *Symbol* | *Factor* | *Prefix* | *Symbol* |
|---|---|---|---|---|---|
| $10^{-24}$ | yocto | y | $10^{1}$ | deka | da |
| $10^{-21}$ | zepto | z | $10^{2}$ | hecto | h |
| $10^{-18}$ | atto | a | $10^{3}$ | kilo | k |
| $10^{-15}$ | femto | f | $10^{6}$ | mega | M |
| $10^{-12}$ | pico | p | $10^{9}$ | giga | G |
| $10^{-9}$ | nano | n | $10^{12}$ | tera | T |
| $10^{-6}$ | micro | μ | $10^{15}$ | peta | P |
| $10^{-3}$ | milli | m | $10^{18}$ | exa | E |
| $10^{-2}$ | centi | c | $10^{21}$ | zetta | Z |
| $10^{-1}$ | deci | d | $10^{24}$ | yotta | Y |

Note: Any of these prefixes may be combined with any of the symbols permitted within the SI. Thus, kPa and GPa will both be common combinations in measurements of pressure, as will mL and cm for measurements of volume and length, respectively. As a general rule, however, the prefix chosen should be 10 raised to that multiple of 3 that will bring the numerical value of the quantity to a positive value less than 1000.

## Table 11A-3. SI-Derived Units

| *Name* | *Symbol* | *Quantity* | *In Terms of Other Units* | *In Terms of SI Base Units* |
|---|---|---|---|---|
| becquerel | Bq | activity (of a radionuclide) | — | $s^{-1}$ |
| coulomb | C | quantity of electricity, electric charge | — | s·A |
| farad | F | capacitance | C/V | $m^{-2} \cdot kg^{-1} \cdot s^{4} \cdot A^{2}$ |
| gray | Gy | absorbed dose, kerma, specific energy imparted | J/kg | $m^{2} \cdot s^{-2}$ |
| henry | H | inductance | Wb/A | $m^{2} \cdot kg \cdot s^{-2} \cdot A^{-2}$ |
| hertz | Hz | frequency | — | $s^{-1}$ |
| joule | J | energy, work, quantity of heat | N·m | $m^{2} \cdot kg \cdot s^{-2}$ |
| lumen | lm | luminous flux | cd·sr | $m^{2} \cdot m^{-2} \cdot cd = cd$ |
| lux | lx | illuminance | $lm/m^{2}$ | $m^{2} \cdot m^{-4} \cdot cd = m^{-2} \cdot cd$ |
| newton | N | force | — | $m \cdot kg \cdot s^{-2}$ |
| ohm | Ω | electric resistance | V/A | $m^{2} \cdot kg \cdot s^{-3} \cdot A^{-2}$ |
| pascal | Pa | pressure, stress | $N/m^{2}$ | $m^{-1} \cdot kg \cdot s^{-2}$ |
| siemens | S | conductance | A/V | $m^{-2} \cdot kg^{-1} \cdot s^{3} \cdot A^{2}$ |
| sievert | Sv | dose equivalent | J/kg | $m^{2} \cdot s^{-2}$ |
| tesla | T | magnetic flux density | $Wb/m^{2}$ | $kg \cdot s^{-2} \cdot A^{-1}$ |
| volt | V | electric potential, potential difference, electromotive force | W/A | $m^{2} \cdot kg \cdot s^{-3} \cdot A^{-1}$ |
| watt | W | power, radiant flux | J/s | $m^{2} \cdot kg \cdot s^{-3}$ |
| weber | Wb | magnetic flux | V·s | $m^{2} \cdot kg \cdot s^{-2} \cdot A^{-1}$ |

**Table 11A-4.** SI-Derived Compound Units

| *Name* | *Symbol* | *Quantity* | *In Terms of Other Units* |
|---|---|---|---|
| ampere per meter | A/m | magnetic field strength | — |
| ampere per square meter | $A/m^2$ | current density | — |
| candela per square meter | $cd/m^2$ | luminance | — |
| coulomb per cubic meter | $C/m^3$ | electric charge density | $m^{-3}{\cdot}s{\cdot}A$ |
| coulomb per kilogram | C/kg | exposure (X-rays and $\gamma$ rays) | — |
| coulomb per square meter | $C/m^2$ | electric flux density | $m^{-2}{\cdot}s{\cdot}A$ |
| cubic meter | $m^3$ | volume | — |
| cubic meter per kilogram | $m^3/kg$ | specific volume | — |
| farad per meter | F/m | permittivity | $m^{-3}{\cdot}kg^{-1}{\cdot}s^4{\cdot}A^2$ |
| henry per meter | H/m | permeability | $m{\cdot}kg{\cdot}s^{-2}{\cdot}A^{-2}$ |
| joule per cubic meter | $J/m^3$ | energy density | $m^{-1}{\cdot}kg{\cdot}s^{-2}$ |
| joule per kelvin | J/K | heat capacity, entropy | $m^2{\cdot}kg{\cdot}s^{-2}{\cdot}K^{-1}$ |
| joule per kilogram | J/kg | specific energy | $m^2{\cdot}s^{-2}$ |
| joule per kilogram kelvin | J/(kg K) | specific heat capacity, specific entropy | $m^2{\cdot}s^{-2}{\cdot}K^{-1}$ |
| joule per mole | J/mol | molar energy | $m^2{\cdot}kg{\cdot}s^{-2}{\cdot}mol^{-1}$ |
| joule per mole kelvin | J/(mol K) | molar entropy, molar heat capacity | $m^2{\cdot}kg{\cdot}s^{-2}{\cdot}K^{-1}{\cdot}mol^{-1}$ |
| kilogram per cubic meter | $kg/m^3$ | density, mass density | — |
| meter per second | m/s | speed, velocity | — |
| meter per second squared | $m/s^2$ | acceleration | — |
| mole per cubic meter[a] | $mol/m^3$ | concentration (amount of substance per volume) | — |
| newton-meter | N·m | moment of force | $m^2{\cdot}kg{\cdot}s^{-2}$ |
| newton per meter | N/m | surface tension | $kg{\cdot}s^{-2}$ |
| pascal second | Pa·s | dynamic viscosity | $m^{-1}{\cdot}kg{\cdot}s^{-1}$ |
| radian per second | rad/s | angular velocity | — |
| radian per second squared | $rad/s^2$ | angular acceleration | — |
| reciprocal meter | $m^{-1}$ | wavenumber | — |
| reciprocal second | $s^{-1}$ | frequency | — |
| square meter | $m^2$ | area | — |
| square meter per second | $m^2/s$ | kinematic viscosity | — |
| volt per meter | V/m | electric field strength | $m{\cdot}kg{\cdot}s^{-3}{\cdot}A^{-1}$ |
| watt per meter kelvin | W/(m K) | thermal conductivity | $m{\cdot}kg{\cdot}s^{-3}{\cdot}K^{-1}$ |
| watt per square meter | $W/m^2$ | heat flux density, irradiance | $kg{\cdot}s^{-3}$ |
| watt per square meter steradian | $W/(m^2\ sr)$ | radiance | — |
| watt per steradian | W/sr | radiant intensity | — |

[a]Liter (L) is a special name for cubic decimeter. The symbol M is not an SI unit, but expressions such as 0.1 M, meaning a solution with concentration of 0.1 mol/L, are acceptable.

**Table 11A-5.** Other Units

| *Name* | *Symbol* | *Quantity* | *Value in SI Units* |
|---|---|---|---|
| angstrom | Å | distance | 1 Å = $10^{-10}$ m = 0.1 nm |
| bar | bar | pressure | 1 bar = $10^5$ Pa = 100 kPa = 0.1 MPa |
| barn | b | area, cross section | 1 b = $10^{-28}$ $m^2$ = 100 $fm^2$ |
| bohr | b, $a_0$ | length | 1 b ≈ 5.291 77 × $10^{-11}$ m |
| curie[a] | Ci | activity | 1 Ci = 3.7 × $10^{10}$ Bq |
| dalton | Da | atomic mass | 1 Da = 1.660 540 × $10^{-27}$ kg |
| darcy[b] | darcy | permeability | — |
| day | day | time | 1 day = 24 h = 86 400 s |
| debye[c] | D | electric dipole moment | — |
| degree | ° | plane angle | 1° = (π/180) rad |
| degree Celsius[d] | °C | temperature | — |
| dyne[e] | dyn | force | — |
| einstein[f] | einstein | light energy | — |
| electronvolt[g] | eV | — | 1 eV = 1.602 19 × $10^{-19}$ J |
| erg[h] | erg | energy or work | — |
| faraday | faraday | electric charge | 1 faraday = 96 485.31 C |
| fermi | f | length | 1 f = $10^{-15}$ m |
| franklin | Fr | electric charge | 1 Fr = 3.335 64 × $10^{-10}$ C |
| galileo | Gal | acceleration | 1 Gal = $10^{-2}$ m $s^{-2}$ |
| gauss | G | magnetic induction | 1 G = $10^{-4}$ Wb/$m^2$ |
| gilbert[i] | Gi | magnetomotive force | — |
| hartree | hartree, $E_h$ | energy | 1 hartree = 4.359 75 × $10^{-18}$ J |
| hectare | ha | area | 1 ha = 1 $hm^2$ = $10^4$ $m^2$ |
| hour | h | time | 1 h = 60 min = 3600 s |
| liter | L | volume | 1 L = 1 $dm^3$ = $10^{-3}$ $m^3$ |
| metric ton | t | mass | 1 t = $10^3$ kg |
| minute | min | time | 1 min = 60 s |
| minute | ′ | plane angle | 1′ = (1/60)° = (π/10 800) rad |
| parsec | pc | length | 1 pc ≈ 3.085 68 × $10^{16}$ m |
| poise[j] | P | dynamic viscosity | — |
| rad | rad, rd[k] | absorbed dose | 1 rad = 0.01 Gy = 1 cGy = 100 erg·$g^{-1}$ |
| roentgen | R | exposure | 1 R = 2.58 × $10^{-4}$ C·$kg^{-1}$ |
| roentgen equivalent man[l] | rem | weighted absorbed dose | 1 rem = 0.01 Sv |
| second | ″ | plane angle | 1″ + (1/60)′ = (π/648 000) rad |
| stokes[m] | St | kinematic viscosity | — |
| svedberg | Sv | time | 1 Sv = $10^{-13}$ s |
| unified atomic mass unit[n] | u | — | 1 u = 1.660 540 × $10^{-27}$ kg |

[a]1 Ci = 2.2 × $10^{12}$ disintegrations per minute.

[b]1 darcy is the permeation achieved by the passage of 1 mL of fluid of 1 cP viscosity flowing in 1 s under a pressure of 1 atm (101 kPa) through a porous medium that has a cross-sectional area of 1 $cm^2$ and a length of 1 cm.

[c]1 D = $10^{-18}$ Fr cm.

[d]Temperature intervals in kelvins and degrees Celsius are identical; however, temperature in kelvins equals temperature in degrees Celsius plus 273.15.

[e]1 dyn is equal to the force that imparts an acceleration of 1 cm/$s^2$ to a 1 g mass.

[f]1 einstein equals Avogadro's number times the energy of one photon of light at the frequency in question.

[g]The electronvolt is the kinetic energy acquired by an electron in passing through a potential difference of 1 V in vacuum.

[h]1 erg is the work done by a 1 dyn force when the point at which the force is applied is displaced by 1 cm in the direction of the force.

[i]1 Gi is the magnetomotive force of a closed loop of one turn in which there is a current of (1/4π) × 10 A.

[j]1 P is the dynamic viscosity of a fluid in which there is a tangential force of 1 dyn/$cm^2$ resisting the flow of two parallel fluid layers past each other when their differential velocity is 1 cm/s per centimeter of separation.

[k]When there is a possibility of confusion with the symbol for radian, rd may be used as the symbol for rad.

[l]1 rem has the same biological effect as 1 rad of X-rays.

[m]1 St is the kinematic viscosity of a fluid with a dynamic viscosity of 1 P and a density of 1 g/$cm^3$.

[n]The unified atomic mass unit is equal to 1/12 of the mass of an atom of the nuclide $^{12}C$.

**Table 11A-6.** Non-SI Units That Are Discouraged

| *Discouraged Unit* | *Value in SI Units* |
|---|---|
| calorie (thermochemical) | 4.184 J |
| conventional millimeter of mercury | 133.322 Pa |
| grad | 2π/400 rad |
| kilogram-force | 9.806 65 N |
| metric carat | 0.2 g |
| metric horsepower | 735.499 W |
| mho | 1 S |
| micron | 1 μm |
| standard atmosphere | 101.325 kPa |
| technical atmosphere | 98.066 5 kPa |
| torr | 133.322 Pa |

➤➤➤➤➤ CHAPTER 12

# Names and Numbers for Chemical Compounds

The use of proper chemical nomenclature is essential for effective scientific communication. More than one million new substances are reported each year, each of which must be identified clearly, unambiguously, and completely in the primary literature. Chemical compounds are named according to the rules established by the International Union of Pure and Applied Chemistry (IUPAC), the International Union of Biochemistry and Molecular Biology (IUBMB) [formerly the International Union of Biochemistry (IUB)], the Chemical Abstracts Service (CAS), the Committee on Nomenclature, Terminology, and Symbols of the American Chemical Society, and other authorities as appropriate. For more information on naming chemical compounds, refer to the bibliography in Chapter 18. This chapter gives the editorial conventions and style points for chemical compound names.

## Components of Chemical Names

The names of chemical compounds may consist of one or more words, and they may include locants, descriptors, and syllabic portions. Locants and descriptors can be numerals, element symbols, small capital letters, Greek letters, Latin letters, italic words and letters, and combinations of these. Treat the word or syllabic portions of chemical names just like other common nouns: use roman type, keep them lowercase in text, capitalize them at the beginnings of sentences and in titles, and hyphenate them only when they do not fit completely on one line.

## Box 12-1. Correct Forms for Alcohols

| *Correct forms* | *Incorrect forms* |
|---|---|
| 1-butanol, butyl alcohol | |
| 2-butanol, *sec*-butyl alcohol | |
| 2-methyl-1-propanol, isobutyl alcohol | isobutanol |
| 2-methyl-2-propanol, *tert*-butyl alcohol | |
| 1-propanol, propyl alcohol | |
| 2-propanol, isopropyl alcohol | isopropanol |
| | any combination of *sec* or *tert* with the words *butanol* or *propanol* |

### *Locants and Descriptors*

➤ Numerals used as locants can occur at the beginning of or within a chemical name. They are set off with hyphens. (See Box 12-1.)

6-aminobenzothiazole
di-2-propenylcyanamide
4a,8a-dihydronaphthalene
5,7-dihydroxy-3-(4-hydroxyphenyl)-4*H*-1-benzopyran-4-one
6-hydroxy-2-naphthalenesulfonic acid
3′-methylphthalanilic acid

➤ Use italic type for chemical element symbols that denote attachment to an atom or a site of ligation.

*B*,*B*′-di-3-pinanyldiborane
bis[(ethylthio)acetato-*O*,*S*]platinum
glycinato-*N*
*N*-acetyl group
*N*-ethylaniline
*N*,*N*′-bis(3-aminopropyl)-1,4-butanediamine
*O*,*O*,*S*-triethyl phosphorodithioate
*P*-phenylphosphinimidic acid
*S*-methyl benzenethiosulfonate

➤ When element symbols are used with a type of reaction as a noun or adjective, use roman type for the symbol and hyphenate it to the word that follows it.

| | |
|---|---|
| N-acetylated | O-substituted |
| N-acetylation | O-substitution |
| N-oxidation | S-methylated |
| N-oxidized | S-methylation |

➤ Use italic type for the capital H that denotes indicated or added hydrogen.

1*H*-1,3-diazepine
3*H*-fluorene
2*H*-indene
phosphinin-2(1*H*)-one
2*H*-pyran-3(4*H*)-thione

➤ Use Greek letters, not the spelled-out forms, in chemical names to denote position or stereochemistry. Use a hyphen to separate them from the chemical name.

α-amino acid (*not* alpha amino acid)
β-naphthol (*not* beta naphthol)
5α,10β,15α,20α-tetraphenylporphyrin

Use the Greek letters eta (η) to indicate hapticity and kappa (κ) to designate the ligating atom in complicated formulas.

bis($\eta^6$-benzene)-chromium

bis(η-cyclopentadienyl) iron

[2-(diphenylphosphino-κ*P*)phenyl-κ$C^1$]hydrido(triphenylphosphine-κ*P*)-nickel(II)

*N,N'*-bis(2-amino-κ*N*-ethyl)ethane-1,2-diamine-κ*N*]chloroplatinum(II)

➤ Use italic type for positional, stereochemical, configurational, and descriptive structural prefixes when they appear with the chemical name or formula. Use a hyphen to separate them from the chemical name. Accepted prefixes include the following:

abeo
*ac*
*altro*
*amphi*
*anti*
*antiprismo*
*ar*
*arachno*
*as*
*asym*
*c*
*catena*
*cis*
*cisoid*
*closo*
*cyclo*
*d*
*dodecahedro*
*E*
*endo*
*erythro*
*exo*
*facgem*
*hexahedro*
*hexaprismo*
*hypho*
*icosahedro*
*klado*
*l*
*m*
*M*
*mer*
*meso*
*n*
*nido*
*o*
*octahedro*
*p*
*P*
*pentaprismo*
*quadro*
*r*
*R*
*R**
*rel*
*retro*
*ribo*
*s*
*S*
*S**
*sec*
*sn*
*sym*
*syn*
*t*
*tert*
*tetrahedro*
*threo*
*trans*
*transoid*
*triangulo*
*triprismo*
*uns*
*vic*
*xylo*
*Z*

➤ Do not capitalize prefixes that are shown here as lowercase, even at the beginning of a sentence or in a title (see Tables 12-1 and 12-2), and never use lowercase for those that are written in capital letters. Enclose the prefixes *E*, *R*, *R**, *S*, *S**, and *Z* in parentheses.

*anti*-bicyclo[3.2.1]octan-8-amine
*ar*-chlorotoluene
*as*-trichlorobenzene
*catena*-triphosphoric acid
*cis*-diamminedichloroplatinum
*cis*-[$PtCl_2(NH_3)_2$]
*cyclo*-hexasulfur, *c*-$S_6$
(*E*,*E*)-2,4-hexadienoic acid
(*E*,*Z*)-1,3-di-1-propenylnaphthalene
*erythro*-2,3-dibromosuccinic acid
*exo*-chloro-*p*-menthane
*m*-ethylpropylbenzene
*meso*-tartaric acid
*o*-dibromobenzene
*p*-aminoacetanilide
*s*-triazine
(*S*)-2,3-dihydroxypropanoic acid
5-*sec*-butylnonane
*sym*-dibromoethane
*tert*-pentyl bromide, *t*-$C_5H_{11}Br$
*threo*-2,3-dihydroxy-1,4-dimercaptobutane
*trans*-2,3-dimethylacrylic acid
*uns*-dichloroacetone
*vic*-triazine

➤ Use small capital letters D and L to indicate absolute configuration with amino acids and carbohydrates.

2-(difluoromethyl)-DL-ornithine
L-galactosamine
2-*O*-β-D-glucopyranosyl-α-D-glucose

➤ Use plus and minus signs enclosed in parentheses as stereochemical descriptors.

(±)-2-allylcyclohexanone
(+)-dihydrocinchonine
(–)-3-(3,4-dihydroxyphenyl)-L-alanine

➤ When the structural prefixes cyclo, iso, neo, and spiro are integral parts of chemical names, close them up to the rest of the name (without hyphens) and do not italicize them.

cyclohexane
isopropyl alcohol
neopentane

However, italicize and hyphenate cyclo as a nonintegral structural descriptor.

*cyclo*-octasulfur
*cyclo*-triphosphoric acid

➤ Use numerals separated by periods within square brackets in names of bridged and spiro alicyclic compounds.

bicyclo[3.2.0]heptane
bicyclo[4.4.0]decane
1-methylspiro[3.5]non-5-ene
spiro[4.5]decane

**Table 12-1.** Examples of Multiword Chemical Names

| *In Text* | *At Beginning of Sentence* | *In Titles and Headings* |
|---|---|---|
| ***Acids*** | | |
| benzoic acid | Benzoic acid | Benzoic Acid |
| ethanethioic *S*-acid | Ethanethioic *S*-acid | Ethanethioic *S*-Acid |
| hydrochloric acid | Hydrochloric acid | Hydrochloric Acid |
| ***Alcohols*** | | |
| ethyl alcohol | Ethyl alcohol | Ethyl Alcohol |
| ethylene glycol | Ethylene glycol | Ethylene Glycol |
| ***Ketones*** | | |
| di-2-naphthyl ketone | Di-2-naphthyl ketone | Di-2-naphthyl Ketone |
| methyl phenyl ketone | Methyl phenyl ketone | Methyl Phenyl Ketone |
| ***Ethers*** | | |
| di-*sec*-butyl ether | Di-*sec*-butyl ether | Di-*sec*-butyl Ether |
| methyl propyl ether | Methyl propyl ether | Methyl Propyl Ether |
| ***Anhydrides*** | | |
| acetic anhydride | Acetic anhydride | Acetic Anhydride |
| phthalic anhydride | Phthalic anhydride | Phthalic Anhydride |
| ***Esters*** | | |
| methyl acetate | Methyl acetate | Methyl Acetate |
| phenyl thiocyanate | Phenyl thiocyanate | Phenyl Thiocyanate |
| propyl benzoate | Propyl benzoate | Propyl Benzoate |
| ***Polymer Names*** | | |
| 1,2-polybutadiene | 1,2-Polybutadiene | 1,2-Polybutadiene |
| poly(butyl methacrylate) | Poly(butyl methacrylate) | Poly(butyl methacrylate) |
| poly(ethylene glycol) | Poly(ethylene glycol) | Poly(ethylene glycol) |
| poly(*N,N*-dimethylacrylamide) | Poly(*N,N*-dimethylacrylamide) | Poly(*N,N*-dimethylacrylamide) |
| ***Other Organic Compounds*** | | |
| aniline hydrochloride | Aniline hydrochloride | Aniline Hydrochloride |
| benzyl hydroperoxide | Benzyl hydroperoxide | Benzyl Hydroperoxide |
| butyl chloride | Butyl chloride | Butyl Chloride |
| dicyclohexyl peroxide | Dicyclohexyl peroxide | Dicyclohexyl Peroxide |
| diethyl sulfide | Diethyl sulfide | Diethyl Sulfide |
| methyl iodide | Methyl iodide | Methyl Iodide |
| 2-naphthoyl bromide | 2-Naphthoyl bromide | 2-Naphthoyl Bromide |
| sodium *S*-phenyl thiosulfite | Sodium *S*-phenyl thiosulfite | Sodium *S*-Phenyl Thiosulfite |
| *tert*-butyl fluoride | *tert*-Butyl fluoride | *tert*-Butyl Fluoride |
| ***Inorganic and Coordination Compounds*** | | |
| ammonium hydroxide | Ammonium hydroxide | Ammonium Hydroxide |
| bis(diethyl phosphato)zinc | Bis(diethyl phosphato)zinc | Bis(diethyl phosphato)zinc |
| calcium sulfate | Calcium sulfate | Calcium Sulfate |
| (dimethyl sulfoxide)-cadmium sulfate | (Dimethyl sulfoxide)-cadmium sulfate | (Dimethyl sulfoxide)-cadmium Sulfate |
| magnesium oxide | Magnesium oxide | Magnesium Oxide |
| sodium cyanide | Sodium cyanide | Sodium Cyanide |
| sulfur dioxide | Sulfur dioxide | Sulfur Dioxide |

**Table 12-2.** Locants and Descriptors in Chemical Names

| *In Text* | *At Beginning of Sentence* | *In Titles and Headings* |
|---|---|---|
| ***Numeral Locants*** | | |
| adenosine 5′-triphosphate | Adenosine 5′-triphosphate | Adenosine 5′-Triphosphate |
| 1,3-bis(bromomethyl)-benzene | 1,3-Bis(bromomethyl)-benzene | 1,3-Bis(bromomethyl)-benzene |
| 2-benzoylbenzoic acid | 2-Benzoylbenzoic acid | 2-Benzoylbenzoic Acid |
| 1-bromo-3-chloropropane | 1-Bromo-3-chloropropane | 1-Bromo-3-chloropropane |
| 2-(2-chloroethyl)pentanoic acid | 2-(2-Chloroethyl)pentanoic acid | 2-(2-Chloroethyl)pentanoic Acid |
| 7-(4-chlorophenyl)-1-naphthol | 7-(4-Chlorophenyl)-1-naphthol | 7-(4-Chlorophenyl)-1-naphthol |
| 1,2-dicyanobutane | 1,2-Dicyanobutane | 1,2-Dicyanobutane |
| 4a,8a-dihydronaphthalene | 4a,8a-Dihydronaphthalene | 4a,8a-Dihydronaphthalene |
| ***Element Symbol Locants*** | | |
| (2,3-butanedione dioximato-*O*,*O*′)copper | (2,3-Butanedione dioximato-*O*,*O*′)copper | (2,3-Butanedione dioximato-*O*,*O*′)copper |
| *N*-ethylaniline | *N*-Ethylaniline | *N*-Ethylaniline |
| *N*,2-dihydroxybenzamide | *N*,2-Dihydroxybenzamide | *N*,2-Dihydroxybenzamide |
| *N*,*N*′-dimethylurea | *N*,*N*′-Dimethylurea | *N*,*N*′-Dimethylurea |
| 3*H*-fluorene | 3*H*-Fluorene | 3*H*-Fluorene |
| *O*,*S*,*S*-triethyl phosphorodithioate | *O*,*S*,*S*-Triethyl phosphorodithioate | *O*,*S*,*S*-Triethyl Phosphorodithioate |
| *S*-methyl benzenethiosulfonate | *S*-Methyl benzenethiosulfonate | *S*-Methyl Benzenethiosulfonate |
| ***Greek Letter Locants and Descriptors*** | | |
| α-hydroxy-β-aminobutyric acid | α-Hydroxy-β-aminobutyric acid | α-Hydroxy-β-aminobutyric Acid |
| 17α-hydroxy-5β-pregnane | 17α-Hydroxy-5β-pregnane | 17α-Hydroxy-5β-pregnane |
| 1α-hydroxycholecalciferol | 1α-Hydroxycholecalciferol | 1α-Hydroxycholecalciferol |
| α-methylbenzeneacetic acid | α-Methylbenzeneacetic acid | α-Methylbenzeneacetic Acid |
| $\alpha_1$-sitosterol | $\alpha_1$-Sitosterol | $\alpha_1$-Sitosterol |
| β-chloro-1-naphthalenebutanol | β-Chloro-1-naphthalenebutanol | β-Chloro-1-naphthalenebutanol |
| β,4-dichlorocyclohexane-propionic acid | β,4-Dichlorocyclohexane-propionic acid | β,4-Dichlorocyclohexane-propionic Acid |
| β-endorphin | β-Endorphin | β-Endorphin |
| ω,ω′-dibromopolybutadiene | ω,ω′-Dibromopolybutadiene | ω,ω′-Dibromopolybutadiene |
| tris(β-chloroethyl)amine | Tris(β-chloroethyl)amine | Tris(β-chloroethyl)amine |
| ***Small Capital Letter Descriptors*** | | |
| β-D-arabinose | β-D-Arabinose | β-D-Arabinose |
| D-1,2,4-butanetriol | D-1,2,4-Butanetriol | D-1,2,4-Butanetriol |
| D-serine | D-Serine | D-Serine |
| DL-alanine | DL-Alanine | DL-Alanine |
| $D_S$-threonine | $D_S$-Threonine | $D_S$-Threonine |
| L-methionine | L-Methionine | L-Methionine |
| ***Positional and Structural Descriptors*** | | |
| 7-bromo-*p*-cymene | 7-Bromo-*p*-cymene | 7-Bromo-*p*-cymene |
| 4-chloro-*m*-cresol | 4-Chloro-*m*-cresol | 4-Chloro-*m*-cresol |
| *m*-hydroxybenzyl alcohol | *m*-Hydroxybenzyl alcohol | *m*-Hydroxybenzyl Alcohol |
| *n*-butyl iodide | *n*-Butyl iodide | *n*-Butyl Iodide |

*Continued on next page*

**Table 12-2.** Locants and Descriptors in Chemical Names—*Continued*

| *In Text* | *At Beginning of Sentence* | *In Titles and Headings* |
|---|---|---|
| ***Positional and Structural Descriptors—Continued*** | | |
| 2-(*o*-chlorophenyl)-1-naphthol | 2-(*o*-Chlorophenyl)-1-naphthol | 2-(*o*-Chlorophenyl)-1-naphthol |
| *o*-dibromobenzene | *o*-Dibromobenzene | *o*-Dibromobenzene |
| *p*-benzenediacetic acid | *p*-Benzenediacetic acid | *p*-Benzenediacetic Acid |
| *p*-*tert*-butylphenol | *p*-*tert*-Butylphenol | *p*-*tert*-Butylphenol |
| *s*-triazine | *s*-Triazine | *s*-Triazine |
| *sec*-butyl alcohol | *sec*-Butyl alcohol | *sec*-Butyl Alcohol |
| *sym*-dibromoethane | *sym*-Dibromoethane | *sym*-Dibromoethane |
| *tert*-pentyl isovalerate | *tert*-Pentyl isovalerate | *tert*-Pentyl Isovalerate |
| 1-(*trans*-1-propenyl)-3-(*cis*-1-propenyl)naphthalene | 1-(*trans*-1-Propenyl)-3-(*cis*-1-propenyl)naphthalene | 1-(*trans*-1-Propenyl)-3-(*cis*-1-propenyl)naphthalene |
| ***Stereochemical Descriptors*** | | |
| *anti*-bicyclo[3.2.1]octan-8-amine | *anti*-Bicyclo[3.2.1]octan-8-amine | *anti*-Bicyclo[3.2.1]octan-8-amine |
| *cis*-1,2-dichloroethene | *cis*-1,2-Dichloroethene | *cis*-1,2-Dichloroethene |
| *d*-camphor | *d*-Camphor | *d*-Camphor |
| *dl*-2-aminopropanoic acid | *dl*-2-Aminopropanoic acid | *dl*-2-Aminopropanoic Acid |
| (*E*)-diphenyldiazene | (*E*)-Diphenyldiazene | (*E*)-Diphenyldiazene |
| *endo*-2-chlorobicyclo[2.2.1]-heptane | *endo*-2-Chlorobicyclo[2.2.1]-heptane | *endo*-2-Chlorobicyclo[2.2.1]-heptane |
| (+)-*erythro*-2-amino-3-methylpentanoic acid | (+)-*erythro*-2-Amino-3-methylpentanoic acid | (+)-*erythro*-2-Amino-3-methylpentanoic Acid |
| *erythro*-β-hydroxyaspartic acid | *erythro*-β-Hydroxyaspartic acid | *erythro*-β-Hydroxyaspartic Acid |
| *exo*-bicyclo[2.2.2]oct-5-en-2-ol | *exo*-Bicyclo[2.2.2]oct-5-en-2-ol | *exo*-Bicyclo[2.2.2]oct-5-en-2-ol |
| *exo*-5,6-dimethyl-*endo*-bicyclo[2.2.2]octan-2-ol | *exo*-5,6-Dimethyl-*endo*-bicyclo[2.2.2]octan-2-ol | *exo*-5,6-Dimethyl-*endo*-bicyclo[2.2.2]octan-2-ol |
| L-*threo*-2,3-dichlorobutyric acid | L-*threo*-2,3-Dichlorobutyric acid | L-*threo*-2,3-Dichlorobutyric Acid |
| *meso*-tartaric acid | *meso*-Tartaric acid | *meso*-Tartaric Acid |
| (1*R**,3*S**)-1-bromo-3-chlorocyclohexane | (1*R**,3*S**)-1-Bromo-3-chlorocyclohexane | (1*R**,3*S**)-1-Bromo-3-chlorocyclohexane |
| *rel*-(1*R*,3*R*)-1-bromo-3-chlorocyclohexane | *rel*-(1*R*,3*R*)-1-Bromo-3-chlorocyclohexane | *rel*-(1*R*,3*R*)-1-Bromo-3-chlorocyclohexane |
| (*S*)-2,3-dihydroxypropanoic acid | (*S*)-2,3-Dihydroxypropanoic acid | (*S*)-2,3-Dihydroxypropanoic Acid |
| *sn*-glycerol 1-(dihydrogen phosphate) | *sn*-Glycerol 1-(dihydrogen phosphate) | *sn*-Glycerol 1-(Dihydrogen phosphate) |
| *syn*-7-methylbicyclo[2.2.1]-heptene | *syn*-7-Methylbicyclo[2.2.1]-heptene | *syn*-7-Methylbicyclo[2.2.1]-heptene |
| *trans*-*cisoid*-*trans*-perhydrophenanthrene | *trans*-*cisoid*-*trans*-Perhydrophenanthrene | *trans*-*cisoid*-*trans*-Perhydrophenanthrene |
| (*Z*)-5-chloro-4-pentenoic acid | (*Z*)-5-Chloro-4-pentenoic acid | (*Z*)-5-Chloro-4-pentenoic Acid |
| (1*Z*,4*E*)-1,2,4,5-tetrachloro-1,4-pentadiene | (1*Z*,4*E*)-1,2,4,5-Tetrachloro-1,4-pentadiene | (1*Z*,4*E*)-1,2,4,5-Tetrachloro-1,4-pentadiene |

➤ Use italic letters within square brackets in names of polycyclic aromatic compounds.

dibenz[*a,j*]anthracene
dibenzo[*c,g*]phenanthrene
dicyclobuta[*de,ij*]naphthalene
1*H*-benzo[*de*]naphthacene
indeno[1,2-*a*]indene

### *Syllabic Portion of Chemical Names*

Multiplying affixes are integral parts of the chemical name; they are set in roman type and are always closed up to the rest of the name (without hyphens). Use hyphens only to set off intervening locants or descriptors. Use enclosing marks (parentheses, brackets, or braces) to ensure clarity or to observe other recommended nomenclature conventions. Multiplying prefixes include the following:

- hemi, mono, di, tri, tetra, penta, hexa, hepta, octa, ennea, nona, deca, deka, undeca, dodeca, etc.
- semi, uni, sesqui, bi, ter, quadri, quater, quinque, sexi, septi, octi, novi, deci, etc.
- bis, tris, tetrakis, pentakis, hexakis, heptakis, octakis, nonakis, decakis, etc.

3,4′-bi-2-naphthol
2,2′-bipyridine
bis(benzene)chromium(0)
1,4-bis(3-bromo-1-oxopropyl)piperazine
1,3-bis(diethylamino)propane
di-*tert*-butyl malonate
dichloride
1,2-ethanediol
hemihydrate
hexachlorobenzene
2,4,6,8-nonanetetrone
pentachloroethane
3,4,5,6-tetrabromo-*o*-cresol
tetrakis(hydroxymethyl)methane
tri-*sec*-butylamine
triamine
triethyl phosphate
tris(amine)
2,3,5-tris(aziridin-1-yl)-*p*-benzoquinone
tris(ethylenediamine)cadmium dihydroxide

## Capitalization of Chemical Names

Chemical names are not capitalized unless they are the first word of a sentence or are part of a title or heading. Then, the first letter of the syllabic portion is capitalized, not the locant, stereoisomer descriptor, or positional prefix. Table 12-1 presents examples of simple chemical names and their capitalization. Table 12-2 presents chemical names that include locants and descriptors.

➤ Some reaction names are preceded by element symbols; they may be used as nouns or adjectives. When they are the first word of a sentence or appear in titles and headings, capitalize the first letter of the word. Do not italicize the element symbol.

N-Oxidation of the starting compounds yielded compounds **3–10.**

N-Benzoylated amines undergo hydroxylation when incubated with yeast.

Preparation of S-Methylated Derivatives

O-Substituted Structural and Functional Analogues

## Punctuation in Chemical Names

➤ Use commas between numeral locants, chemical element symbol locants, and Greek locants, with no space after the comma. When a single locant consists of a numeral and a Greek letter together with no space or punctuation, the numeral precedes the Greek letter. When the Greek letter precedes the numeral, they indicate two different locants and should be separated by a comma. For example, α,2 denotes two locants; 1α is viewed as one locant.

(6α,11β,16α)-6-fluoro-16-methylpregna-1,4-diene
β,4-dichlorocyclohexanepropionic acid
1,2-dinitrobutane
*N*,*N*-dimethylacetamide
2,3,3a,4-tetrahydro-1*H*-indole

➤ Use hyphens to separate locants and configurational descriptors from each other and from the syllabic portion of the name.

α-ketoglutaric acid
2-benzoylbenzoic acid
1,4-bis(2-ethylhexyl) sulfosuccinate
3-chloro-4-methylbenzoic acid
*cis*-dichloroethylene
D-arabinose
(1,4-dioxaspiro[4.5]dec-2-ylmethyl) guanidine
(*E*)-2-(3,7-dimethyl-2,6-octadienyl)-1,4-benzenediol
*N*-hydroxy-*N*-nitrosobenzeneamine
*N*-methylmethanamine
4-*O*-β-D-galactopyranosyl-D-fructose
tetrahydro-3,4-dipiperonyl-2-furanol
*trans*-2-bromocyclopentanol

➤ Do not use hyphens to separate the syllables of a chemical name unless the name is too long to fit on one line. Appendix 12-1 is a list of prefixes, suffixes, roots, and some complete words hyphenated as they would be at the end of a line.

# Specialized Groups of Chemicals

## *Polymers*

Polymer names are often one or two words in parentheses following the prefix “poly”. “Poly” is a syllabic prefix, not a descriptor, and thus is set in roman type. Here is a short list of correctly formatted names of frequently cited polymers. (These names are not necessarily IUPAC or CA index preferences.)

nylon-6
nylon-6,6
polyacrylamide
poly(acrylic acid)
polyacrylonitrile
polyamide
poly(aryl sulfone)
polybutadiene
1,2-polybutadiene
1,4-polybutadiene
poly(butyl acrylate)
poly(butyl methacrylate)
poly(*n*-butyl methacrylate)
poly(butylene terephthalate)
polycarbonate
polychloroprene
poly(*N*,*N*-dimethylacrylamide)
poly(dimethylsiloxane)
polyester
polyether
poly(ether imide)
poly(ether ketone)
poly(ether sulfone)
poly(ethyl acrylate)
poly(ethyl methacrylate)
polyethylene
poly(ethylene adipate)
poly(ethylene glycol)
poly(ethylene oxide)
poly(ethylene terephthalate)
polyformaldehyde
polyimidazole
polyimide
poly(isobutyl methacrylate)
polyisobutylene
polyisoprene
poly(methacrylic acid)
poly(methyl acrylate)
poly(methyl methacrylate)
poly(methylene)
poly(*N*,*N*′-hexamethyleneadipamide)
poly(oxy-1,4-phenylene)
poly(oxyethylene)
poly(oxymethylene)
poly(phenylene ether)
poly(phenylene oxide)
poly(phenylene sulfide)
polypropylene
poly(propylene glycol)
polystyrene
polysulfide
polysulfone
poly(tetrafluoroethylene)
poly(tetramethylene oxide)
polythiazole
poly(thiocarbonate)
polyurethane
poly(vinyl acetate)
poly(vinyl alcohol)
poly(vinyl butyral)
poly(vinyl chloride)
poly(vinyl ether)
poly(vinyl trichloroacetate)
poly(vinylidene chloride)
poly(vinylpyrrolidone)
povidone

➤ In text, keep polymer names lowercase. As the first word of a sentence and in titles or headings, capitalize only the first letter of the polymer name.

New Uses for Poly(ethylene terephthalate)

Poly(vinyl chloride) is a less useful polymer than poly(ethylene glycol).

Reactions of Poly(methyl methacrylate)

➤ In copolymer nomenclature, descriptive lowercase italic infixes may be used. These include alt, blend, block (or b), co, cross, graft (or g), inter, per, stat, and ran.

polybutadiene-*graft*-[polystyrene:poly(methyl methacrylate)]
poly(*cross*-butadiene)
poly[*cross*-(ethyl acrylate)]-*inter*-polybutadiene
poly(ethylene-*alt*-carbon monoxide)
polyisoprene-*blend*-polystyrene
poly[(methyl methacrylate)-*b*-(styrene-*co*-butadiene)]
poly[(methyl methacrylate)-*co*-styrene]
polystyrene-*block*-polybutadiene
poly(styrene-*co*-butadiene)
poly(styrene-*g*-acrylonitrile)
poly(vinyl trichloroacetate)-*cross*-polystyrene

## *Saccharides*

Abbreviations for the major monosaccharides are presented in Table 12-3.

➤ Designation of the bond between two monosaccharides should specify the first residue, its anomeric configuration (α or β), the position of attachment on both sugar residues, and the second residue. Additional locants and configurational descriptors may be present but are not required. A number of variations may be used.

Galβ1→4Glc　　Galβ1–4Glc　　Galβ1,4Glc

Galβ(1→4)Glc　　Galβ(1–4)Glc　　Galβ(1,4)Glc

**Table 12-3.** Abbreviations for Major Monosaccharides

| *Saccharide* | *Abbreviation* |
|---|---|
| arabinose | Ara |
| fucose | Fuc |
| galactose | Gal |
| *N*-acetylgalactosamine | GalNAc |
| glucose | Glc |
| *N*-acetylglucosamine | GlcNAc |
| glucuronic acid | GlcA |
| mannose | Man |
| *N*-acetylneuraminic acid | Neu5Ac |
| *N*-glycoloylneuraminic acid | Neu5Gc |
| rhamnose | Rha |
| xylose | Xyl |

➤ For larger oligosaccharides, use parentheses or brackets for branched residues:

Manα1→6(Manα1→3)Manβ1→4GlcNAcβ1→4(±Fucα1→6)GlcNAc

Manα(1–6)[Manα(1–3)]Manβ(1–4)GlcNAcβ(1–4)[±Fucα(1–6)]GlcNAc

## Nucleic Acids

Table 12-4 presents the standard abbreviations for nucleic acids.

➤ By convention, numbering of the bases is unprimed and numbering of the sugars is primed. A nucleic acid polymer typically has the phosphate group attached to the 5′ position of the first nucleotide (the 5′ end), and the other end terminates in a hydroxyl group at the 3′ position of the sugar of the last nucleotide (the 3′ end). By convention, nucleotide sequences are almost always read and presented in the 5′ to 3′ direction. The sequence may be presented in unbroken form or in evenly spaced blocks.

5′-TAGCTAACCCGTTTTAGCGTCGTC-3′

5′-TAGCT AACCC GTTTT AGCGT CGTC-3′

➤ Complementary base pairs are joined by hydrogen bonds and are best represented by a centered dot: dA·dT and dG·dC are the canonical pairings in double-stranded DNA, although alternate pairings, DNA·RNA hybrids, triplexes, and other variations may occur.

**Table 12-4.** Abbreviations for Nucleic Acids

| | | | *Abbreviation for* | | |
|---|---|---|---|---|---|
| *Base* | *Sugar* | *Nucleoside*[a] *(base + sugar)* | *Nucleoside Monophosphate*[b] | *Nucleoside Diphosphate* | *Nucleoside Triphosphate* |
| ***Naturally occurring in RNA, ribonucleic acid*** | | | | | |
| adenine | ribose | adenosine, A | AMP | ADP | ATP |
| cytosine | ribose | cytidine, C | CMP | CDP | CTP |
| guanine | ribose | guanosine, G | GMP | GDP | GTP |
| uracil | ribose | uridine, U | UMP | UDP | UTP |
| ***Naturally occurring in DNA, deoxyribonucleic acid*** | | | | | |
| adenine | deoxyribose | deoxyadenosine, dA | dAMP | dADP | dATP |
| cytosine | deoxyribose | deoxycytidine, dC | dCMP | dCDP | dCTP |
| guanine | deoxyribose | deoxyguanosine, dG | dGMP | dGDP | dGTP |
| thymine | deoxyribose | thymidine, dT[c] | dTMP | dTDP | dTTP |

[a]Other nucleosides may occur (e.g., inosine, I, or xanthosine, X), but those listed are the canonical building blocks of nucleic acids and need never be defined. N may be used for an unspecified nucleoside or a mixture of all four (i.e., dNTPs were added to the reaction mixture).

[b]A nucleoside monophosphate is also known as a nucleotide.

[c]Thymidine consists of thymine + deoxyribose by definition and is abbreviated dT by definition; deoxythymidine would imply a different compound lacking another oxygen.

## *Amino Acids*

➤ The abbreviations of the essential 20 amino acids are listed in Table 12-5; these abbreviations do not need to be defined. In analytical situations, undifferentiated mixtures of aspartic acid/asparagine (Asx, B) or glutamic acid/glutamine (Glx, Z) may occur.

➤ For amino acids other than the essential 20, three-letter abbreviations may be used and defined at their first appearance.

| | |
|---|---|
| homocysteine | Hcy |
| hydroxyproline | Hyp |
| norvaline | Nva |
| ornithine | Orn |

Three-letter amino acid abbreviations may be preceded by a configurational designator, D or L, set in small capital letters.

➤ Always capitalize the three-letter and one-letter abbreviations for amino acids.

➤ In sequences of amino acids, separate the three-letter abbreviations with hyphens.

Pro-Gln-Ile-Ala

**Table 12-5.** Abbreviations for Amino Acids

| *Amino Acid* | *Three-Letter Abbreviation* | *One-Letter Abbreviation* |
|---|---|---|
| alanine | Ala | A |
| arginine | Arg | R |
| asparagine | Asn | N |
| aspartic acid | Asp | D |
| cysteine | Cys | C |
| glutamic acid | Glu | E |
| glutamine | Gln | Q |
| glycine | Gly | G |
| histidine | His | H |
| isoleucine | Ile | I |
| leucine | Leu | L |
| lysine | Lys | K |
| methionine | Met | M |
| phenylalanine | Phe | F |
| proline | Pro | P |
| serine | Ser | S |
| threonine | Thr | T |
| tryptophan | Trp | W |
| tyrosine | Tyr | Y |
| valine | Val | V |

➤ Do not use abbreviations for individual amino acids in running text.

Selective labeling of arginine and serine made it possible to monitor the kinetics of folding of the individual residues.

➤ Position numbers may follow one- or three-letter abbreviations or spelled-out names and may be closed up, hyphenated, spaced, or superscripted. Designation of a mutation includes the original amino acid, the position number, and the new amino acid:

A134V
Ala134Val
alanine-134 → valine

## Combinatorial Compounds

➤ Combinatorial libraries are described by generic representations, which consist of a generic structure together with one or more lists of substituents. The position of substituents is indicated by superatoms in the generic structure.

➤ Superatoms are designated by roman letters: R for any set of substituents or residues, for example, or Ar for a list of aromatic substituents. Superatoms may be distinguished by the addition of designation digits, typically a subscript following the superatom: $R_A$, $R_B$, and $R_C$ could specify the residue of reagents A, B, and C.

➤ Subscript numbers are generally used to indicate the order in which residues were introduced to the reaction scheme: $R_1$, $R_2$, $R_3$, and so on.

➤ The *Journal of Combinatorial Chemistry* recommends the use of ChemSet notation, in which a structure number is followed by the reagent sets associated with it. The structure number is typeset in bold, and the reagent sets are set in italics and enclosed in curly brackets.

**1**{*1–5*} + **2**{*1–6*} + **4**{*1–4*} → **5**{*1–5,1–4,1–6*}

**4**{*1–5,1–4,1–6*}

➤ The composition of a final library can be described in terms of mixtures or separate products. An uppercase X is used to indicate products mixed; an uppercase O is used to indicate products separate.

**4**{X,X,O} **4**{X *1–5*,X *1–4*,O *1–6*}

**4**{O,O,O} **4**{O *1–5*,O *1–4*,O *1–6*}

Additional information on the representation of combinatorial chemistry is given in Appendix 12-2.

➤➤➤➤➤

APPENDIX 12-1

# End-of-Line Hyphenation of Chemical Names

This appendix contains a list of prefixes, suffixes, roots, and some complete words hyphenated as they would be at the end of a line. To hyphenate a chemical name such as

5-(2-chloroethyl)-9-(diaminomethyl)-2-anthracenol

look up each syllable that is to be hyphenated in the list. Also, chemical names can be broken after hyphens that are integral in their names. Follow other standard rules for hyphenation of regular words; for example, try to leave at least three characters on each line. Thus, the example given could be hyphenated as follows:

5-(2-chlo-ro-eth-yl)-9-(di-ami-no-meth-yl)-2-anth-ra-cenol

Most desk dictionaries contain the names of common chemicals; they also give end-of-line hyphenation.

ace-naph-tho
ace-tal
acet-al-de-hyde
acet-amide
acet-ami-do
acet-amin-o-phen
acet-an-i-lide
ace-tate
ac-et-azol-amide
ace-tic
ace-to
ace-to-ace-tic
ace-tone
ace-to-ni-trile
ace-tyl
acet-y-late
acet-y-lene
acro-le-in
ac-ryl-am-ide
ac-ry-late
acryl-ic
ac-ry-lo
ad-i-po-yl
al-kyl
al-lyl
ami-di-no
amide
ami-do
amine
ami-no
am-mine
am-mo-nio
am-mo-ni-um
an-thra
an-thra-cene
an-thra-ce-no
an-thryl
ar-se-nate
ar-si-no
aryl
az-i-do
azi-no
azo
benz-ami-do
ben-zene
benz-hy-dryl
ben-zo-yl
ben-zyl
ben-zyl-i-dene
bi-cy-clo
bo-ryl
bro-mide
bro-mo
bu-tane
bu-ten-yl
bu-tyl
bu-tyl-ene
bu-tyl-i-dene
car-ba-mate
car-bam-ic
car-ba-mide
carb-an-ion
car-ba-ryl

car-ba-zole
car-bi-nol
car-bol-ic
car-bon-ate
car-bon-ic
car-bo-ni-um
car-bon-yl
car-box-ami-do
car-boxy
car-box-yl
car-byl-a-mi-no
chlo-ride
chlo-ro
chlo-ro-syl
chlo-ryl
cu-mene
cy-a-nate
cy-a-nide
cy-a-na-to
cy-a-no
cy-clo
cy-clo-hex-ane
cy-clo-hex-yl
di-azo
di-bo-ran-yl
di-car-bon-yl
di-im-ino
di-oxy
di-oyl
diyl
do-de-cyl
ep-oxy
eth-ane
eth-a-no
eth-a-nol
eth-a-no-yl
eth-en-yl
eth-yl
eth-yl-ene
eth-yl-i-dene
eth-yn-yl
fluo-res-cence
fluo-ride
fluo-ro
form-al-de-hyde
form-ami-do
for-mic
form-imi-do-yl
for-myl
fu-ran
ger-myl
gua-ni-di-no
gua-nyl
halo
hep-tane
hep-tyl
hex-ane
hex-yl
hy-dra-zide
hy-dra-zine
hy-dra-zi-no
hy-dra-zo
hy-dra-zo-ic
hy-dric
hy-dride
hy-dri-od-ic
hy-dro
hy-dro-chlo-ric
hy-dro-chlo-ride
hy-dro-chlo-ro
hy-drox-ide
hy-droxy
hy-drox-yl
imi-da-zole
imide
imi-do
imi-do-yl
imi-no
in-da-mine
in-da-zole
in-dene
in-de-no
in-dole
io-date
io-dide
iodo
io-do-syl
io-dyl
iso-cy-a-na-to
iso-cy-a-nate
iso-cy-a-nide
iso-pro-pen-yl
iso-pro-pyl
mer-cap-to
mer-cu-ric
meth-an-ami-do
meth-ane
meth-ano
meth-yl
meth-yl-ate
meth-yl-ene
meth-yl-i-dene
mono
mono-ac-id
mono-amine
naph-tha-lene
naph-tho
naph-thyl
neo-pen-tyl
ni-trate
ni-tric
ni-trile
ni-trilo
ni-trite
ni-tro
ni-troso
oc-tane
oc-tyl
ox-idase
ox-ide
ox-ido
ox-ime
oxo
ox-o-nio
oxy
palm-i-toyl
pen-tane
pen-tyl
pen-tyl-i-dene
per-chlo-rate
per-chlo-ride
per-chlo-ryl
per-man-ga-nate
per-ox-idase
per-ox-ide
per-oxy
phen-ac-e-tin
phen-an-threne

phen-an-thro
phen-an-thryl
phen-a-zine
phe-no
phe-nol
phe-none
phen-ox-ide
phen-oxy
phen-yl
phen-yl-ene
phos-phate
phos-phide
phos-phine
phos-phi-no
phos-phin-yl
phos-phite
phos-pho
phos-pho-nio
phos-pho-no
phos-phor-anyl
phos-pho-li-pase
phos-pho-lip-id
phos-pho-ni-um
phos-pho-ric
phos-pho-rus
phos-pho-ryl
plum-byl
pro-pane
pro-pen-yl
pro-pen-yl-ene
pro-pyl
pro-pyl-ene
pro-pyl-i-dene
pu-rine
py-ran
pyr-a-zine
pyr-a-zole
pyr-i-dine
pyr-id-a-zine
pyr-role
quin-o-line
qui-none
sel-e-nate
se-le-nic
sel-e-nide
sel-e-nite
se-le-no
si-lane
sil-anyl
sil-ox-anyl
sil-ox-yl
si-lyl
spi-ro
stan-nic
stan-nite
stan-nous
stan-nyl
stib-ino
sty-rene
sty-ryl
sul-fa-mo-yl
sul-fate
sul-fe-no
sul-fe-nyl
sul-fide
sul-fi-do
sul-fi-no
sul-fi-nyl
sul-fite
sul-fo
sul-fon-ami-do
sul-fo-nate
sul-fone
sul-fon-ic
sul-fo-nio
sul-fo-nyl
sulf-ox-ide
sul-fu-ric
sul-fu-rous
sul-fu-ryl
tet-ra
thio
thio-nyl
thio-phene
thi-oxo
thi-oyl
tol-u-ene
tol-u-ide
tol-yl
tri-a-zine
tri-a-zole
tri-yl
urea
ure-ide
ure-ido
uric
vi-nyl
vi-nyl-i-dene
xan-thene
xan-tho
xy-lene
xy-li-dine
xy-lyl
xy-li-din-yl
yl-i-dene

➤➤➤➤➤

# APPENDIX 12-2

# Representation of Combinatorial Chemistry

Derek Maclean

Combinatorial chemistry entails the reaction of *sets* of reagents to produce sets or *libraries* of products in numbers up to the size of each reagent set multiplied together. Thus 5 reagents A, plus 4 reagents B, plus 6 reagents C could prepare $5 \times 4 \times 6$, or 120 products ABC. The unique feature of this strategy is the ability to efficiently prepare very large numbers of compounds—a so-called *combinatorial library*. The challenges in describing combinatorial chemistry consist of concisely but accurately reporting the constituents and form of such a compound collection.

The individual products of combinatorial chemistry are called *members*. The distinction between a member and a *compound* is important and is based on the respective level of characterization. A compound will meet the typical standards for reporting new chemical structures; a member will fall short of that standard. In fact, a member may, in principle, simply be *expected* to be present in a library, especially for large libraries, and those that consist of mixtures rather than discrete samples of members.

The structural representation of combinatorial chemistry consists of a *generic structure* plus a list of *substituents* that may be present in that structure.

The generic structure is very similar to a typical structural formula with the addition of special notation to indicate the potential for variable substitution at certain parts of the molecule.

The position of variable substituents on a generic structure is indicated by *superatoms*, such as the R designation. In a combinatorial library, R need not simply designate an alkyl radical but is conventionally used to represent any set of substituents or *residues*. The residues R typically define those portions of reagents or building blocks that are found in the final product of the synthesis and that vary among library members. Particular superatoms may more precisely define the composition of a library, such as Ar for a list of aromatic substituents.

To distinguish superatoms within a generic structure, it is typical to use additional designation digits. Thus $R_A$, $R_B$, and $R_C$ could specify the residue of reagents A, B, and C in our example. The position of the additional digit has not been standardized but may be typically added as a subscript following the superatom.

The designation digit may usefully convey additional information beyond the simple differentiation of superatoms. For instance, $R_1$, $R_2$, $R_3$, and so on may indicate the order in which these residues were introduced to the reaction scheme. Alternatively, $R_X$ or $R_O$ make use of the pool designations X and O (see the section "ChemSet Notation") to concisely describe library composition in a generic structure.

It is often possible to draw multiple generic structures for a combinatorial library. Care should be taken in the choice of the appropriate structure, taking into account the purpose for the structure. Often the maximum common substructure is a good choice, but other structures may better illustrate structure–activity relationships or the synthetic potential of a combinatorial reaction strategy. The relationship between residue and reagent or product should be defined. This may be conveniently shown in the generic reaction scheme; see Figure 12A-1.

A typical feature of combinatorial synthesis is the use of techniques that facilitate the isolation of products and intermediates. The most common example is the attachment to a solid support, such as a polymer bead, to allow isolation by simple filtration (*solid-phase chemistry*). A variety of solid and soluble supports have been developed. These may be attached to either the products or the reagents of a library.

Such supports can be dealt with in generic reaction schemes as a special type of superatom. Often, a pictorial representation is used. A polymer bead may be conveniently represented by an s orbital in common chemistry software packages. Few standards have been developed for the representation of other supports, but a filled structure designates a solid support and an open structure may be used for a soluble support.

## ChemSet Notation

A convenient descriptive notation for combinatorial libraries (recommended by the *Journal of Combinatorial Chemistry*) is the ChemSet terminology.

A ChemSet is denoted by a structure number followed by the reagent sets associated with that structure. Thus the combinatorial library with 120 products could be described as follows, where the numbers in curly brackets define the reagents which were used to prepare that library; see Figure 12A-1.

**5**{*1–5,1–6,1–4*}

The synthetic scheme to prepare such a library may also be described in terms of ChemSets:

**1**{*1–5*} + **2**{*1–6*} + **4**{*1–4*} → **5**{*1–5,1–6,1–4*}

The composition of the final library may consist of a mixture of all members, or as a collection of discrete product samples, or as a set of smaller *pools*. An exten-

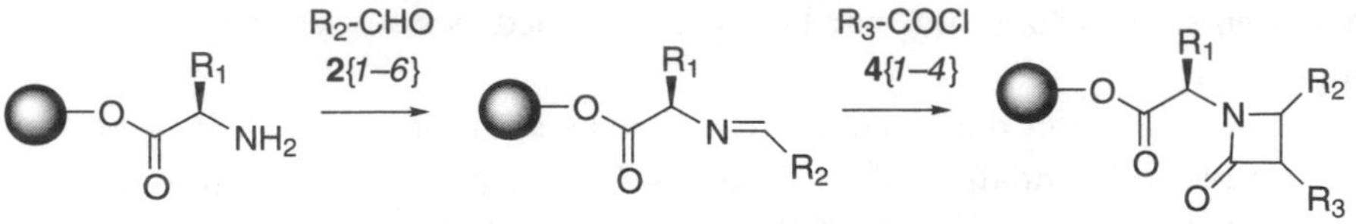

**Figure 12A-1.** Combinatorial reaction scheme and ChemSet notation.

*Source:* Reprinted from *J. Am. Chem. Soc.* **1996,** *118,* 253–254. Copyright 1996 American Chemical Society.

sion to the ChemSet notation allows these situations to be precisely defined. The useful designations X and O indicate respectively, "products mixed" and "products separate". Thus, the library **5**{X *1–5*,X *1–6*,O *1–4*} is a set of 6 pools, each with 20 members deriving from a single reagent from set **4**, since the O indicates that the six individual product mixtures were kept separate from one another after adding reagents **4**. In contrast, **5**{O *1–5*,O *1–4*,O *1–6*} represents 120 discrete product samples, also called *parallel synthesis,* since the O for each set indicates that all products were kept separate at each stage.

If all of a given set of reagents have been used in a given pool, then the number for that set may be omitted. Thus, **5**{X,X,O} and **5**{O,O,O} accurately describe the composition of the above libraries. However, if a subset of possible reagents have been used, this must be indicated using the numerical list format, e.g., **5**{X *1–6*,O *1–2,3*}.

➤➤➤➤➤

APPENDIX 12-3

# CAS Registry Numbers

Chemical substances, their syntheses, the determination of their properties, and their applications are the core of chemistry and the main occupation of chemists. In their communications, chemists represent chemical substances by structural diagrams, names, molecular formulas, codes, and identification numbers. One of the most frequently used identification numbers is the CAS Registry Number. Today, CAS Registry Numbers are often used to identify chemical substances in handbooks, indexes, databases, and inventories, and even on many commercial product labels.

The CAS Chemical Registry System is a computer-based system that uniquely identifies chemical substances on the basis of their molecular structures. Begun originally in 1965 to support indexing for *Chemical Abstracts* (*CA*), the CAS Chemical Registry System now serves not only as a support system for identifying substances within Chemical Abstracts Service operations but also as an international resource for chemical substance identification by scientists, industry, and regulatory bodies. The CAS Registry provides a means of bridging the many differences in systematic, generic, proprietary, and trivial substance names that may be used to identify a single substance.

The CAS Chemical Registry System database is the largest collection of information on naturally occurring and synthetic chemical substances in the world, including organic compounds, inorganic compounds, organometallics, metals, polymers, coordination compounds, alloys, elements, isotopes, nuclear particles, proteins, nucleic acids, and minerals. By the end of 2005, the CAS Chemical Registry System contained records for more than 27 million organic and inorganic substances, with new records added at the rate of some 5000 per day. A running total of registered substances can be found on the CAS Web site at http://www.cas.org/cgi-bin/regreport.pl.

The database contains CAS Registry Numbers, structures, and names for substances reported in the chemical literature covered in *CA,* in addition to substances registered from special collections, for governmental and industrial organizations, and for individual requesters. CAS Registry Numbers are also assigned to sequences such as DNA and proteins.

CAS Registry Numbers are assigned in sequential order as substances are entered into the CAS Chemical Registry System database for the first time; the numbers have no chemical significance. CAS Registry Numbers link the molecular structure diagram, systematic *CA* index name, synonyms, molecular formula, and other identifying information for each substance. Because CAS Registry Numbers are independent of the many different systems of chemical nomenclature, they can bridge these systems and link often unrecognized synonymous names.

A format was developed using hyphens to make the numbers easier to read and to recognize. A CAS Registry Number includes up to nine digits that are separated into three parts by hyphens. The first part, starting from the left, has up to six digits, the second part has two digits, and the final part is a single check digit to verify the validity of the total number (e.g., 7732-18-5 for water).

Within the registry system, each substance is assigned a separate CAS Registry Number. For example, each salt of an acid receives a distinct number, and an ion receives a number different from that of the neutral compound.

CAS Registry Numbers are included in the printed *Chemical Abstracts* chemical substance and formula indexes and in the CAS databases. The full set of CAS Chemical Registry System database information—structures, names, formulas, and ring data—is available for search and display through STN International, SciFinder, and other CAS search services. CAS Registry information is also available in CAS databases offered by other online system vendors.

In addition to their inclusion in the CAS databases, CAS Registry Numbers are used in many public and private databases. Many handbooks, guides, and other reference works include CAS Registry Numbers and provide special indexes that allow the reader to find the proper place in the text without first having to identify the full name of the substance. The reader benefits because the full name may differ from handbook to handbook.

CAS Registry Numbers are also widely used as standard identifiers for chemical substances in many of the commercial chemical inventories of governmental regulatory agencies, such as the Toxic Substances Control Act (TSCA) Inventory in the United States, the European Inventory of Existing Commercial Chemical Substances (EINECS), and the Canadian Domestic and Non-Domestic Substance Lists (DSL/NDSL).

Whenever a chemical substance is sold, transported, imported, exported, reported to a regulatory agency, or disposed of, a CAS Registry Number is probably involved.

➤➤➤➤➤ CHAPTER 13

# Conventions in Chemistry

This chapter presents a quick reference guide for the use of typefaces (roman, italic, and bold), Greek letters, superscripts and subscripts, and special symbols that are commonly used in chemistry. Appendix 13-1 presents the symbols for commonly used physical quantities.

Detailed recommendations from the International Union of Pure and Applied Chemistry (IUPAC, http://www.iupac.org) are given in the book titled *Quantities, Units and Symbols in Physical Chemistry*, 2nd edition, nicknamed the "green book", published by Blackwell Science, Oxford, U.K., 1993. Updates are published as articles in the journal *Pure and Applied Chemistry*.

Detailed recommendations from the International Organization for Standardization (ISO, http://www.iso.org) are given in the ISO Standards Handbook 2, *Quantities and Units*, published by ISO, Geneva, Switzerland, 1993. Some individual standards have been amended, and their updates are available at http://www.iso.org/iso/en/prods-service/ISOstore/store.html. The *National Institute of Standards (NIST) Special Publication 330, 2001 Edition*, available at http://physics.nist.gov/Pubs/SP330/sp330.pdf, is the U.S. updated edition of the English version of the *Bureau International des Poids et Mesures*. The booklet *The International System of Units (SI)*, 7th ed., published by the International Bureau of Weights and Measures (BIPM), Sèvres, France, 1998, is the definitive reference on SI units.

Some books and journals follow IUPAC recommendations for representations of various chemical conventions. Some specify the use of ISO standards. Others are less stringent as long as the manuscript is consistent in usage within itself. Always consult the author guidelines.

## Subatomic Particles and Quanta

➤ Use lowercase Latin or Greek letters for abbreviations for subatomic particles.

| | | | |
|---|---|---|---|
| alpha particle | $\alpha$ | neutrino | $\nu_e$ |
| beta particle | $\beta$ | neutron | n |
| deuteron | d | photon | $\gamma$ |
| electron | e | pion | $\pi$ |
| helion | h | proton | p |
| muon | $\mu^{\pm}$ | triton | t |

➤ Indicate electric charges with the appropriate superscript (+, –, or 0).

$n^0$
$e^+$
$e^-$
$\pi^{\pm}$

If the symbols p and e are used without indication of charge, they refer to positive proton and negative electron, respectively.

## Electronic Configuration

➤ Denote electron shells with the uppercase roman letters K, L, M, and N.

➤ Name electron subshells and atomic orbitals with the lowercase roman letters s, p, d, and f. Write principal energy levels 1–7 on the line and to the left of the letter; give the number of electrons in the orbital as a superscript to the right of the letter. Specify orbital axes with italic subscripts.

| | |
|---|---|
| 7s electron | $3d^44s4p^2$ configuration |
| $5f^2$ ions | $p_xp_yp_z$ |
| 5f orbital | $d_{xz}d_{yz}d_{xy}$ |
| 6d orbital | $d_{z^2}$ |
| $sp^3$ hybrid orbital | $d_{x^2-y^2}$ |
| $f^{n-3}ds^2$ configuration | |

The ground state of boron is $1s^22s^22p_x{}^12p_y{}^02p_z{}^0$.

The valence-shell configuration of nitrogen is $2s^22p_x{}^12p_y{}^12p_z{}^1$.

The electronic configuration of potassium is $1s^22s^22p^63s^23p^64s^1$.

The valence-electron configuration is described by $5d^{10}6s^1$.

➤ Use Greek letters for some bonding orbitals and the bonds they generate.

$\pi$ bond
$\sigma$ orbital
$\sigma*$ orbital

➤ Name the *electronic states of atoms* with the uppercase roman letters S, P, D, F, G, H, I, and K, corresponding to quantum numbers $l = 0–7$. Use the corresponding lowercase letters to indicate the orbital angular momentum of a single electron. The left superscript is the spin multiplicity; the right subscript is the total angular momentum quantum number *J*.

| | | | |
|---|---|---|---|
| $^{2}S_{0}$ | $^{4}P_{1/2}$ | $^{7}F_{0}$ | $^{2}p_{3/2}$ |
| $^{2}s_{0}$ | $^{4}p_{1/2}$ | $^{7}f_{0}$ | $^{7}D_{1}$ |
| $^{8}F_{1/2}$ | $^{8}G_{1/2}$ | $^{2}P_{3/2}$ | $^{7}d_{1}$ |
| $^{8}f_{1/2}$ | $^{8}g_{1/2}$ | | |

➤ Name the *electronic states of molecules* with the uppercase roman letters A, B, E, and T; the ground state is X. Use the corresponding lowercase letters for one-electron orbitals. A tilde (~) is added for polyatomic molecules. The subscripts describe the symmetry of the orbital.

| | | | |
|---|---|---|---|
| $\tilde{A}$ | $^{2}a_{1g}$ | $E_{g}$ | $e_{2g}$ |
| $^{2}A_{1g}$ | $a_{2g}$ | $E_{2g}$ | $T_{2g}$ |
| $A_{2g}$ | $^{3}B_{1}$ | $e_{g}$ | $t_{2g}$ |
| $\tilde{a}$ | $^{3}b_{1}$ | | |

## Chemical Elements and Formulas

➤ Write the names of the chemical elements in roman type and treat them as common nouns.

| | | |
|---|---|---|
| calcium | francium | oxygen |
| californium | helium | seaborgium |
| carbon | hydrogen | uranium |
| einsteinium | | |

➤ Write the symbols for the chemical elements in roman type with an initial capital letter.

| | | | |
|---|---|---|---|
| Ca | Es | H | Sg |
| Cf | Fr | O | U |
| C | He | | |

The complete list of chemical elements and symbols is given in Table 13-1.

➤ Even when symbols are used, the element's name is pronounced. Therefore, choose the article (a or an) preceding the element symbol to accommodate the pronunciation of the element name. (This usage does not apply to isotopes, as described in the section on isotopes.)

a Au electrode (pronounced "a gold electrode")
a N-containing compound (pronounced "a nitrogen-containing compound")
a He–Ne laser (pronounced "a helium–neon laser")

**Table 13-1.** Atomic Weights of the Elements 2001

| *Name* | *Sym.* | *No.* | *Atomic Wt.* | *Notes* |
|---|---|---|---|---|
| Actinium | Ac | 89 | * | |
| Aluminum | Al | 13 | 26.981538(2) | |
| Americium | Am | 95 | * | |
| Antimony | Sb | 51 | 121.760(1) | *g* |
| Argon | Ar | 18 | 39.948(1) | *g, r* |
| Arsenic | As | 33 | 74.92160(2) | |
| Astatine | At | 85 | * | |
| Barium | Ba | 56 | 137.327(7) | |
| Berkelium | Bk | 97 | * | |
| Beryllium | Be | 4 | 9.012182(3) | |
| Bismuth | Bi | 83 | 208.98038(2) | |
| Bohrium | Bh | 107 | * | |
| Boron | B | 5 | 10.811(7) | *g, m, r* |
| Bromine | Br | 35 | 79.904(1) | |
| Cadmium | Cd | 48 | 112.411(8) | *g* |
| Calcium | Ca | 20 | 40.078(4) | *g* |
| Californium | Cf | 98 | * | |
| Carbon | C | 6 | 12.0107(8) | *g, r* |
| Cerium | Ce | 58 | 140.116(1) | *g* |
| Cesium | Cs | 55 | 132.90545(2) | |
| Chlorine | Cl | 17 | 35.453(2) | *g, m, r* |
| Chromium | Cr | 24 | 51.9961(6) | |
| Cobalt | Co | 27 | 58.933200(9) | |
| Copper | Cu | 29 | 63.546(3) | *r* |
| Curium | Cm | 96 | * | |
| Dubnium | Db | 105 | * | |
| Dysprosium | Dy | 66 | 162.500(1) | *g* |
| Einsteinium | Es | 99 | * | |
| Erbium | Er | 68 | 167.259(3) | *g* |
| Europium | Eu | 63 | 151.964(1) | *g* |
| Fermium | Fm | 100 | * | |
| Fluorine | F | 9 | 18.9984032(5) | |
| Francium | Fr | 87 | * | |
| Gadolinium | Gd | 64 | 157.25(3) | *g* |
| Gallium | Ga | 31 | 69.723(1) | |
| Germanium | Ge | 32 | 72.64(1) | |
| Gold | Au | 79 | 196.96655(2) | |
| Hafnium | Hf | 72 | 178.49(2) | |
| Hassium | Hs | 108 | * | |
| Helium | He | 2 | 4.002602(2) | *g, m* |
| Holmium | Ho | 67 | 164.93032(2) | |
| Hydrogen | H | 1 | 1.00794(7) | *g, m, r* |
| Indium | In | 49 | 114.818(3) | |
| Iodine | I | 53 | 126.90447(3) | |
| Iridium | Ir | 77 | 192.217(3) | |
| Iron | Fe | 26 | 55.845(2) | |
| Krypton | Kr | 36 | 83.798(2) | *g, m* |
| Lanthanum | La | 57 | 138.9055(2) | *g* |
| Lawrencium | Lr | 103 | * | |
| Lead | Pb | 82 | 207.2(1) | *g, r* |
| Lithium | Li | 3 | 6.941(2)† | *g, m, r* |
| Lutetium | Lu | 71 | 174.967(1) | *g* |
| Magnesium | Mg | 12 | 24.3050(6) | |
| Manganese | Mn | 25 | 54.938049(9) | |
| Meitnerium | Mt | 109 | * | |
| Mendelevium | Md | 101 | * | |
| Mercury | Hg | 80 | 200.59(2) | |
| Molybdenum | Mo | 42 | 95.94(2) | *g* |
| Neodymium | Nd | 60 | 144.24(3) | *g* |
| Neon | Ne | 10 | 20.1797(6) | *g, m* |
| Neptunium | Np | 93 | * | |
| Nickel | Ni | 28 | 58.6934(2) | |
| Niobium | Nb | 41 | 92.90638(2) | |
| Nitrogen | N | 7 | 14.0067(2) | *g, r* |
| Nobelium | No | 102 | * | |
| Osmium | Os | 76 | 190.23(3) | *g* |
| Oxygen | O | 8 | 15.9994(3) | *g, r* |
| Palladium | Pd | 46 | 106.42(1) | *g* |
| Phosphorus | P | 15 | 30.973761(2) | |
| Platinum | Pt | 78 | 195.078(2) | |
| Plutonium | Pu | 94 | * | |
| Polonium | Po | 84 | * | |
| Potassium | K | 19 | 39.0983(1) | |
| Praseodymium | Pr | 59 | 140.90765(2) | |
| Promethium | Pm | 61 | * | |
| Protactinium | Pa | 91 | 231.03588(2)* | |
| Radium | Ra | 88 | * | |
| Radon | Rn | 86 | * | |
| Rhenium | Re | 75 | 186.207(1) | |
| Rhodium | Rh | 45 | 102.90550(2) | |
| Rubidium | Rb | 37 | 85.4678(3) | *g* |
| Ruthenium | Ru | 44 | 101.07(2) | *g* |
| Rutherfordium | Rf | 104 | * | |
| Samarium | Sm | 62 | 150.36(3) | *g* |
| Scandium | Sc | 21 | 44.955910(8) | |
| Seaborgium | Sg | 106 | * | |
| Selenium | Se | 34 | 78.96(3) | *r* |
| Silicon | Si | 14 | 28.0855(3) | *r* |
| Silver | Ag | 47 | 107.8682(2) | *g* |
| Sodium | Na | 11 | 22.989770(2) | |
| Strontium | Sr | 38 | 87.62(1) | *g, r* |
| Sulfur | S | 16 | 32.065(5) | *g, r* |
| Tantalum | Ta | 73 | 180.9479(1) | |
| Technetium | Tc | 43 | * | |
| Tellurium | Te | 52 | 127.60(3) | *g* |

*Continued on next page*

**Table 13-1.** Atomic Weights of the Elements 2001—*Continued*

| *Name* | *Sym.* | *No.* | *Atomic Wt.* | *Notes* | *Name* | *Sym.* | *No.* | *Atomic Wt.* | *Notes* |
|---|---|---|---|---|---|---|---|---|---|
| Terbium | Tb | 65 | 158.92534(2) | | Ununquadium | Uuq | 114 | * | |
| Thallium | Tl | 81 | 204.3833(2) | | Unununium | Uuu | 111 | * | |
| Thorium | Th | 90 | 232.0381(1)* | *g* | Uranium | U | 92 | 238.02891(3)* | *g, m* |
| Thulium | Tm | 69 | 168.93421(2) | | Vanadium | V | 23 | 50.9415(1) | |
| Tin | Sn | 50 | 118.710(7) | *g* | Xenon | Xe | 54 | 131.293(6) | *g, m* |
| Titanium | Ti | 22 | 47.867(1) | | Ytterbium | Yb | 70 | 173.04(3) | *g* |
| Tungsten | W | 74 | 183.84(1) | | Yttrium | Y | 39 | 88.90585(2) | |
| Ununbium | Uub | 112 | * | | Zinc | Zn | 30 | 65.409(4) | |
| Ununhexium | Uuh | 116 | * | | Zirconium | Zr | 40 | 91.224(2) | *g* |
| Ununnilium | Uun | 110 | * | | | | | | |

*Notes:* Scaled to the relative atomic mass, $A_r(^{12}C) = 12$, where $^{12}C$ is a neutral atom in its nuclear and electronic ground state.

The atomic weights of many elements are not invariant but depend on the origin and treatment of the material. The standard values of $A_r(E)$ and the uncertainties (in parentheses following the last significant figure to which they are attributed) apply to elements of natural terrestrial origin. The footnotes to this table elaborate the types of variation that may occur for individual elements and that may be larger than the listed uncertainties of values of $A_r(E)$. Names of elements with atomic numbers 110 to 116 are provisional.

*Element has no stable nuclides. However, three such elements (Pa, Th, and U) do have a characteristic terrestrial isotopic composition, and for these an atomic weight is tabulated.

†Commercially available Li materials have atomic weights that range between 6.939 and 6.996; if a more accurate value is required, it must be determined for the specific material.

*g* Geological specimens are known in which the element has an isotopic composition outside the limits for normal material. The difference between the atomic weight of the element in such specimens and that given in the table may exceed the stated uncertainty.

*m* Modified isotopic compositions may be found in commercially available material because it has been subjected to an undisclosed or inadvertent isotopic fractionation. Substantial deviations in atomic weight of the element from that given in the table can occur.

*r* Range in isotopic composition of normal terrestrial material prevents a more precise $A_r(E)$ being given; the tabulated $A_r(E)$ value should be applicable to any normal material.



➤ Write the names of chemical compounds in roman type and treat them as common nouns. (Names for chemical compounds are discussed further in Chapter 12.)

benzaldehyde
calcium carbonate
chlorobenzene
ethanol
hydrochloric acid
iron(III) nitrate
isopropyl iodide
magnesium sulfate
mercuric sulfate
methyl salicylate
phenol
sodium hydroxide

➤ Use roman type for the symbols for chemical compounds.

$BaSO_4$
$C_2H_5OH$
$C_6H_5Cl$
$C_6H_5OH$
$CaCO_3$
$CH_3COOH$
$Fe(NO_3)_3$
$H_3PO_4$
$HCl$
$HgSO_4$
$NaOH$
$Ni_3P_2O_8$
$P_2S_5$
$VF_5$
$Zn(C_2H_3O_2)_2$

➤ You may use both chemical symbols and element names in text, but it is best to use one or the other consistently. Do not mix symbols and words within a name.

NaCl *or* sodium chloride, *not* Na chloride

➤ Unnamed elements may be designated by using the atomic number (for example, element 125). They may also be designated by using the systematic name or symbol devised by IUPAC for elements of atomic number greater than 100 that have not yet received trivial names. In this system, an element name consists of a series of numerical roots corresponding to the numerals in the atomic number of the element, followed by "ium". The roots are as follows:

| | | | | | |
|---|---|---|---|---|---|
| 0 | nil | 4 | quad | 7 | sept |
| 1 | un | 5 | pent | 8 | oct |
| 2 | bi | 6 | hex | 9 | enn |
| 3 | tri | | | | |

The symbols consist of the first letters of the numerical roots.

EXAMPLES

| | | |
|---|---|---|
| element 146 | unquadhexium | Uqh |
| element 187 | unoctseptium | Uos |
| element 209 | binilennium | Bne |
| element 290 | biennilium | Ben |
| element 501 | pentnilunium | Pnu |
| element 502 | pentnilbium | Pnb |
| element 503 | pentniltrium | Pnt |
| element 900 | ennilnilium | Enn |

Drop the final "n" in "enn" when it is followed by "nil" (see elements 290 and 900) and the final "i" in "bi" and "tri" when they are followed by "ium" (see elements 502 and 503).

➤ You may use common abbreviations for organic groups in formulas and structures, but not in text. These (and only these) abbreviations need not be defined.

| | | | |
|---|---|---|---|
| Ac | acetyl | Bn, Bzl | benzyl |
| Ar | aryl | Et | ethyl |
| Bu | butyl | Me | methyl |
| *i*-Bu | isobutyl | Ph | phenyl |
| *sec*-Bu | *sec*-butyl | Pr | propyl |
| *t*-Bu | *tert*-butyl | *i*-Pr | isopropyl |
| Bz | benzoyl | R, R′ | alkyl |

➤ Use square brackets in formulas for *coordination entities.*

$[Cr(C_6H_6)_2]$
$K[PtCl_3(C_2H_4)]$

➤ In the formula for an *addition compound,* use a centered dot, closed up on each side. (Although the IUPAC books show a space on each side, this spacing would wreak havoc with many typesetting systems.)

$BH_3{\cdot}NH_3$
$Ni(NO_3)_2{\cdot}2Ni(OH)_2$

*Water of hydration* follows a centered dot, closed up on each side.

$Na_2SO_4{\cdot}10H_2O$
$Zn(NO_3)_2{\cdot}H_2O$

➤ Use either a slash or an en dash between components of a mixture, but not a colon.

dissolved in 5:1 glycerin/water
dissolved in 5:1 glycerin–water

the metal/ligand (1:1) reaction mixture
the metal–ligand (1:1) reaction mixture
the metal–ligand (1/1) reaction mixture

the methane/oxygen/argon (1/50/450) matrix
the methane/oxygen/argon (1:50:450) matrix

## Reference to the Periodic Table

➤ Always use lowercase for the word "group", even with a specific number.

group 15 elements
group IVB elements

➤ Always use lowercase for the words "periodic table".

The elements in group 8 of the periodic table include Fe, Ru, and Os.

## Atoms and Molecules

Nuclide descriptors are specified with superscripts and subscripts to the element symbol, as follows.

### *Use the Left Superscript for Mass Number*

➤ The mass number of an atom is usually shown only for isotopes or in discussions of isotopes.

$^{12}C$
$^{35}Cl$
$^{32}S$

### Use the Left Subscript for Atomic Number

➤ The atomic number of an atom is usually used only in discussions of nuclear chemistry.

$_{6}C$
$_{16}S$

### Use the Right Superscript for Ionic Charge

➤ The charge number is followed by the sign of the ionic charge. When the charge number is 1, only the sign is used.

$Ca^{2+}$
$Na^{+}$
$NO_3^{-}$

➤ Stagger the subscript and superscript; do not align them. The subscript comes first with ionic charge.

$PO_4{}^{3-}$

➤ Do not use multiple plus or minus signs, and do not circle the charge.

$Hg^{2+}$ (*not* $Hg^{++}$)

### Use the Right Asterisk for Excited Electronic State

He*
NO*

### Use the Right Superscript for Oxidation Number

➤ You may use superscript roman numerals for oxidation numbers. In formulas, do not use numbers on the line to avoid confusion with the symbols for iodine or vanadium.

$Co^{III}$
$Fe^{II}Cl_2$
$Mn^{III/IV}$
$Mn^{III}–Mn^{IV}$
$Mn^{III}/Mn^{IV}$
$Mn^{IV}O_2$
$(NH_3)_2Pt^{II}$
$Ni^{0}$
$O^{-II}$
$Pb^{IV}O_2$
$Ru^{II}–Ru^{III}$
$Ru^{II}/Ru^{III}$

➤ Stagger the subscript and superscript; do not align them. The subscript follows the superscript with oxidation number.

$Pb^{II}{}_2$

➤ You may also write oxidation numbers on the line in parentheses closed up to the element name or symbol.

cobalt(III) *or* Co(III)
copper(II) *or* Cu(II)
diammineplatinum(II)
ferrate(VI) ion
iron(II) *or* Fe(II)
iron(II) chloride
manganese(IV) oxide
Mn(III)–Mn(IV) complex
Mn(III)/Mn(IV) complex
potassium tetracyanonickelate(0)

### *Use the Right Subscript for Number of Atoms*

➤ With an element symbol, use a subscript to indicate the number of atoms, whether in formulas or in narrative text.

| | | |
|---|---|---|
| $Al_2O_3$ | $(CH_3)_4C$ | $H_2S$ |
| $C_6$ | $Fe_3$ | $NH_4$ |
| $C_6H_5CH_3$ | $FeSi_2$ | |

The $C_{60}$ fullerene molecule is shaped like a soccer ball.

➤ With an element name, follow the usual conventions for numbers in text.

Molecules composed of 60 carbon atoms are shaped like soccer balls.

In this reaction, three hydrogen atoms are lost.

### *Atom in a Specific Position*

➤ Use either words or symbols and numbers on the line to refer to an atom in a specific position.

at the carbon in the 6-position *or* at C6 *or* at C-6

the atom in the β-position *or* the β atom

## Isotopes

➤ Specify the isotope of an element by a mass number written as a left superscript to the element symbol.

$^{13}C$
$^{15}N$
$^{32}S$
$^{29}Si$
$^{235}U$

➤ Alternatively, indicate an isotope by using the spelled-out element name hyphenated to its mass number.

carbon-14
uranium-235

➤ In either case, the isotope name or symbol is pronounced first, then the number. Thus, $^{14}C$ is pronounced "c fourteen". Consequently, choose the article (a or an) preceding the isotope to accommodate the pronunciation of the element name or symbol, not the number.

| | |
|---|---|
| a carbon-14 isotope | a $^{14}C$ isotope (pronounced "c fourteen") |
| a hydrogen-3 isotope | an $^{3}H$ isotope (pronounced "aitch three") |
| a nitrogen-15 isotope | an $^{15}N$ isotope (pronounced "en fifteen") |

➤ Use the symbols $^{2}H$ or D for deuterium and $^{3}H$ or T for tritium when no other nuclides are present.

| | | |
|---|---|---|
| $D_2O$ | $CH_3O^3H$ | $^2H_2S$ |
| $CD_2H_2$ | $(T_2N)_2CO$ | $CH_2TOH$ |
| $^3H_2$ | $HDSO_4$ | $^3H_2$ |

An *isotopically unmodified* compound is one whose isotopic nuclides are present in the proportions that occur in nature. An *isotopically modified* compound has a nuclide composition that deviates measurably from that occurring in nature.

An *isotopically substituted* compound has a composition such that all of the molecules of the compound have only the indicated nuclides at the designated positions. To indicate isotopic substitution in formulas, the nuclide symbols are incorporated into the formulas. To indicate isotopic substitution in spelled-out compound names, the number and symbol (and locants if needed) are placed in parentheses closed up to the name.

| | | |
|---|---|---|
| $N^{15}NF_2$ | $^{238}UCl_3$ | ($^{15}N$)ammonia |
| $^{24}NaCl$ | $^{32}PO_4{}^{3-}$ | ($^{14}C_6$)glucose |
| $H_2N^{14}CONH_2$ | $Mo(^{12}CO)_6$ | (1,3-$^3H_2$)benzene |
| $^{14}CH_4$ | $Na_2{}^{35}S$ | 1-chloro(2-$^3H$)benzene |

An *isotopically labeled* compound is a mixture of an isotopically unmodified compound with an analogous isotopically substituted compound or compounds. Isotopically labeled compounds may be specifically labeled or selectively labeled. To indicate isotopic labeling, the number and symbol (and locants if needed) are enclosed in square brackets closed up to the compound name or formula.

SPECIFICALLY LABELED:

$[^{14}C]H_4$
$CH_2[^2H_2]$
$CH_3CH_2[^{18}O]H$

SELECTIVELY LABELED:

[$^{2}$H]$CH_4$
[$^{2}$H]$PH_3$
[$^{36}$Cl]$SOCl_2$

[6,7-$^{15}$N]adenosine
[$^{15}$N]alanine
[$^{15}$N]ammonium chloride
[1,3-$^{3}H_2$]benzene
[$^{57}$Co]cyanocobalamin
2,4-diamino[$^{18}$O]phenol
[2,8-$^{3}$H]inosine
[2-$^{14}$C]leucine

➤ When the isotope position is specified by a group name that is part of the parent compound, italicize the group name.

[*methyl*-$^{14}$C]toluene

➤ Isotopically labeled compounds may also be described by inserting the symbol in brackets into the name of the compound.

hydrogen [$^{36}$Cl]chloride
[$^{35}$S]sulfuric [$^{2}$H]acid

➤ Do not use the left superscript within an abbreviation.

[$^{32}$P]CMP (*not* CM$^{32}$P)

➤ To indicate *general labeling*, use the symbol G in the names of selectively labeled compounds in which all positions of the designated element are labeled, but not necessarily in the same isotopic ratio.

D-[G-$^{14}$C]glucose

➤ To indicate *uniform labeling*, use the symbol U in the names of selectively labeled compounds in which all positions of the designated element are labeled in the same isotopic ratio.

D-[U-$^{14}$C]glucose

➤ When it is unknown or irrelevant whether the compound is isotopically labeled or isotopically substituted, simply hyphenate the isotope symbol to the compound name and do not use square brackets or parentheses.

$^{14}$C-glucose
$^{3}$H-benzene
$^{15}$N-adenosine

The *Boughton system*, used in *Chemical Abstracts*, does not distinguish between labeling and substitution. The isotopic variation is shown by the symbol for the isotope (with a subscript numeral to indicate the number of isotopic atoms) placed after the name or relevant portion of the name; locants are cited if necessary. The locants and symbols are in italics, except subscripts and Greek letters, and hyphens are used to separate them.

acetamide-$1$-$^{13}C$-$^{15}N$
acetic-$^{17}O_2$ acid
benzeneacetic-*carboxy*,α-$^{14}C_2$ acid
benzoic-$^{18}O$ acid
4-(2-propenyl-$3$-$^{13}C$-oxy)benzoic acid
toluene-*methyl*-$^{14}C$

➤ In this system, deuterium and tritium are represented by italic lowercase letters *d* and *t*, respectively.

acetic-$t_3$ acid-*t*
alanine-*N*,*N*,*1*-$d_3$
ammonia-*d*-*t*
ethane-*1*-*d*-*2*-*t*
1-(ethyl-*2*,*2*,*2*-$d_3$)-4-(methyl-$d_3$)benzene
methan-*t*-ol
methane-$d_4$
tri(silyl-$d_3$)phosphine
urea-$t_4$

# Radicals

➤ In the formula of a free radical, indicate the unshared electron by a superscript or centered dot closed up to the element symbol or formula. The superscript dot comes after the symbol or formula; centered dots come before or after the symbol or formula.

| | | |
|---|---|---|
| $Br^{\cdot}$ | $H^{\cdot}$ | $\cdot SH$ |
| $Br\cdot$ | $HO^{\cdot}$ | $(SiH_3)^{\cdot}$ |
| $\cdot CH_3$ | $\cdot NH_2$ | $\cdot SnH_3$ |
| $C_6H_5\cdot$ | | |

➤ Charged radical cations and anions are often indicated by the symbol, formula, or structure with a superscript dot followed by a plus or minus sign. However, in mass spectrometry, the reverse is used. Therefore, use the order of dots and signs for charges that is appropriate for the context.

| | | |
|---|---|---|
| $(Ag_2)^{\cdot+}$ | $R^{\cdot-}$ | $R^{(\cdot)(2-)}$ |
| $C_6H_5NO^{\cdot 3-}$ | $R_2^{\cdot+}$ | $(SO_2)^{\cdot-}$ |
| $HCO^{\cdot+}$ | | |

MASS SPECTROMETRY

$C_3H_6^{+\cdot}$
$R^{+\cdot}$

# Bonds

➤ For linear formulas in text, do not show single bonds unless the bonds are the subject of the discussion.

$C_6H_5CH_3$
$C_6H_5COOCOCH_3$
$CH_3CHOHCH_3$
$CH_3COOH$
$H_2SO_4$

➤ When necessary for the discussion, indicate bonds by en dashes.

the $–CH_2–$ segment
the C–H distances
the C–C–C angle
$(–CH_2–CH(CH_3)–O–)_n$

➤ When necessary, show double and triple bonds in linear formulas.

$CH_3C{\equiv}CH$

$CH_2{=}CH_2$

R–C(=O)–OH (O double-bonded to C, drawn below) *is better as* RCOOH *or* $RCO_2H$ *or* RC(=O)OH

➤ Use three centered dots to indicate association of an unspecified type (e.g., hydrogen bonding, bond formation, or bond breaking).

C···Pt
F···H–$NH_3$
$H_2O$···π aromatic hydrogen bonding
$Mg^{2+}$···O–
Ni···Al

# Crystallography

## *Planes and Directions in Crystals*

➤ Miller indices of a crystal face or a single net plane are enclosed in parentheses. (123) or (*hkl*) is a plane or set of planes that describe crystal faces; $(h_1h_2h_3)$ is a single net plane.

➤ Laue indices are not enclosed. 123 or *hkl* is the Bragg reflection from the set of net planes (123) or (*hkl*), respectively.

➤ Indices of a set of all symmetrically equivalent crystal faces or net planes are enclosed in braces. {*hkl*} is a form.

➤ Indices of a zone axis or lattice direction are enclosed in square brackets. [123] or [*uvw*] is a direction.

➤ Indices of a set of symmetrically equivalent lattice directions are enclosed in angle brackets. <*uvw*> represents all crystallographically equivalent directions of the type [*uvw*].

$1\bar{2}0$
1,10,1
11,0,1
1,–2,0 reflections

the (111) face
the (120) face
the [001] axis
the [010] direction
the [101] direction

*h*00 diffraction lines
the *hk*0 zone
the 002 reflection
the 00*l* class of reflections

➤ When indices are used with spelled-out element names, separate the name of the element and the index with a space.

copper (111)
rhenium (010)
a gold (111) substrate
on silicon (111) surfaces
the silver (110) surface

➤ However, when indices are used with element symbols, close up the element symbol to the index.

Au(210)
CdTe(100)
Cu(111)
GaAs(100)
Rh(010)
Si(400)

an iodine-modified Ag(111) electrode

the Ag(110) surface

## Types of Crystal Lattices

| | |
|---|---|
| bcc | body-centered cubic |
| ccp | cubic close-packed |
| fcc | face-centered cubic |
| hcp | hexagonal close-packed |

## Symmetry Operations and Structural Point Groups

➤ Use italic type for the letters in symmetry operations and structural point groups. The symbols (Schoenflies) are as follows: *E*, identity; *C*, cyclic; *D*, dihedral; *T*, tetrahedral; *O*, octahedral; *I*, icosahedral; *S*, rotation–reflection; and σ, mirror plane. Align subscripts and superscripts.

| | | |
|---|---|---|
| $C_1$ | $C_{2h}$ | $S_3$ |
| $C_i$ | $C_{3h}$ | $S_4$ |
| $C_s$ | $D_{2d}$ $(V_d)$ | $2S_6$ |
| $C_2$ | $D_{3d}$ | $S_8$ |
| $C_2^3$ | $D_{4d}$ | $\sigma$ |
| $C_2^4$ | $D_{2h}$ $(V_h)$ | $2\sigma_v$ |
| $C_\infty$ | $D_{3h}$ | $3\sigma_v$ |
| $C_{2v}$ | $D_{4h}$ | $4\sigma_v$ |
| $C_{3v}$ | $D_{\infty h}$ | $3\sigma_d$ |
| $C_{4v}$ | $I_h$ | $\sigma_h$ |
| $C_{\infty v}$ | $O_h$ $(K_h)$ | $T_d$ |

## Crystallographic Point Groups

➤ Use arabic numerals or combinations of numerals and the italic letter *m* to designate the 32 crystallographic point groups (Hermann–Mauguin). The number is the degree of the rotation, and *m* stands for mirror plane. Use an overbar to indicate rotation inversion.

| | | |
|---|---|---|
| 1 | $mmm$ | $6/mmm$ |
| $m$ | $4mm$ | $\bar{4}3m$ |
| $2/m$ | 32 | $m\bar{3}m$ |
| $mm2$ | 622 | |

## Space Groups

➤ Designate space groups by a combination of unit cell type and point group symbol, modified to include screw axes and glide planes (Hermann–Mauguin); 230 space groups are possible. Use italic type for conventional types of unit cells (or Bravais lattices): *P*, primitive; *I*, body-centered; *A*, A-face-centered; *B*, B-face-centered; *C*, C-face-centered; *F*, all faces centered; and *R*, rhombohedral.

| | | |
|---|---|---|
| *Pnma* | *Pnn*2 | *P*1 |
| *C*2/*c* | $P4_32_12$ | *Fdd*2 |
| *Pbcn* | $Fm\bar{3}m$ | *Aba*2 |
| $I4_1/a$ | *R*3*m* | *I*4/*mmm* |
| *Fd*3*m* | $Cmc2_1$ | |

### *Crystallographic Information File*

A description of the Crystallographic Information File, CIF, is included in Appendix 13-2.

## Chirality

➤ Use italic type for certain chirality symbols and symmetry site terms.

| | | | |
|---|---|---|---|
| *A* | anticlockwise | *OC* | octahedron |
| *C* | clockwise | *TP* | trigonal phase |
| *CU* | cube | *TPR* | trigonal prism |
| *DD* | dodecahedron | *TPY* | trigonal pyramid |

These symbols are often combined with coordination numbers and position designations for stereochemical descriptors (e.g., *OC*-6-11′).

➤ In chemical names, use (*R*) and (*S*), with designated locants when applicable, as prefixes to designate absolute configuration.

(*R*)-hydroxyphenylacetic acid
(*S*)-2,3-dihydroxypropanoic acid
(1*S*,2*S*,4*R*)-trichloro-1,2,4-trimethylcyclohexane

➤ Indicate optical rotation by plus and minus signs in parentheses and hyphenate them to the chemical name.

(±)-4-(2-aminopropyl)phenol
(+)-glucose
(−)-tartaric acid

➤ Use small capital letters D and L for absolute configuration with amino acids and carbohydrates.

| | |
|---|---|
| β-D-cellotetraose | 5-hydroxy-L-lysine |
| D-allothreonine | hydroxy-L-proline |
| D-glucose | L-alanine |
| D-valine | L-alloisoleucine |
| DL-leucine | L-ascorbic acid |
| hydroxy-DL-glutamic acid | L-phenylalanine |

➤ Use a hyphen between (+) or (–) and D or L.

(–)-D-fructose
(–)-D-glyceraldehyde
(+)-L-phosphoglycerol

# Concentration

➤ Use square brackets enclosing an element symbol or formula to indicate its concentration in reactions and equations, but not in narrative text.

CORRECT

$[Mg^{2+}] = 3\times10^{-2}$ M

The Mg concentration decreased with repeated washings.

INCORRECT

The [Mg] was found to be greater in the unwashed samples.

➤ Do not use square brackets to indicate concentration with a spelled-out name.

$[Ca^{2+}]$ (*not* [calcium])

[NaCl] (*not* [sodium chloride])

➤ Do not use italic type for the chemical concentration unit M (molar, moles per cubic decimeter, moles per liter) or the unit N (normal). Use italic type for the unit *m* (molal, moles per kilogram). Use a space between the number and these abbreviations, that is, on each side of these abbreviations.

8 M urea
1 mM EDTA
6 N HCl
2.0 *m* NaOH

➤ When concentration is given as percentage, use the percent sign closed up to the number.

20% $H_2SO_4$
90% acetonitrile/10% water

➤ Generally, the negative logarithm of the hydrogen ion concentration is denoted by pH; the negative logarithm of the hydroxide ion concentration is denoted by pOH. Use a space to separate pH or pOH and the number. Use roman type for pH and pOH; always use lowercase for "p"; always capitalize "H" and "OH".

> Solutions were titrated to pH >11.
>
> The UV spectra were measured at pH 6.
>
> A pOH of <12 was acceptable.

## Chemical Reactions

➤ Short chemical reactions may be run into text or they may be displayed and numbered, if numbering is needed. Long chemical reactions should be displayed separately from the text. The sequential numbering system used may integrate both chemical and mathematical equations, or separate sequences using different notations may be used for different types of equations (e.g., eqs 1–3 could be used for a set of chemical reactions and eqs I–III could be used for a set of mathematical equations). The use of lettering, rather than numbering, sequences is also acceptable.

$$\mathrm{Cr(CO)_4 + CO \rightarrow Cr(CO)_5} \qquad (1)$$

$$\mathrm{NH_3 + HCOOH \rightarrow NH_2CHO + H_2O} \qquad (2)$$

$$\mathrm{(C_6H_5)_2P{-}P(C_6H_5)_2 \rightarrow 2(C_6H_5)_2P^{\bullet}} \qquad (3)$$

$$\mathrm{Fe(CO)_5 + OCH_3^- \rightarrow Fe(CO)_4(CO_2CH_3)^-} \qquad (4)$$

➤ Many kinds and combinations of arrows can be used. For example, two full arrows in opposite directions ($\leftrightarrows$) indicate a reaction that is proceeding in both directions. Two arrows with half heads in opposite directions ($\rightleftharpoons$) indicate a reaction in equilibrium. A single arrow with heads on both sides ($\leftrightarrow$) indicates resonance structures, not a reaction.

➤ Specify the number of each species (molecules, atoms, ions, etc.) of reactants and products by a numeral written on the line and closed up to the symbol.

$$\mathrm{2Al + 6NaOH \rightarrow 2Na_3AlO_3 + 3H_2} \qquad (5)$$

➤ To indicate the aqueous, solid, liquid, or gas state, use the appropriate abbreviations on the line, in parentheses, and with no space preceding them.

$$\mathrm{Ag(s) + H^+(aq) + Cl^-(aq) \rightarrow AgCl(s) + \tfrac{1}{2}H_2(g)} \qquad (6)$$

$$\mathrm{4FeS(s) + 7O_2(g) \rightarrow 2Fe_2O_3(s) + 4SO_2(g)} \qquad (7)$$

➤ Indicate reaction conditions and catalysts over and under the arrow in a smaller type size. The Greek capital letter delta indicates heat; $h\nu$ indicates light, where $h$ is Planck's constant and the Greek letter nu is the photon frequency.

$$PhS^- \xrightarrow{h\nu} PhS^\bullet + e^-$$

$$C_3H_8(g) + 5O_2(g) \xrightarrow{\Delta} 3CO_2(g) + 4H_2O(g)$$

$$RC{\equiv}N + 2H_2 \xrightarrow{\text{Pt or Pd}} RCH_2NH_2$$

$$2H_2O + CH_3C(CH_3){=}C(CH_3)CH_3 \xrightarrow[H_2SO_4,\Delta]{KMnO_4} 2CH_3C({=}O)CH_3 + 2H_2$$

$$(C_2H_5)_2C{=}O + 2CH_3OH \xrightarrow[125\ ^\circ C]{H^+} (C_2H_5)_2C(OCH_3)_2 + H_2O$$

➤ Specify *nuclear reactions* according to the following scheme:

$$\text{initial nuclide} \begin{pmatrix} \text{incoming} & & \text{outgoing} \\ \text{particle(s)} & , & \text{particle(s)} \\ \text{or quanta} & & \text{or quanta} \end{pmatrix} \text{final nuclide}$$

EXAMPLES

$^{14}N(\alpha,p)^{17}O$
$^{59}Co(n,\gamma)^{60}Co$
$^{23}Na(\gamma,3n)^{20}Na$
$^{31}P(\gamma,pn)^{29}Si$

➤ Treat *chemical equations* that include structures with rings as illustrations. They are discussed in Chapter 17.

➤ Abbreviate reaction types with capital roman letters and arabic numerals.

| | | |
|---|---|---|
| $S_N1$ | $S_N2$ | first- and second-order nucleophilic substitution, respectively |
| E1 | E2 | first- and second-order elimination, respectively |
| $S_{RN}1$ | $S_{RN}2$ | first- and second-order radical nucleophilic substitution, respectively |

➤ Subscripts denote a chemical process or reaction.

| | |
|---|---|
| ads | adsorption |
| at | atomization |
| c | combustion |
| dil | dilution |
| dpl | displacement |
| f | formation |
| fus | fusion |
| imm | immersion |
| mix | mixing |
| r | reaction in general |
| sol | solution |
| sub | sublimation |
| trs | transition |
| vap | vaporization |

➤ Certain superscripts are recommended.

| | |
|---|---|
| ‡ | activated complex, transition state |
| ′ | apparent |
| E | excess quantity |
| id | ideal |
| ∞ | infinite dilution |
| * | pure substance |
| °, ⊖ | standard state |

# Reporting Analytical Data

There is no best way to present data. A presentation that is suitable for one paper or publication may be unsuitable for another. The following are examples of acceptable presentations of analytical data. These are not necessarily real examples; they may be combinations of data from two or more samples, intended to show various style possibilities. You need not define the abbreviations and symbols in the paper.

## *Melting and Boiling Points*

mp 175.5 °C (lit.[25] mp 175–176 °C)
mp 225 °C dec
bp 127 °C

Abbreviations: mp, melting point; bp, boiling point; lit., literature value; and dec, decomposition. A full space is used between the number and the unit °C; the degree symbol is closed up to the C. A superscript number after "lit." denotes the number of the reference.

## *Specific Rotation*

$[\alpha]_{\mathrm{D}}^{20}$ + 25.4 (*c* 1.00, $CHCl_3$)

Abbreviations: α, specific rotation; D, the sodium D line or wavelength of light used for the determination; and the superscript number, temperature (°C) at which the determination was made. In parentheses: *c* stands for concentration; the number following *c* is the concentration in grams per 100 mL of solution; and last is the solvent name or formula.

## *NMR Spectroscopy*

$^1$H NMR (400 MHz, $CD_3OD$, δ): 8.73 (s, 3H, $-OCH_3$), 7.50 (s, 1H, CH), 7.15 (d, *J* = 8.2 Hz, 1H, Ar H), 6–3 (br s, 5H, NH and $NH_2$).

Compound **5:** $^1$H NMR (500 MHz, $CDCl_3$, δ) 1.12 (t, *J* = 7.1 Hz, $-CH_2\mathit{CH}_3$, 3H), 3.34 (q, *J* = 7.1 Hz, $-\mathit{CH}_2CH_3$, 2H), 3.38 (t, *J* = 6.0 Hz, $-\mathit{CH}_2CH_2OH$, 2H), 3.72 (t, *J* = 6.0 Hz, $-CH_2\mathit{CH}_2OH$, 2H), 6.57 (dd, *J* = 8.7 Hz, Ar H, 2H).

$^{13}$C NMR (DMSO-$d_6$, δ): 175.4 (C=O), 156.5 ($C_4$), 147.4 ($C_6$), 138.3 ($C_2$), 110.5 (d, *J* = 11.3 Hz, $C_5$), 52.3 ($CH_3$), 28.4 and 28.8 ($C_7$).

$^{13}$C NMR (DMSO-$d_6$, δ): 0.43 (2C), 27.56 (4C), 131.8 (1C), 161.9 (2C).

$^{13}$C NMR ($CDCl_3$, 75.4 MHz): δ 213.50 (s, C-21), 178.27 (s, C-2), 168.69 (s, C-8), 164.61 (d, C-10), 119.67 (d, C-7), 52.45 (t, C-22), 38.95 (q, C-25).

$^{13}$C NMR ($C_6D_{12}$, δ): 6.51 (s, $C_5Me_5$), 14.41 (d, $J$ = 157 Hz, $PMe_3$), 28.68 (s, Me), 105.1 (t, $J$ = 3.7 Hz, $C_5Me_5$), 128.52 (s), 135.19 (br s), 212.56 (C–O).

If the experimental conditions have already been described elsewhere in the paper, they need not be repeated.

NMR: 3.81, 2.56, and 2.12 ppm.

Compound **27:** NMR 5.14, 3.90, 2.67, and 1.88 ppm.

Abbreviations: δ, chemical shift in parts per million (ppm) downfield from the standard; *J*, coupling constant in hertz; and the multiplicities s, singlet; d, doublet; t, triplet; q, quartet; and br, broadened. Italicized elements or groups are those that are responsible for the shifts.

## IR Spectroscopy

IR (KBr) $\bar{\nu}_{max}$: 967 (Ti=O), 3270 cm$^{-1}$ (NH).

IR (KBr, thin film) $\bar{\nu}_{max}$ (cm$^{-1}$): 3017, 2953 (s, OH), 2855 (s), 2192, 1512, 1360, 1082, 887.

IR (dry film) $\bar{\nu}_{max}$: 3324 (OH), 2973–2872 (CH, aliphatic), 1706 (C=O, ketone), 1595, 1437, 1289, 1184, 1048, 870, 756, 677 cm$^{-1}$.

IR: 2000, 2030, 2040, 2050 cm$^{-1}$.

IR (cm$^{-1}$): 3130, 3066, 2964, 1654, 1500, 1371.

Compound **6:** IR 2910, 2487, 1972, 1564, 1190 cm$^{-1}$.

GC–FTIR $\bar{\nu}_{max}$ (cm$^{-1}$): 2979 (w), 1400 (m), 1264 (s), 827 (vs).

Abbreviations: $\bar{\nu}_{max}$ is the wavenumber of maximum absorption peaks in reciprocal centimeters, and the absorptions are w, weak; m, medium; s, strong; vw, very weak; vs, very strong; and br, broad.

## Mass Spectrometry

MS *m/z* (relative intensity): 238.2058 (44.8%), 195.1487 (100%), 153.1034 (21.2%).

GC–MS *m/z* (% relative intensity, ion): 202 (9, M + 4), 200 (32, M + 2), 198 (23, M$^+$), 142 (35, M – 2CO), 321 (95, M – Me), 415 (M$^+$ – Cl).

HRMS–FAB (*m/z*): [M + H]$^+$ calcd for $C_{21}H_{38}N_4O_6S$, 475.259; found, 475.256.

EIMS (70 eV) *m/z:* M$^+$ 420 (15), 241 (15), 201 (59), 135 (14), 69 (23).

Abbreviations: *m/z* is the mass-to-charge ratio, M is the molecular weight of the molecule itself, M$^+$ is the molecular ion, HRMS is high-resolution mass spectrometry, FAB is fast atom bombardment, and EIMS is electron-impact mass spectrometry.

## UV–Visible Spectroscopy

UV (hexanes) $\lambda_{max}$, nm ($\varepsilon$): 250 (1070).

UV ($CH_3OH$) $\lambda_{max}$ (log $\varepsilon$) 210 (3.33), 242 (sh, 3.02), 288 (sh, 2.21), 421 nm (3.16).

Abbreviations: $\lambda_{max}$ is the wavelength of maximum absorption in nanometers; $\varepsilon$ is the molar absorption coefficient or molar absorptivity; and sh is the shoulder. The wavenumber, $\overline{\nu}$, in reciprocal micrometers, might also be given.

## Quantitative Analysis

Anal. Calcd for $C_{45}H_{28}N_4O_7$: C, 62.47; H, 3.41; N, 6.78. Found: C, 61.80; H, 3.55; N, 6.56.

All values are given as percentages.

## X-ray Diffraction Spectroscopy

➤ Separate the element and its spectral line by a space.

| | | |
|---|---|---|
| Au $L_I$ | Cu Kα | Cu Kα radiation |
| Au $L_{III}$ | Cu Kβ | the Co K absorption edge |
| Cr K | Mo Kα | the Ni K edge |

## Citing ASTM, ANSI, and ISO Standards

ASTM International, originally known as the American Society for Testing and Materials; ANSI, the American National Standards Institute; and ISO, the International Organization for Standardization, are organizations that set standards in a variety of areas.

➤ For ASTM standards, separate the letter and the number of the standard by a space. For ANSI standards, close up the letter and the number.

ASTM

| | | |
|---|---|---|
| D 3137 | DS 4B | STP 1249 |
| D 573 | DS 64 | STP 169C |
| D 130 | DS 48A | STP 315I |
| D 1660 | E 380-93 | |

as described in ASTM Standard D 1223

ANSI

| | |
|---|---|
| Z358.1-1990 | Z88.2-1992 |

as noted in ANSI Standard H35.1M-1993

ISO

| | | |
|---|---|---|
| 14020 | 9000 | 71.040 |

ISO/DIS 14010

according to ISO Standard 11634

➤➤➤➤➤

APPENDIX 13-1

# Symbols for Commonly Used Physical Quantities

## Atoms and Molecules

| NAME | SYMBOL | SI UNIT |
|---|---|---|
| Atomic mass | $m_a$ | kg |
| Atomic mass constant | $m_u$ | kg |
| Atomic number | $Z$ | dimensionless |
| Decay constant | $\lambda$ | $s^{-1}$ |
| Electron rest mass | $m_e$ | kg |
| Electronic term | $T_e$ | $m^{-1}$ |
| Elementary charge (of a proton) | $e$ | C |
| *g* factor, *g* value | $g$ | dimensionless |
| Ionization energy | $E_i$, $I$ | J |
| Magnetogyric ratio | $\gamma$ | $s^{-1}{\cdot}T^{-1}$ |
| Mass number | $A$ | dimensionless |
| Neutron number | $N$ | dimensionless |
| Nucleon number | $A$ | dimensionless |
| Planck constant | $h$ | J·s |
| Planck constant/2π | $\hbar$ | J·s |
| Proton number | $Z$ | dimensionless |
| Proton rest mass | $m_p$ | kg |
| Rotational constants | $A$, $B$, $C$ | $m^{-1}$ |
| Rotational term | $F$ | $m^{-1}$ |
| Total angular momentum component | $m_j$, $m_J$ | dimensionless |
| Total term | $T$ | $m^{-1}$ |
| Unified atomic mass unit | $m_u$ | kg |
| Vibrational quantum number | $v$ | dimensionless |
| Vibrational term | $G$ | $m^{-1}$ |

## Chemical Kinetics

| NAME | SYMBOL | SI UNIT |
|---|---|---|
| Boltzmann constant | $k$, $k_B$ | J/K, $J{\cdot}K^{-1}$ |
| Energy of activation; Arrhenius or activation energy | $E$, $E_a$, $E_A$ | J/mol, $J{\cdot}mol^{-1}$ |
| Half-life | $t_{1/2}$ | s |
| Photochemical yield | $\phi$ | dimensionless |
| Quantum yield | $\phi$ | dimensionless |

### Chemical Kinetics—*Continued*

| NAME | SYMBOL | SI UNIT |
|---|---|---|
| Rate constant, first order | $k$ | $s^{-1}$ |
| Rate constant, second order | $k$ | $mol^{-1}{\cdot}s^{-1}$ |
| Rate of concentration change of substance B | $r_B$ | $mol/(m^3{\cdot}s)$, $mol{\cdot}m^{-3}{\cdot}s^{-1}$ |
| Rate of conversion | $\dot{\zeta}$ | mol/s, $mol{\cdot}s^{-1}$ |
| Rate of reaction | $v$ | $mol/(m^3{\cdot}s)$, $mol{\cdot}m^{-3}{\cdot}s^{-1}$ |
| Relaxation time | $\tau$ | s |
| Scattering angle | $\theta$ | rad |
| Standard enthalpy of activation | $\Delta^{\ddagger}H^{\circ}$, $\Delta H^{\ddagger}$ | J/mol, $J{\cdot}mol^{-1}$ |
| Standard entropy of activation | $\Delta^{\ddagger}S^{\circ}$, $\Delta S^{\ddagger}$ | J/(mol·K), $J{\cdot}mol^{-1}{\cdot}K^{-1}$ |
| Standard Gibbs energy of activation | $\Delta^{\ddagger}G^{\circ}$, $\Delta G^{\ddagger}$ | J/mol, $J{\cdot}mol^{-1}$ |
| Standard internal energy of activation | $\Delta^{\ddagger}U^{\circ}$, $\Delta U^{\ddagger}$ | J/mol, $J{\cdot}mol^{-1}$ |
| Temperature, absolute | $T$ | K |
| Thermal energy | $kT$ | J |
| Volume of activation | $\Delta^{\ddagger}V$, $\Delta V^{\ddagger}$ | J/mol, $J{\cdot}mol^{-1}$ |

## Electricity and Magnetism

| NAME | SYMBOL | SI UNIT |
|---|---|---|
| Capacitance | $C$ | C/V, $C{\cdot}V^{-1}$, F |
| Charge density | $\rho$ | $C/m^3$, $C{\cdot}m^{-3}$ |
| Conductance | $G$ | S |
| Conductivity | $\kappa$ | S/m, $S{\cdot}m^{-1}$ |
| Dielectric polarization | $\boldsymbol{P}$ | $C/m^2$, $C{\cdot}m^{-2}$ |
| Electric charge | $Q$ | C |
| Electric current | $I$ | A |
| Electric current density | $j$ | $A/m^2$, $A{\cdot}m^{-2}$ |
| Electric dipole moment | $\boldsymbol{p}$, $\mu$ | C·m |
| Electric displacement | $\boldsymbol{D}$ | $C/m^2$, $C{\cdot}m^{-2}$ |
| Electric field strength | $\boldsymbol{E}$ | V/m, $V{\cdot}m^{-1}$ |
| Electric potential | $V$, $\phi$ | V, J/C, $J{\cdot}C^{-1}$ |
| Electric potential difference | $U$, $\Delta V$, $\Delta\phi$ | V |
| Electric resistance | $R$ | Ω |
| Electric susceptibility | $\chi_e$ | dimensionless |
| Electromagnetic moment | $\mu$, $\boldsymbol{m}$ | $A{\cdot}m^2$, J/T, $J{\cdot}T^{-1}$ |
| Impedance | $Z$ | Ω |
| Inductance | $\boldsymbol{H}$ | A/m, $A{\cdot}m^{-1}$, H |
| Magnetic field strength | $\boldsymbol{H}$ | A/m, $A{\cdot}m^{-1}$, H |
| Magnetic flux | $\Phi$ | Wb |
| Magnetic flux density | $\boldsymbol{B}$ | T |
| Magnetic induction | $\boldsymbol{B}$ | T |
| Magnetic moment | $\boldsymbol{M}$ | A/m, $A{\cdot}m^{-1}$ |
| Magnetic susceptibility | $\chi$ | dimensionless |

### Electricity and Magnetism—*Continued*

| NAME | SYMBOL | SI UNIT |
|---|---|---|
| Magnetization | $\boldsymbol{M}$ | A/m, $A \cdot m^{-1}$ |
| Permeability | $\mu$ | H/m, $H \cdot m^{-1}$ |
| Permittivity | $\varepsilon$ | F/m, $F \cdot m^{-1}$ |
| Polarization (of a particle) | $\alpha$ | $m^2 \cdot C \cdot V^{-1}$ |
| Relative permeability | $\mu_r$ | dimensionless |
| Relative permittivity | $\varepsilon_r$ | dimensionless |
| Resistance | $R$ | Ω |
| Resistivity | $\rho$ | Ω·m |
| Self-inductance | $L$ | H |
| Voltage | $U, \Delta V, \Delta\phi$ | V |

# Electrochemistry

| NAME | SYMBOL | SI UNIT |
|---|---|---|
| Charge number of an ion | $z$ | dimensionless |
| Conductivity | $\kappa$ | S/m, $S \cdot m^{-1}$ |
| Diffusion rate constant | $k_d$ | m/s, $m \cdot s^{-1}$ |
| Electric current | $I$ | A |
| Electric current density | $j$ | $A/m^2$, $A \cdot m^{-2}$ |
| Electric mobility | $u$ | $m^2/(V \cdot s)$, $m^2 \cdot V^{-1} \cdot s^{-1}$ |
| Electrode potential | $E$ | V |
| Electrolytic conductivity | $\kappa$ | S/m, $S \cdot m^{-1}$ |
| Electromotive force (emf) | $E$ | V |
| Elementary charge | $e$ | C |
| Faraday constant | $F$ | C/mol, $C \cdot mol^{-1}$ |
| Half-wave potential | $E_{1/2}$ | V |
| Ionic strength | | |
| concentration basis | $I_c, I$ | $mol/m^3$, $mol \cdot m^{-3}$ |
| molality basis | $I_m, I$ | mol/kg, $mol \cdot kg^{-1}$ |
| Mass-transfer coefficient | $k_d$ | m/s, $m \cdot s^{-1}$ |
| Molar conductivity | $\Lambda$ | $S \cdot m^2/mol$, $S \cdot m^2 \cdot mol^{-1}$ |
| Reduction potential | $E°$ | V |
| Standard electrode potential | $E°$ | V |
| Standard electromotive force (emf) | $E°$ | V |
| Surface charge density | $\sigma$ | $C/m^2$, $C \cdot m^{-2}$ |
| Transport number | $t$ | dimensionless |

# General Chemistry

| NAME | SYMBOL | SI UNIT |
|---|---|---|
| Amount concentration | $c$ | $mol/m^3$, $mol \cdot m^{-3}$ |
| Amount of substance | $n$ | mol |
| Atomic weight | $A_r$ | dimensionless |
| Concentration | $c$ | $mol/m^3$, $mol \cdot m^{-3}$ |
| Degree of dissociation | $\alpha$ | dimensionless |

### General Chemistry—*Continued*

| NAME | SYMBOL | SI UNIT |
|---|---|---|
| Extent of reaction | $\zeta$ | mol |
| Mass fraction | $w$ | dimensionless |
| Molality | $m$, $b$ | mol/kg, mol·kg$^{-1}$ |
| Molar mass | $M$ | kg/mol, kg·mol$^{-1}$ |
| Molar volume | $V_m$ | m$^3$/mol, m$^3$·mol$^{-1}$ |
| Molarity | M | mol/L, mol·L$^{-1}$ |
| Mole fraction | $x$ | dimensionless |
| Molecular weight | $M_r$ | dimensionless |
| Number concentration | $C$, $n$ | m$^{-3}$ |
| Number density of entities | $C$, $n$ | m$^{-3}$ |
| Number of entities | $N$ | dimensionless |
| Partial pressure of substance B | $p_B$ | Pa |
| Relative atomic mass | $A_r$ | dimensionless |
| Relative molecular mass | $M_r$ | dimensionless |
| Stoichiometric coefficient | $\nu$ | dimensionless |
| Surface concentration | $\Gamma$ | mol/m$^2$, mol·m$^{-2}$ |
| Volume fraction | $\phi$ | dimensionless |

# Mechanics

| NAME | SYMBOL | SI UNIT |
|---|---|---|
| Angular momentum | $\boldsymbol{L}$ | J·s |
| Density | $\rho$ | kg/m$^3$, kg·m$^{-3}$ |
| Energy | $E$ | J |
| Force | $\boldsymbol{F}$ | N |
| Gravitational constant | $G$ | N·m$^2$/kg$^2$, N·m$^2$·kg$^{-2}$ |
| Hamiltonian function | $H$ | J |
| Kinetic energy | $E_k$, $K$ | J |
| Lagrange function | $L$ | J |
| Mass | $m$ | kg |
| Moment of force | $M$ | N·m |
| Moment of inertia | $I$ | kg·m$^2$ |
| Momentum | $\boldsymbol{p}$ | kg·m/s, kg·m·s$^{-1}$ |
| Potential energy | $E_p$, $V$ | J |
| Power | $P$ | W |
| Pressure | $P$, $p$ | Pa, N/m$^2$, N·m$^{-2}$ |
| Reduced mass | $\mu$ | kg |
| Relative density | $d$ | dimensionless |
| Specific volume | $v$ | m$^3$/kg, m$^3$·kg$^{-1}$ |
| Surface tension | $\gamma$ | N/m, N·m$^{-1}$, J/m$^2$, J·m$^{-2}$ |
| Torque | $\boldsymbol{T}$ | N·m |
| Weight | $\boldsymbol{G}$, $\boldsymbol{W}$ | N |
| Work | $w$, $W$ | J |

# NMR Spectroscopy

| NAME | SYMBOL | SI UNIT |
|---|---|---|
| Bohr magneton | $\mu_B$, $\beta$ | J/T, $J \cdot T^{-1}$ |
| Bohr radius | $a_0$ | m |
| Chemical shift, δ scale | $\delta$ | dimensionless |
| Coupling constant | | |
| (indirect) spin–spin | $J_{AB}$ | Hz |
| direct (dipolar) | $D_{AB}$ | Hz |
| reduced spin–spin | $K_{AB}$ | $T^2 \cdot J^{-1}$, $N \cdot A^{-2} \cdot m^{-3}$ |
| Delay time | $\tau$ | s |
| Electron spin quantum component | $m_s$, $m_S$ | dimensionless |
| Electron spin quantum number | $s$, $S$ | dimensionless |
| Hyperfine coupling constant | $a$, $A$, $\boldsymbol{T}$ | Hz |
| Larmor angular frequency | $\omega_L$ | $s^{-1}$ |
| Larmor frequency | $\nu_L$ | Hz |
| Magnetogyric ratio | $\gamma$ | $s^{-1} \cdot T^{-1}$ |
| Nuclear magneton | $\mu_N$ | J/T, $J \cdot T^{-1}$ |
| Nuclear spin quantum component | $M_I$ | dimensionless |
| Nuclear spin quantum number | $I$ | dimensionless |
| Orbital quantum number | $l$, $L$ | dimensionless |
| Orbital quantum number component | $m_l$, $m_L$ | dimensionless |
| Principal quantum number | $n$ | dimensionless |
| Quadrupole moment | $\boldsymbol{Q}$, $\Theta$ | $C \cdot m^2$ |
| Relaxation time | | |
| longitudinal | $T_1$ | s |
| transverse | $T_2$ | s |
| Shielding constant | $\sigma$ | dimensionless |

# Polymer Chemistry

| NAME | SYMBOL | SI UNIT |
|---|---|---|
| Bulk modulus | $K$ | Pa |
| Complex permittivity | $\varepsilon^*$ | F/m, $F \cdot m^{-1}$ |
| Crack-tip radius | $\rho_c$ | m |
| Electrophoretic mobility | $\mu$ | $m^2 \cdot V^{-1} \cdot s^{-1}$ |
| Flory–Huggins interaction parameter | $\chi$ | |
| Fracture strain | $\gamma_f$, $\varepsilon_f$ | dimensionless |
| Fracture stress | $\sigma_f$ | Pa |
| Glass-transition temperature | $T_g$ | K |
| Modulus of elasticity | $E$ | Pa |
| Tensile strength | $\sigma$ | Pa |
| Viscosity | $\nu$ | Pa·s |
| Volume fraction | $V_f$ | dimensionless |
| Yield stress | $\sigma_y$ | Pa |
| Young's modulus | $E$ | Pa |

# Radiation

| NAME | SYMBOL | SI UNIT |
|---|---|---|
| Absorbance | $A$ | dimensionless |
| Absorption factor | $\alpha$ | dimensionless |
| Angle of optical rotation | $\alpha$ | dimensionless, rad |
| Angular frequency | $\omega$ | $s^{-1}$, rad/s, rad·$s^{-1}$ |
| Emissivity, emittance | $\varepsilon$ | dimensionless |
| Frequency | $\nu$ | Hz |
| Linear decadic absorption coefficient | $a$ | $m^{-1}$ |
| Molar decadic absorption coefficient | $\varepsilon$ | $m^2$/mol, $m^2$·$mol^{-1}$ |
| Molar refraction | $R_m$ | $m^3$/mol, $m^3$·$mol^{-1}$ |
| Radiant energy | $Q$, $W$ | J |
| Radiant intensity | $I$ | W/sr, W·$sr^{-1}$ |
| Radiant power | $P$ | W |
| Refractive index | $n$ | dimensionless |
| Speed of light | $c$ | m/s, m·$s^{-1}$ |
| Stefan–Boltzmann constant | $\sigma$ | W/($m^2$·$K^4$), W·$m^{-2}$·$K^{-4}$ |
| Transmittance | $T$ | dimensionless |
| Wavelength | $\lambda$ | m |
| Wavenumber (in a vacuum) | $\bar{\nu}$ | $m^{-1}$ |

# Space and Time

| NAME | SYMBOL | SI UNIT |
|---|---|---|
| Acceleration | $\boldsymbol{a}$ | $m/s^2$, m·$s^{-2}$ |
| Area | $A$, $S$, $A_S$ | $m^2$ |
| Cartesian space coordinates | $x$, $y$, $z$ | m |
| Characteristic time interval | $\tau$ | s |
| Circular frequency | $\omega$ | $s^{-1}$, rad/s, rad·$s^{-1}$ |
| Diameter | $d$ | m |
| Frequency | $\nu$, $f$ | Hz |
| Height | $h$ | m |
| Length | $l$ | m |
| Position vector | $\boldsymbol{r}$ | m |
| Radius | $r$ | m |
| Speed | $v$, $u$, $w$, $c$ | m/s, m·$s^{-1}$ |
| Thickness, distance | $d$, $\delta$ | m |
| Time | $t$ | s |
| Time constant | $\tau$ | s |
| Velocity | $\boldsymbol{v}$, $\boldsymbol{u}$, $\boldsymbol{w}$, $\boldsymbol{c}$ | m/s, m·$s^{-1}$ |
| Volume | $V$ | $m^3$ |

# Thermodynamics

| NAME | SYMBOL | SI UNIT |
|---|---|---|
| Absolute activity | $\lambda$ | dimensionless |
| Affinity of a reaction | $A$ | J/mol, J·mol$^{-1}$ |
| Chemical potential | $\mu$ | J/mol, J·mol$^{-1}$ |
| Cubic expansion coefficient | $\alpha$ | K$^{-1}$ |
| Energy | $E$ | J |
| Enthalpy | $H$ | J |
| Entropy | $S$ | J/K, J·K$^{-1}$ |
| Fugacity | $f$ | Pa |
| Gas constant | $R$ | J/(K·mol), J·K$^{-1}$·mol$^{-1}$ |
| Gibbs energy | $G$ | J |
| Heat | $q$, $Q$ | J |
| Heat capacity, molar | $C_m$ | J/(K·mol), J·K$^{-1}$·mol$^{-1}$ |
| Heat capacity at constant pressure | $C_p$ | J/K, J·K$^{-1}$ |
| Heat capacity at constant volume | $C_v$ | J/K, J·K$^{-1}$ |
| Helmholtz energy | $A$ | J |
| Internal energy | $U$ | J |
| Isothermal compressibility | $\kappa$ | Pa$^{-1}$ |
| Joule–Thomson coefficient | $\mu$ | K/Pa, K·Pa$^{-1}$ |
| Pressure, osmotic | $\Pi$ | Pa |
| Pressure coefficient | $\beta$ | Pa/K, Pa·K$^{-1}$ |
| Specific heat capacity | $c$ | J/(K·kg), J·K$^{-1}$·kg$^{-1}$ |
| Surface tension | $\gamma$, $\sigma$ | J/m$^2$, J·m$^{-2}$, N/m, N·m$^{-1}$ |
| Temperature | | |
| Celsius | $t$, $\theta$ | °C |
| thermodynamic | $T$ | K |
| Viscosity | $\eta$ | Pa·s |
| Work | $w$, $W$ | J |

# Transport Properties

| NAME | SYMBOL | SI UNIT |
|---|---|---|
| Coefficient of heat transfer | $h$ | W/(m$^2$·K), W·m$^{-2}$·K$^{-1}$ |
| Diffusion coefficient | $D$ | m$^2$/s, m$^2$·s$^{-1}$ |
| Flux of a quantity $x$ | $J_x$, $J$ | varies |
| Heat flow rate | $\phi$ | W |
| Kinematic viscosity | $\nu$ | m$^2$/s, m$^2$·s$^{-1}$ |
| Mass flow rate | $q_m$ | kg/s, kg·s$^{-1}$ |
| Mass-transfer coefficient | $k_d$ | m/s, m·s$^{-1}$ |
| Thermal conductivity | $\lambda$, $k$ | W/(m·K), W·m$^{-1}$·K$^{-1}$ |
| Thermal diffusion coefficient | $D_T$ | m$^2$/s, m$^2$·s$^{-1}$ |
| Thermal diffusivity | $a$ | m$^2$/s, m$^2$·s$^{-1}$ |
| Viscosity | $\eta$ | Pa·s |
| Volume flow rate | $q_v$, $V$ | m$^3$/s, m$^3$·s$^{-1}$ |

➤➤➤➤➤

APPENDIX 13-2

# The Crystallographic Information File

Frank H. Allen

The Crystallographic Information File (CIF) is the internationally agreed on standard file format for the exchange of crystallographic information. Most major journals now require that crystal structure data for small molecules and inorganic compounds, obtained by single-crystal and powder diffraction analyses, be deposited as electronic CIFs as part of the manuscript submission process. A related format, the macromolecular CIF, or mmCIF, has been developed for data deposition with the Protein Data Bank. The CIF (and mmCIF) standard is maintained by the International Union of Crystallography (IUCr), and full information and leading references can be found at www.iucr.org/iucr-top/cif/.

Important sources of three-dimensional crystal structure data, including data in CIF or mmCIF format, are the following:

- The Cambridge Crystallographic Data Centre produces the Cambridge Structural Database (http://www.ccdc.cam.ac.uk/products/csd/), which covers organic and metal-organic small-molecule crystal structures.
- CRYSTMET (http://www.tothcanada.com/) contains data for metals, intermetallics, and alloys.
- FIZ Karlsruhe produces the ICSD (http://www.fiz-informationsdienste.de/en/FG/Kristall/index.html), which contains information on inorganic crystal structures.
- The Nucleic Acid Database (http://ndbserver.rutgers.edu/) contains data on oligonucleotides.
- The Protein Data Bank (http://www.rcsb.org/pdb/) contains biological macromolecular structure data.

CIF principles are simple. Every *data item* is represented by a unique *data name* followed by its associated *data value*. Data items are described in an electronic dictionary that defines meaning, usage, and (where appropriate) permitted values or ranges of values. Data names start with an underscore (underline) character, and data values can be any type of string (text, numeric, or mixed), ranging from a single character to many lines of text. Data values are delimited by spaces, double or single quotes, or pairs of lines beginning with a semicolon (to delimit multiline data items). Related data items, such as those that relate to an individual crystal structure, are grouped together in a *data block*. The start of a block is designated by the string `data_` prefixing the name of the block, and the end of a block is recognized by another `data_` record introducing a new block, or by the end of the file. A complete CIF may contain any number of data blocks, each reporting an individual crystal structure, and these may be preceded by a data block containing items (such as author names and contact details) that are common for all structures in the complete CIF.

These basic principles, together with the data names in the CIF standard, make the CIF human readable, as illustrated below:

```
data_structure_1
_cell_length_a                5.959(1)
_chemical_formula_moiety      'C23 H36 O7'
_publ_contact_author
;
Dr J. Smith
Department of Chemistry
University of Nowhere
Nowhere
Anystate 20761
USA
;
```

If a data name is preceded by `loop_`, then a series of values can be associated with that name, e.g., `_symmetry_equiv_pos_as_xyz`. A series of data values can also be grouped together under different data names using the `loop_` construct, as in the example below, which shows a loop containing the atom labels and *x,y,z*-coordinates for four atoms. Here, the data names can be regarded as the column headings in a conventional printed table.

```
loop_
_atom_site_label
_atom_site_fract_x
_atom_site_fract_y
_atom_site_fract_z
I1 0.26639(7) 0.61557(3) 0.94292(3)
I2 0.64548(7) 0.364889(3) 0.56299(3)
P3 0.0438(2) 0.27607(11) 0.74432(13)
O1 -0.0989(7) 0.2488(3) 0.6619(4)
```

The crystallographic part of a CIF is generated automatically by the software package used for structure refinement. However, the software can only output data items that it knows about, i.e., those items that are input to or are generated by the refinement process, or which can be generated reliably from these data.

This is seldom sufficient for journal submission, where additional data, such as author names and contact details, and other chemical or crystal property information are almost always required. Because these additional requirements will vary from journal to journal, authors should consult the author guidelines for the journal of their choice to ensure that they submit files that conform to the publisher's requirements.

Although the CIF can be edited using standard text editors, this is not recommended because the strict syntax, if broken in any small way, renders the file unreadable by application software. To solve the editing issue, a number of stand-alone CIF editors have been written. For example, the enCIFer program is available for free download from the Cambridge Crystallographic Data Centre Web site (http://www.ccdc.cam.ac.uk). The program will handle single-block or multiblock CIFs and has a graphical interface that permits the following:

- location, display, and correction of syntax or format violations using the current CIF dictionary;
- spreadsheet representation of looped data items;
- editing and/or addition of individual or looped data items;
- addition of certain standard information (basic bibliographic and/or property information) via two data entry wizards; and
- three-dimensional visualization of the structure(s) contained in the CIF.

The IUCr Web site (http://www.iucr.org/iucr-top/cif/) lists other software that can be used to manipulate or visualize CIF content. Most importantly, they maintain a Web-based service that provides a classified validation report on any format-compliant CIF. This service can be accessed at http://checkcif.iucr.org/. CheckCIF reports can be lengthy, but it is wise to scan them carefully, particularly the most serious Class A alerts. Journal editors and their referees will often indicate the types of alerts that they would normally expect to be fixed by the authors before publication.

➤➤➤➤➤ CHAPTER 14

# References

Janet S. Dodd, Leah Solla,
and Paula M. Bérard

This chapter presents style conventions for citing references within a manuscript and for listing complete reference citations. Many of the references in the examples were created to illustrate a style point under discussion; they may not be real references.

## Citing References in Text

In ACS publications, you may cite references in text in three ways:

1. By superscript numbers, which appear outside the punctuation if the citation applies to a whole sentence or clause.

    Oscillation in the reaction of benzaldehyde with oxygen was reported previously.[3]

2. By italic numbers in parentheses on the line of text and inside the punctuation.

    The mineralization of TCE by a pure culture of a methane-oxidizing organism has been reported (*6*).

3. By author name and year of publication in parentheses inside the punctuation (known as author–date).

    The primary structure of this enzyme has also been determined (Finnegan et al., 2004).

In ACS books, all three of these systems are used, depending on the subject matter and series. Table 14-1 lists the referencing systems used by the ACS journals currently in print.

**Table 14-1.** ACS Periodicals, with Referencing Style, *CASSI* Abbreviation, and 2006 Volume Number

| *Name as Registered in the U.S. Patent and Trademark Office* | *Referencing Style*[a] | CASSI *Abbreviation* | *2006 Vol.* |
|---|---|---|---|
| *Accounts of Chemical Research* | 1 | *Acc. Chem. Res.* | 39 |
| *ACS Chemical Biology* | 2 | *ACS Chem. Biol.* | 1 |
| *Analytical Chemistry* | 1 | *Anal. Chem.* | 78 |
| *review issues* | 2 | | |
| *Biochemistry* | 2 | *Biochemistry* | 45 |
| *Bioconjugate Chemistry* | 2 | *Bioconjugate Chem.* | 17 |
| *Biomacromolecules* | 1 | *Biomacromolecules* | 7 |
| *Biotechnology Progress* | 2 | *Biotechnol. Prog.* | 22 |
| *Chemical & Engineering News* | | *Chem. Eng. News* | 84 |
| *Chemical Research in Toxicology* | 2 | *Chem. Res. Toxicol.* | 19 |
| *Chemical Reviews* | 1 | *Chem. Rev.* | 106 |
| *Chemistry of Materials* | 1 | *Chem. Mater.* | 18 |
| *Crystal Growth & Design* | 1 | *Cryst. Growth Des.* | 6 |
| *Energy & Fuels* | 1 | *Energy Fuels* | 20 |
| *Environmental Science & Technology* | 2 | *Environ. Sci. Technol.* | 40 |
| *Industrial & Engineering Chemistry Research* | 1 | *Ind. Eng. Chem. Res.* | 45 |
| *Inorganic Chemistry* | 1 | *Inorg. Chem.* | 45 |
| *Journal of Agricultural and Food Chemistry* | 2 | *J. Agric. Food Chem.* | 54 |
| *Journal of the American Chemical Society* | 1 | *J. Am. Chem. Soc.* | 128 |
| *Journal of Chemical and Engineering Data* | 1 | *J. Chem. Eng. Data* | 51 |
| *Journal of Chemical Information and Modeling* | 1 | *J. Chem. Inf. Model.* | 46 |
| *Journal of Chemical Theory and Computation* | 1 | *J. Chem. Theory Comput.* | 2 |
| *Journal of Combinatorial Chemistry* | 1 | *J. Comb. Chem.* | 8 |
| *Journal of Medicinal Chemistry* | 1 | *J. Med. Chem.* | 49 |
| *Journal of Natural Products* | 1 | *J. Nat. Prod.* | 69 |
| *The Journal of Organic Chemistry* | 1 | *J. Org. Chem.* | 71 |
| *The Journal of Physical Chemistry A* | 1 | *J. Phys. Chem. A* | 110 |
| *The Journal of Physical Chemistry B* | 1 | *J. Phys. Chem. B* | 110 |
| *Journal of Proteome Research* | 1 | *J. Proteome Res.* | 5 |
| *Langmuir* | 1 | *Langmuir* | 22 |
| *Macromolecules* | 1 | *Macromolecules* | 39 |
| *Molecular Pharmaceutics* | 1 | *Mol. Pharm.* | 3 |
| *Nano Letters* | 1 | *Nano Lett.* | 6 |
| *Organic Letters* | 1 | *Org. Lett.* | 8 |
| *Organic Process Research & Development* | 1 | *Org. Process Res. Dev.* | 10 |
| *Organometallics* | 1 | *Organometallics* | 25 |

[a]Reference style 1 uses superscript numbers, and 2 uses italic numbers in parentheses on the line of the text.

➤ In all three systems, the author's name may be made part of the sentence. In such cases, in the author–date system, place only the year in parentheses.

> The syntheses described by Fraser[8] take advantage of carbohydrate topology.
>
> Jensen (*3*) reported oscillation in the reaction of benzaldehyde with oxygen.
>
> According to Harris (2003), drug release is controlled by varying the hydrolytic stability of the ester bond.

➤ With numerical reference citations, start with 1 and number consecutively throughout the paper, including references in text and those in tables, figures, and other nontext components. If a reference is repeated, do not give it a new number; use the original reference number.

➤ Whenever authors are named, if a reference has two authors, give both names joined by the word "and". If a reference has more than two authors, give only the first name listed, followed by "et al." Do not use a comma before et al.; always use a period after al.

> Allison and Perez[12]
> Johnson et al. (*12*)
> (O'Brien and Alemmo, 2005)
> (Bachrach et al., 2004)

➤ To cite more than one reference by the same principal author and various coauthors in one of the numerical citation systems, use the principal author's name followed by "and co-workers" or "and colleagues".

> Pauling and co-workers[10,11]
> Cram and colleagues (*27–29*)

➤ When citing more than one reference at one place by number in one of the numerical systems, list the numbers in ascending order and separate them by commas (*without* spaces as superscripts, *with* spaces on line), or if they are part of a consecutive series, use an en dash to indicate a range of three or more.

> in the literature[2,5,8]
> were reported[3–5,10]
>
> in the literature (*2, 5, 8*)
> were reported (*3–5, 10*)

➤ When citing more than one reference at one place by the author–date system, list them alphabetically according to the first author's name, followed by a comma and the year. Use a semicolon to separate individual references.

> (Axelrod, 2003; Cobbs and Stolman, 2005; Gerson et al., 2001)

➤ When citing more than one reference by the same author at one place by the author–date system, do not repeat the name. List the name followed by the year of each of the references in ascending order; separate the years by commas. If an author has more than one reference in the same year, add lowercase letters to the years to differentiate them. Add letters to all of the years, for example, 2005a, 2005b, etc., not 2005, 2005a, etc. (The references in the list will need to be listed the same way, for example, 2005a, 2005b.

(Trapani, 2003, 2005; Zillman, 2004)
(Knauth, 2005a, 2005b)
(Fordham, 2004; Fordham and Rizzo, 2004)

➤ Cite the reference in a logical place in the sentence.

recent investigations (cite)
other developments (cite)
was reported (cite)
as described previously (cite)
previous results (cite)
were demonstrated (cite)
a molecular mechanics study (cite)
Marshall and Levitt's approach (cite)
the procedure of Lucas et al. (cite)

## Style for Reference Lists

***Authors are responsible for the accuracy and completeness of all references.*** Authors should check all parts of each reference listing against the original document.

A reference must include certain minimum data:

- Periodical references must include the author names, abbreviated journal title, year of publication, volume number (if any), and initial page of cited article (the complete span is better).
- Book references must include the author or editor names, book title, publisher, city of publication, and year of publication.
- For material other than books and journals, sufficient information must be provided so that the source can be identified and located.

In lists, references always end with a period.

Table 14-2 provides sample references for common reference types.

## Periodicals

RECOMMENDED FORMATS

Author 1; Author 2; Author 3; etc. Title of Article. *Journal Abbreviation* **Year,** *Volume,* Inclusive Pagination.

Author 1; Author 2; Author 3; etc. *Journal Abbreviation* **Year,** *Volume,* Inclusive Pagination.

The journal *Biochemistry* is an exception. Consult this journal's instructions to authors for the correct format.

### *Author Name Field*

Include all author names in a reference citation. With multiple authors, separate the names from one another by semicolons. Always end the author field with a period (exception: *Biochemistry*). List the names in inverted form: surname first, then first initial, middle initial, and qualifiers (Jr., II). Some publications list the first 10 authors followed by a semicolon and et al.; check the guidelines.

Cotton, F. A.
Basconi, J.; Lin, P. B.
Chandler, J. P., III; Levine, S. M.
Schafer, F. W., Jr.
Fishman, W., II.

Farhataziz. (a single name is uncommon, but does occur; no period in *Biochemistry*)

Inderjit; Fontana, M. J. (the first author has a single name)

### *Article Title Field*

Article titles are not essential in reference citations, but they are considered desirable to highlight the contents of a paper and facilitate location in reference libraries. Some ACS publications include the article title in journal references, and some do not; check the publication itself. Article titles are set in roman type without quotation marks and end with a period (or a question mark if that is part of the title). In ACS journals, capitalization follows that of the original publication; in other publications, the main words are capitalized.

Caruso, R. A.; Susha, A.; Caruso, F. Multilayered Titania, Silica, and Laponite Nanoparticle Coatings on Polystyrene Colloidal Templates and Resulting Inorganic Hollow Spheres. *Chem. Mater.* **2001**, *13*, 400–409.

### *Journal Abbreviation Field*

The journal name is an essential component of a periodical reference citation. Abbreviate the name according to the *Chemical Abstracts Service Source Index*

**Table 14-2.** Common Types of References with Examples

| *Reference Type* | *See Pages* | *Example* |
|---|---|---|
| ***Print Sources*** | | |
| Journal article | | |
| with article title | 291 | Klingler, J. Influence of Pretreatment on Sodium Powder. *Chem. Mater.* **2005,** *17,* 2755–2768. |
| without article title | 291 | Klingler, J. *Chem. Mater.* **2005,** *17,* 2755–2768. |
| Nonscientific magazines and newspapers | 299 | Squires, S. Falling Short on Nutrients. *The Washington Post,* Oct 4, 2005, p H1. |
| Books | | |
| without editors | 300 | Le Couteur, P.; Burreson, J. *Napoleon's Buttons: How 17 Molecules Changed History;* Jeremy P. Tarcher/Putnam: New York, 2003; pp 32–47. |
| with editors | 300 | Almlof, J.; Gropen, O. Relativistic Effects in Chemistry. In *Reviews in Computational Chemistry;* Lipkowitz, K. B., Boyd, D. B., Eds.; VCH: New York, 1996; Vol. 8, pp 206–210. |
| Series publication | | |
| cited as a book | 306 | Puls, J.; Saake, B. Industrially Isolated Hemicelluloses. In *Hemicelluloses: Science and Technology;* Gatenholm, P., Tenkanen, M., Eds.; ACS Symposium Series 864; American Chemical Society: Washington, DC, 2004; pp 24–37. |
| cited as a journal | 306 | Puls, J.; Saake, B. *ACS Symp. Ser.* **2004,** *864,* 24–37. |
| Meeting or conference, full citation | 307–308 | Garrone, E.; Ugliengo, P. In *Structure and Reactivity of Surfaces,* Proceedings of the European Conference, Trieste, Italy, Sept 13–20, 1988; Zecchina, A., Cost, G., Morterra, C., Eds.; Elsevier: Amsterdam, 1988. |
| Theses | 309–310 | Mäckel, H. Capturing the Spectra of Silicon Solar Cells. Ph.D. Thesis, The Australian National University, December 2004. |
| Patents | 310–311 | Sheem, S. K. Low-Cost Fiber Optic Pressure Sensor. U.S. Patent 6,738,537, May 18, 2004. |
| Government publications, U.S. | 311–314 | *Agriculture Fact Book 2000;* U.S. Department of Agriculture, U.S. Government Printing Office: Washington, DC, 2000. |
| Technical reports and bulletins | 314 | Tschantz, B. A.; Moran, B. M. *Modeling of the Hydrologic Transport of Mercury in the Upper East Fork Poplar Creek (UEFPC) Watershed;* Technical Report for Lockheed Martin Energy Systems: Bethesda, MD, September 2004. |
| Material Safety Data Sheets | 315 | *Titanium Dioxide;* MSDS No. T3627; Mallinckrodt Baker: Phillipsburg, NJ, Nov 12, 2003. |
| Personal communications | 315–316 | Henscher, L. X. University of Minnesota, Minneapolis, MN. Personal communication, 2001. |
| ***Online Periodicals*** | | |
| Based on print editions | 318 | Fine, L. Einstein Revisited. *J. Chem. Educ.* [Online] **2005,** *82,* 1601 ff. http://jchemed.chem.wisc.edu/Journal/Issues/2005/Nov/abs1601.html (accessed Oct 15, 2005). |
| Published in advance of print issue | 318–319 | Pratt, D. A.; van der Donk, W. A. Theoretical Investigations into the Intermediacy of Chlorinated Vinylcobalamins in the Reductive Dehalogenation of Chlorinated Ethylenes. *J. Am. Chem. Soc.* [Online early access]. DOI: 10.1021/ja047915o. Published Online: Dec 8, 2004. http://pubs.acs.org/cgi-bin/asap.cgi/jacsat/asap/html/ja047915o.html (accessed Dec 8, 2004). |

*Continued on next page*

**Table 14-2.** Common Types of References with Examples—*Continued*

| *Reference Type* | *See Pages* | *Example* |
|---|---|---|
| Retrieved from a database provider | 318 | Hallet, V. Scanning the Globe for Organic Chemistry. *U.S. News and World Report* [Online], April 19, 2004, p 59. Business Source Premier. http://www.epnet.com/academic/bussourceprem.asp (accessed April 24, 2005). |
| Published only electronically | 319 | Zloh, M.; Esposito, D.; Gibbons, W. A. Helical Net Plots and Lipid Favourable Surface Mapping of Transmembrane Helices of Integral Membrane Proteins: Aids to Structure Determination of Integral Membrane Proteins. *Internet J. Chem.* [Online] **2003,** *6,* Article 2. http://www.ijc.com/articles/2003v6/2/ (accessed Oct 13, 2004). |
| From preprint servers | 319 | Ward, D. W.; Nelson, K. A. Finite Difference Time Domain (FDTD) Simulations of Electromagnetic Wave Propagation Using a Spreadsheet. 2004, arXiv:physics/0402096. arXiv.org e-Print archive. http://arxiv.org/abs/physics/0402096 (accessed Oct 13, 2004). |
| Online books | | |
| without editors | 319–320 | Tour, J. M. *Molecular Electronics: Commercial Insights, Chemistry, Devices, Architecture and Programming* [Online]; World Scientific: River Edge, NJ, 2003; pp 177–180. http://legacy.netlibrary.com/ebook_info.asp?product_id=91422&piclist=19799,20141,20153 (accessed Nov 7, 2004). |
| with editors | 320 | Oleksyn, B. J.; Stadnicka, K.; Sliwinski, J. Structural Chemistry of Enamines: A Statistical Approach. In *The Chemistry of Enamines* [Online]; Rappoport, Z., Ed.; The Chemistry of Functional Groups; Patai, S., Rappoport, Z., Series Eds.; Wiley & Sons: New York, 1994; Chapter 2, pp 87–218. http://www3.interscience.wiley.com/cgi-bin/summary/109560980/SUMMARY (accessed April 24, 2005). |
| Online encyclopedias | 320 | Alkanolamines from Nitro Alcohols. *Kirk-Othmer Encyclopedia of Chemical Technology* [Online]; Wiley & Sons, Posted March 14, 2003. http://www.mrw.interscience.wiley.com/kirk/articles/alkaboll.a01/frame.html (accessed Nov 7, 2004). |
| ***Other Online Sources*** | | |
| General Web sites | 320–321 | ACS Publications Division Home Page. http://pubs.acs.org (accessed Nov 7, 2004). |
| Electronic lists and newsgroups | 322 | Chemical Information List Server, CHMINF-L@iubvm.ucs.indiana.edu (accessed Oct 13, 2004).<br>Computational Chemistry List, solvent discussion in archived messages of September 2003, chemistry@ccl.net (accessed Nov 10, 2004). |
| Electronic mail messages | 322 | Solla, L. R. Cornell University, Ithaca, NY. Personal communication, 2005. |
| CD-ROMs and DVDs | | |
| periodicals | 322 | Fleming, S. A.; Jensen, A. W. Substitutent Effects on the Photocleavage of Benzyl–Sulfur Bonds. Observation of the "*Meta*" Effect. *J. Org. Chem.* [CD-ROM] **1996,** *61,* 7044. |
| books | 322–323 | *Green Chemistry: Meeting Global Challenges* [DVD]; American Chemical Society: Washington, DC, 2003. |

(*CASSI*), and italicize it. One-word journal names are not abbreviated (e.g., *Biochemistry, Macromolecules, Nature, Science*). No punctuation is added to end this field; thus, a period will be there with an abbreviation but not with a spelled-out word.

*CASSI* and its quarterly supplements provide an extensive list of recommended journal abbreviations. Appendix 14-1 lists *CASSI* abbreviations for more than 1000 of the most commonly cited journals. ACS publication names, their abbreviations, and their volume numbers for 2006 are given in Table 14-1. Note that, in some cases, the word "the" is part of the title.

Sometimes journal names change. Authors should use the abbreviation of the journal title that was in use at the time the article was published. *CASSI* lists the journal titles and the range of years during which the title was being used.

### Information Found in *CASSI*

Entries are arranged in *CASSI* alphabetically according to the abbreviated form of the title. Abbreviations are based on the standards of the International Organization for Standardization (ISO). Recommended abbreviations are indicated in boldface type. See Appendix 14-2 for a sample *CASSI* entry with a description of each element in an entry.

### Using *CASSI* Abbreviations

➤ The boldface components of the publication title form the abbreviated title. Use a period after each abbreviation, and maintain the punctuation shown in *CASSI*.

> **J***ournal of* **Polym***er* **Sci***ence*, **Part A: Polym***er* **Sci***ence*
> *J. Polym. Sci., Part A: Polym. Sci.*

➤ Maintain the word spacing shown in *CASSI*, except for D.C., N.Y., U.K., U.S., and U.S.A.

> *Analyst (Cambridge, U.K.)*
> *Anesth. Analg. (Hagerstown, MD, U.S.)*
> *Ann. N.Y. Acad. Sci.*
> *Proc. Natl. Acad. Sci. U.S.A.*
> *Science (Washington, DC, U.S.)*

➤ Use a terminal period only if the last word of the periodical title is abbreviated.

> **Int***ernational* **J***ournal of* **Nanosci***ence*
> *Int. J. Nanosci.* (last word is abbreviated; period is used)

> **J***ournal of* **Controlled Release**
> *J. Controlled Release* (last word is not abbreviated; no period is used)

➤ If the periodical abbreviation in *CASSI* shows a hyphen with spaces on both sides, change the hyphen to an em dash closed up on each side.

> **Annu***al* **Tech***nical* **Conf***erence* - **Soc***iety of* **Plast***ics* **Eng***ineers*
> *Annu. Tech. Conf.—Soc. Plast. Eng.*

➤ If a boldface **n** precedes the volume number in *CASSI*, use the abbreviation "No." before the volume number in italics in the entry.

> **Br***itish* **Med***ical* **J***ournal* ... **n6372 1983**
> *Br. Med. J.* **1983,** *No. 6372.*

Include all the information shown for volume in italics, especially for references to government publications and reports.

> **Los Alamos Nat***ional* **Lab***oratory*, [**Rep***ort*] **LA** (**U***nited* **S***tates*) ... **LA-14240-SR 2005**
> *Los Alamos Natl. Lab., [Rep.] LA (U.S.)* **2005,** *LA-14240-SR.*

### Exceptions to the Rules of *CASSI* Abbreviations

➤ Strict rules for *CASSI* abbreviations can be modified for periodicals whose titles include multiple parts, sections, and series.

ABBREVIATION

> *Acta Crystallogr., Sect. C: Cryst. Struct. Commun.* **2005,** *61,* 99–102.

ACCEPTABLE VARIATION The section title need not be named:

> *Acta Crystallogr., Sect. C* **2005,** *61,* 99–102.

ACCEPTABLE VARIATION The section can be indicated by the volume number:

> *Acta Crystallogr.* **2005,** *C61,* 99–102.

➤ For some periodicals whose *CASSI* abbreviation includes a place of publication, you need not add the place of publication unless its omission would create ambiguity. If *CASSI* lists only one journal with a given main title, there is no ambiguity in omitting the place of publication.

| USE | NOT NECESSARILY |
|---|---|
| *Clin. Chem.* | *Clin. Chem. (Washington, DC, U.S.)* |
| *Nature* | *Nature (London, U.K.)* |
| *Science* | *Science (Washington, DC, U.S.)* |

In contrast, omission of the place of publication would create ambiguity for different journals having the same main title.

> *Transition Met. Chem. (Dordrecht, Neth.)*
> *Transition Met. Chem. (N.Y.)*

### *Year of Publication Field*

The year of publication is essential information in a periodical citation. The year is set in boldface type, followed by a comma in boldface type.

### *Publication Volume Field*

The volume number is important information and is recommended for all periodical citations; it is essential for publications having more than one volume per year (such as the *Journal of Chemical Physics*). The volume number is set in italic type and is separated from the pagination information by a comma, which is set in italic type.

➤ For periodicals in which each issue begins with page 1, include issue information (either the number or the date) in the publication volume field. Issue information is set in roman type, enclosed in parentheses, and spaced from the volume number, which it directly follows.

ISSUE NUMBER

Mullin, R. *Chem. Eng. News* **2005,** *83* (42), 7.

DATE OF ISSUE

Mullin, R. *Chem. Eng. News* **2005,** *83* (Oct 17), 7.

➤ For publications that have supplements, the following form is recommended.

Taylor, C. W.; Kumar, S. *Eur. J. Cancer* **2005,** *40* (Suppl. 1), 781.
*Eur. J. Anaesthesiol.* **2005,** *22* (Suppl. S36), 1–35.

➤ For journals that have no volume numbers, include issue numbers, especially when the pagination of each issue begins with page 1. Use the following form. Note that the issue number is not italicized.

Wills, M. R.; Savory, J. *Lancet* **1983,** No. 2, 29.

### *Pagination Field*

Pagination is an essential element of a reference citation. The complete page range is preferable, but initial page numbers are acceptable.

➤ In page spans, use all digits, closed up, with no commas or spaces.

2–15
44–49
103–107

1376–1382
2022–2134
11771–11779

➤ You may also indicate pagination in reference citations by "f" or "ff", which mean "and following" page or pages, respectively. The f or ff is set in roman type and is spaced from the preceding number:

60 f (indicates page 60 and the page following—pages 60 and 61)

60 ff (indicates page 60 and pages following)

58–60 ff (indicates pages 58 through 60 and pages following—essentially the same as 58 ff except that the three pages enumerated contain the most pertinent information and other relevant information is scattered on the rest of the pages)

➤ The pagination field may also include terms such as "and references therein" and similar expressions (especially in references to review articles). This phrase follows the page numbers and is not separated by a comma.

Puskas, J. E.; Chan, S. W. P.; McAuley, K. B.; Shaikh, S.; Kaszas, G. *J. Polym. Sci., Part A: Polym. Chem.* **2005,** *43,* 5394–5413 and references therein.

➤ Some publications use article numbering, rather than page numbering, where each article starts on page 1. Use the article number in the pagination field.

Brosset, C. *Ark. Kemi, Mineral. Geol.* **1945,** *20A,* No. 7.

## Use of Punctuation To Indicate Repeating Fields of Information

The choice of what punctuation to use to indicate repeating fields of information depends on whether the publication will appear strictly in print or on the Web. For publications that will appear in both print and on the Web, use the rules for Web publications.

➤ In references that will appear only in print publications, use a semicolon, a comma, or a period to indicate repeating information.

1. Same authors in multiple publications:

   Chauvin, Y.; Gilbert, B.; Guibard, I. *Vib. Spectrosc.* **1991,** *1,* 299–304; *J. Chem. Soc., Chem. Commun.* **1990,** 1715–1716.

2. Same authors in multiple publications, but with letters to separate the references (the semicolon from the previous example is changed to a period):

   (a) Schrock, R. R. *Chem. Commun.* **2003,** 2389. (b) *J. Mol. Catal. A* **2004,** *213,* 21–30.

3. Same authors of multiple articles in the same journal:

   Lu, Y.; Pignatello, J. J. *Environ. Sci. Technol.* **2002,** *36,* 4553–4561; **2004,** *38,* 5853–5862.

When the year and volume are the same:

Clay, S. A.; Koskinen, W. C. *Weed Sci.* **1990,** *38,* 74–80, 262–266.

When the year is the same but the volumes are different:

> Badyal, R.; Fleissner, A. *Chem. Phys.* **2005,** *317,* 73–86; **2005,** *316,* 201–215.

➤ In references that will appear only in Web publications, provide complete references so that the references can be properly linked. If two or more references with the same authors are cited, it is not acceptable to combine them into a single reference.

1. Same authors in multiple publications:

   > Chauvin, Y.; Gilbert, B.; Guibard, I. *Vib. Spectrosc.* **1991,** *1,* 299–304; Chauvin, Y.; Gilbert, B.; Guibard, I. *J. Chem. Soc., Chem. Commun.* **1990,** 1715–1716.

2. Same authors in multiple publications, but with letters to separate the references:

   > (a) Schrock, R. R. *Chem. Commun.* **2003,** 2389. (b) Schrock, R. R. *J. Mol. Catal. A* **2004,** *213,* 21–30.

3. Same authors of multiple articles in the same journal:

   > Lu, Y.; Pignatello, J. J. *Environ. Sci. Technol.* **2002,** *36,* 4553–4561; Lu, Y.; Pignatello, J. J. *Environ. Sci. Technol.* **2004,** *38,* 5853–5862.

The same principle holds no matter what information is being repeated: provide each reference in its entirety. Do not use the Latin terms ibid. (in the same place) or idem (the same).

## *References to* Chemical Abstracts

Use a semicolon to separate the periodical citation from a reference to its abstract (*Chemical Abstracts*).

> Mohamed, A. M.; Hawata, A.; El-Torgoman, M.; El-Kousy, S. M.; Ismail, A. E.; Øgaard Madsen, J.; Søtofte, I.; Senning, A. *Eur. J. Org. Chem.* **2002,** 2039–2045; *Chem. Abstr.* **2003,** *138,* 4195.
>
> Mloston, G.; Majchrazak, A.; Senning, A.; Søtofte, I. *J. Org. Chem.* **2002,** *67,* 5690–5695; *Chem. Abstr.* **2002,** *137,* 201289.

*Chemical Abstracts* routinely contains more than one abstract per page. The method of distinguishing which abstract was being cited has changed over the years. Three variations are worth noting.

1. The column (two columns per page) in which the abstract occurs followed by a superscript number:

   > *Chem. Abstr.* **1946,** *40,* $4463^8$. (This is the eighth abstract in column 4463.)

2. The column (two columns per page) in which the abstract occurs followed by a letter, either on the line or superscript (generally italic):

*Chem. Abstr.* **1953,** *47,* 1167*f.* (This is abstract f in column 1167.)
*Chem. Abstr.* **1947,** *41,* 571$^{d}$. (This is abstract d in column 571.)

3. The abstract number itself followed by an online letter (roman), often a computer check character:

   *Chem. Abstr.* **1989,** *110,* 8215j. (This is abstract number 8215.)

## *Special Situations*

➤ You may treat Beilstein references as periodical references.

*Beilstein, 4th ed.* **1950,** *12,* 237.

➤ Cite journals published in a foreign language either by the actual non-English title or by a translated form.

*Nippon Ishikai Zasshi* or *J. Jpn. Med. Assoc.*
*Nouv. J. Chim.* or *New J. Chem.*

➤ When citing an article printed in the English translation of a foreign-language journal, include reference to the original article, if possible, and use a semicolon to separate the two citations.

Tarasov, Y. I.; Kochikov, I. V.; Kovtun, D. M.; Vogt, N.; Novosadov, B. K.; Saakyan, A. S. *J. Struct. Chem. (Engl. Transl.)* **2004,** *45* (5), 778–785; *Zh. Strukt. Khim.* **2004,** *45* (5), 822–829.

➤ Separate two or more companion publications with a semicolon.

Clear, J. M.; Kelly, J. M.; O'Connell, C. M.; Vos, J. G. *J. Chem. Res., Miniprint* **2005,** 3038; *J. Chem. Res., Synop.* **2005,** 260.

## *Nonscientific Magazines and Newspapers*

RECOMMENDED FORMAT

Author 1; Author 2; Author 3; etc. Title of Article. *Title of Periodical,* Complete Date, Pagination.

For nonscientific magazines and other periodicals that are not abstracted by Chemical Abstracts Service, give the authors' names in inverted form ending with a period, the article title in roman type with main words capitalized and ending with a period, the full magazine title in italic type followed by a comma in italic type, the complete date of the issue (see pp 160–161 about dates) ending with a comma, and the pagination.

Squires, S. Falling Short on Nutrients. *The Washington Post,* Oct 4, 2005, p H1.

## Books

Some ACS publications include the chapter title in book references, and some do not; check the publication itself. Also, consult the instructions to authors in *Biochemistry* for exceptions to the format presented here and elsewhere in this chapter.

RECOMMENDED FORMATS FOR BOOKS WITHOUT EDITORS

Author 1; Author 2; Author 3; etc. Chapter Title. *Book Title,* Edition Number; Series Information (if any); Publisher: Place of Publication, Year; Volume Number, Pagination.

Author 1; Author 2; Author 3; etc. *Book Title;* Series Information (if any); Publisher: Place of Publication, Year; Volume Number, Pagination.

When a book has authors and no editors, it means either that the entire book was written by one author or that two or more authors collaborated on the entire book.

Le Couteur, P.; Burreson, J. *Napoleon's Buttons: How 17 Molecules Changed History;* Jeremy P. Tarcher/Putnam: New York, 2003; pp 32–47.

Morris, R. *The Last Sorcerers: The Path from Alchemy to the Periodic Table;* Joseph Henry Press: Washington, DC, 2003; pp 145–158.

RECOMMENDED FORMATS FOR BOOKS WITH EDITORS

Author 1; Author 2; Author 3; etc. Chapter Title. In *Book Title,* Edition Number; Editor 1, Editor 2, etc., Eds.; Series Information (if any); Publisher: Place of Publication, Year; Volume Number, Pagination.

Author 1; Author 2; Author 3; etc. In *Book Title,* Edition Number; Editor 1, Editor 2, etc., Eds.; Series Information (if any); Publisher: Place of Publication, Year; Volume Number, Pagination.

When a book has editors, it means that different authors wrote various parts of the book independently of each other. The word "In" before the book title indicates that the authors mentioned wrote only a part of the book, not the entire book.

Holbrey, J. D.; Chen, J.; Turner, M. B.; Swatloski, R. P.; Spear, S. K.; Rogers, R. D. Applying Ionic Liquids for Controlled Processing of Polymer Materials. In *Ionic Liquids in Polymer Systems: Solvents, Additives, and Novel Applications;* Brazel, C. S., Rogers, R. D., Eds.; ACS Symposium Series 913; American Chemical Society: Washington, DC, 2005; pp 71–88.

Almlof, J.; Gropen, O. Relativistic Effects in Chemistry. In *Reviews in Computational Chemistry;* Lipkowitz, K. B., Boyd, D. B., Eds.; VCH: New York, 1996; Vol. 8, pp 206–210.

If the book as a whole is being referenced, the author names might not appear.

*Ionic Liquids in Polymer Systems: Solvents, Additives, and Novel Applications;* Brazel, C. S., Rogers, R. D., Eds.; ACS Symposium Series 913; American Chemical Society: Washington, DC, 2005.

*Advances in Inorganic Chemistry and Radiochemistry;* Emeléus, H. J., Sharpe, A. G., Eds.; Academic: New York, 2001.

### Author Name Field

➤ Separate the names of multiple authors by semicolons, and always end the author field with a period (except in *Biochemistry*). List names in inverted form: surname first, then first initial, middle initial, and qualifiers (Jr., II).

➤ If a book has no primary authors because each chapter was written by a different author, you may place the editor names in the author name field (especially for lists in alphabetical order). Separate editor names by commas, and in this case, the period after the abbreviation Ed. or Eds. terminates the field.

Stocker, J. H., Ed. *Chemistry and Science Fiction;* American Chemical Society: Washington, DC, 1998.

➤ A book might have no named authors because it was compiled by a committee or organization. These books are discussed under the section "Works Written by an Organization or a Committee", p 307.

### Chapter Title Field

Chapter titles are not essential, but they are considered desirable components in reference citations because they highlight the contents of a paper and facilitate its location in reference libraries. Chapter titles are set in roman type and end with a period.

Puls, J.; Saake, B. Industrially Isolated Hemicelluloses. In *Hemicelluloses: Science and Technology;* Gatenholm, P., Tenkanen, M., Eds.; ACS Symposium Series 864; American Chemical Society: Washington, DC, 2004; pp 24–37.

### Book Title Field

Book titles are essential elements in book reference citations. In general, book titles should not be abbreviated. They are set in italic type and are separated from the next field of the reference by a semicolon, which is set in italic type.

➤ The edition number (in ordinal form) and the abbreviation "ed." follow the book title, set off by an italic comma; they are set in roman type. The edition information is separated from the next field of the reference by a semicolon.

*Reagent Chemicals,* 10th ed.;

➤ When both authors and editors are given, use the word "In" (set in roman type) immediately before the title of the book to indicate that the cited authors wrote only part of the book.

Hillman, L. W. In *Dye Laser Principles with Applications;* Duarte, F. J., Hillman, L. W., Eds.; Academic: New York, 1990; Chapter 2.

## Editor Name Field

For books with editors, list the names of the editors, after title and edition information, in inverted form as described in the section "Author Name Field", separated from one another by commas. The names are denoted as editors by including the abbreviation "Eds." or "Ed." after the final name. The editor field is set in roman type and ends with a semicolon (unless it is used in the author field location).

*Lignocellulose Biodegradation;* Saha, B. C., Hayashi, K., Eds.; ACS Symposium Series 889; American Chemical Society: Washington, DC, 2004.

*The Chemistry of the Atmosphere: Oxidants and Oxidation in the Earth's Atmosphere;* Bandy, A. R., Ed.; Royal Society of Chemistry: Cambridge, U.K., 1995.

In books that have no primary authors, the names of the editors may appear in either the author name field (especially for lists in alphabetical order) or the editor name field. When the editor names appear in the author name field, they are separated by commas and the field ends with a period.

Saha, B. C., Hayashi, K., Eds.; *Lignocellulose Biodegradation;* ACS Symposium Series 889; American Chemical Society: Washington, DC, 2004.

Bandy, A. R., Ed. *The Chemistry of the Atmosphere: Oxidants and Oxidation in the Earth's Atmosphere;* Royal Society of Chemistry: Cambridge, U.K., 1995.

## Publication Information Field

The name of the publisher, place of publication, and year of publication are essential elements in a book reference.

### Name of Publisher

Check the title page, front and back, for the publisher's name and location. Names and addresses of publishers are also listed in *Chemical Abstracts Service Source Index, 1907–2004 Cumulative,* pp 21I–39I.

➤ Generally, do not abbreviate publishers' names.

American Chemical Society, *not* Am. Chem. Soc. *or* ACS
American Ceramic Society, *not* Am. Ceram. Soc.

EXCEPTION You may use well-known acronyms or abbreviations created by the publishers themselves.

AIChE *or* American Institute of Chemical Engineers
ASTM *or* American Society for Testing and Materials
IUPAC *or* International Union of Pure and Applied Chemistry

➤ In some publisher's names, words such as Co., Inc., Publisher, and Press are not essential.

Academic Press: New York *or* Academic: New York

➤ Expanded names are also not essential.

John Wiley & Sons *or* John Wiley *or* Wiley

➤ It is not necessary to repeat the publisher's name for a book compiled by the organization that published it.

*CRC Handbook of Chemistry and Physics,* 85th ed.; Boca Raton, FL, 2004.

### Place of Publication

For the place of publication, give the city and state for U.S. cities or the city and country for all others. The country or state is not needed if the city is considered a major city in the world and could not be confused easily with other cities of the same name (e.g., London, Paris, New York, and Rome). Use the two-letter postal abbreviations (listed in Chapter 10) for states. Spell out names of countries unless they have standard abbreviations, such as U.K. for United Kingdom.

Birmingham, U.K.
Boca Raton, FL
Cambridge, MA
Cambridge, U.K.
Chichester, U.K.
Dordrecht, Netherlands
Elmsford, NY
Englewood Cliffs, NJ
London
New York
Princeton, NJ
Springfield, IL
Springfield, MA
Washington, DC

### Year of Publication

In book references, the year is set in lightface (not bold) roman type, following the place of publication. Terminate the field with a period or with a semicolon if further information is given.

Gould, S. J. *The Structure of Evolutionary Theory;* Belknap Press: Cambridge, MA, 2002.

Kline, R. B. *Principles and Practice of Structural Equation Modeling,* 2nd ed.; Guilford Press: New York, 2004.

## *Volume and Pagination Field*

### Volume Information

➤ The volume field contains specific information, such as volume number and chapter number. Use the following abbreviations and spelled out forms with the capitalization, spelling, and punctuation shown:

Abstract
Chapter

No.
Paper
Part
Vol. (for specific volumes, Vol. 4; Vols. 1, 2; Vols. 1 and 2; Vols. 3–5)
vols. (for a number of volumes, 4 vols.)

*Annual Review of Physical Chemistry;* Leone, S. R., McDermott, A. E., Paul, A., Eds.; Annual Reviews: Palo Alto, CA, 2005; Vol. 56.

➤ If a volume or part number refers to the volume or part of an entire series of books, this information is placed where a series number would normally appear and not in the volume field for the specific book being cited.

Wiberg, K. In *Investigations of Rates and Mechanisms of Reactions;* Lewis, E. S., Ed.; Techniques of Chemistry, Vol. VI, Part I; Wiley & Sons: New York, 1974; p 764.

➤ If the book or set of books as a whole is the reference, do not include individual volume information.

*McGraw-Hill Encyclopedia of Science and Technology,* 9th ed.; McGraw-Hill: New York, 2002; 20 vols.

### Pagination Information

➤ If you are citing a chapter, the complete page range is best, but initial page numbers are acceptable. Pagination may also be indicated by "f" or "ff" notation (meaning "and following" page or pages, respectively). The f or ff is set in roman type and is spaced from the preceding number. These points are illustrated under the "Pagination Field" heading for periodicals.

➤ Pagination information is set in roman type and ends with a period, except when miscellaneous information follows it, in which case it should end with a semicolon (see the next section). Use the abbreviations "p" and "pp" to indicate single and multiple pages, respectively.

p 57
p 93 f
pp 48–51
pp 30, 52, 76
pp 30, 52, 76 ff

pp 30, 52, and 76
pp 562–569
pp 562–9 (acceptable in journals)
2005; Vol. 2, p 35.
2004; pp 55–61.

➤ If the book as a whole is the reference, page numbers need not be given.

## Miscellaneous Information

If you wish to include additional information about a book that is important for the reader to know, you may add it at the end of the reference with or without parentheses, append it to the title in parentheses before the semicolon, or place it between the title and the publisher.

AOCS. *Official Methods and Recommended Practices of the American Oil Chemists' Society;* Link, W. E., Ed.; Champaign, IL, 1958 (revised 1973).

Brown, H. C. *The Nonclassical Ion Problem;* Plenum: New York, 1977; Chapter 5 (with comments by P. v. R. Schleyer).

Otsu, T.; Kinoshita, M. *Experimental Methods of Polymer Synthesis* (in Japanese); Kagakudojin: Kyoto, Japan, 1972; p 72.

Tessier, J. Structure, Synthesis and Physical–Chemical Properties of Deltamethrin. In *Deltamethrin Monograph;* Tessier, J., Ed.; Roussel-Uclaf: Paris, 1982; pp 37–66; translated by B. V. d. G. Walden.

Tessier, J. Structure, Synthesis and Physical–Chemical Properties of Deltamethrin. In *Deltamethrin Monograph;* Tessier, J., Ed.; Walden, B. V. d. G., Translator; Roussel-Uclaf: Paris, 1982.

*Volatile Compounds in Foods and Beverages;* Maarse, H., Ed.; Marcel Dekker: New York, 1991; see also references therein.

## Special Situations

➤ *Organic Syntheses* collective volumes should be treated as books.

*Organic Syntheses;* Wiley & Sons: New York, Year; Collect. Vol. No., Pagination.

| YEAR | COLLECTIVE VOLUME NO. |
|---|---|
| 1941 | I |
| 1943 | II |
| 1955 | III |
| 1963 | IV |
| 1973 | V |
| 1988 | VI |
| 1990 | VII |
| 1993 | VIII |
| 1998 | IX |
| 2004 | X |

*Organic Syntheses, Cumulative Indices for Collective Volumes I–VIII* was published in 1995. Beginning with Volume 82, each volume of *Organic Syntheses* is planned to be published online on orgsyn.org in installments about every three months, with printed volumes appearing annually.

➤ For references to the *Kirk-Othmer Encyclopedia,* include the article title followed by a period, similar to the citation of a chapter title.

Chloramines and Bromamines. *Kirk-Othmer Encyclopedia of Chemical Technology,* 4th ed.; Wiley & Sons: New York, 1993; Vol. 4, pp 931–932.

## *Series Publications*

Publications such as book series that are periodical in nature but are not journals may be styled as either books or journals. *CASSI* lists every document abstracted and indexed by the Chemical Abstracts Service; hence, book titles are included and abbreviated. Key words to look for with these types of publications include "Advances", "Methods", "Progress", and "Series".

RECOMMENDED FORMAT FOR CITATION AS A BOOK

Author 1; Author 2; Author 3; etc. In *Title;* Editor 1, Editor 2, Eds.; Series Title and Number; Publisher: Place of Publication, Year; Pagination.

➤ In book format, use the regular citation format for a book reference, but include information pertaining to the series. The series title is spelled out and set in roman type.

Kebarle, P. In *Techniques for the Study of Ion–Molecule Reactions;* Saunders, W., Farrar, J. M., Eds.; Techniques of Chemistry Series 20; Wiley & Sons: New York, 1988; p 125.

*Lignocellulose Biodegradation;* Saha, B. C., Hayashi, K., Eds.; ACS Symposium Series 889; American Chemical Society: Washington, DC, 2004.

➤ If a volume or part number is given for a series of books instead of a series number, cite this information where a series number would normally appear.

Wiberg, K. In *Investigations of Rates and Mechanisms of Reactions;* Lewis, E. S., Ed.; Techniques of Chemistry, Vol. VI, Part I; Wiley & Sons: New York, 1974; p 764.

➤ As for any book, you may cite specific chapters.

Puls, J.; Saake, B. Industrially Isolated Hemicelluloses. In *Hemicelluloses: Science and Technology;* Gatenholm, P., Tenkanen, M., Eds.; ACS Symposium Series 864; American Chemical Society: Washington, DC, 2004; pp 24–37.

➤ In journal format, the series title is used as a journal title, abbreviated according to *CASSI* and italicized, and the series number is used as a journal volume number.

RECOMMENDED FORMAT FOR CITATION AS A JOURNAL

Author 1; Author 2; Author 3; etc. *Abbreviation* **Year,** *Volume,* Pagination.

Puls, J.; Saake, B. *ACS Symp. Ser.,* **2004,** *864,* 24–37.
Kebarle, P. *Tech. Chem. (N.Y.)* **1988,** *20,* 125.

## Works Written by an Organization or a Committee

An organization or a committee may be the author of a book or periodical article. Acronyms for very well known organizations may be used. It is not necessary to repeat the publisher's name for a work compiled by the organization that published it.

BOOK FORMAT

American Chemical Society, Committee on Analytical Reagents. *Reagent Chemicals: Specifications and Procedures,* 10th ed.; Washington, DC, 2006.

World Health Organization. *Pathology and Genetics of Tumours of the Head and Neck;* Albany, NY, 2002; Vol. 9.

PERIODICAL FORMAT

International Union of Pure and Applied Chemistry, Physical Chemistry Division, Commission on Molecular Structure and Spectroscopy. Presentation of Molecular Parameter Values for Infrared and Raman Intensity Measurements. *Pure Appl. Chem.* **1988,** *60,* 1385–1388.

IUPAC. Molecular Absorption Spectroscopy, Ultraviolet and Visible (UV/VIS). *Pure Appl. Chem.* **1988,** *60,* 1449–1460.

## Meetings and Conferences

References to work presented at conferences and meetings must be treated on a case-by-case basis. At least three types of citations are possible:

1. Full citations of published abstracts and proceedings. The format resembles that of a book citation.
2. *CASSI* citations of published abstracts and proceedings. The format is that of a periodical citation.
3. References to oral presentations, posters, or demonstrations at technical meetings, possibly accompanied by handouts or brochures. These references contain no publication information.

### *Full Citations*

RECOMMENDED FORMAT

Author 1; Author 2; Author 3; etc. Title of Presentation. In *Title of the Collected Work,* Proceedings of the Name of the Meeting, Location of Meeting, Date of Meeting; Editor 1, Editor 2, etc., Eds.; Publisher: Place of Publication, Year; Abstract Number, Pagination.

The format resembles that of a book citation. The title field, however, includes additional information on the meeting title, location, and dates. The actual title of the book (collected work) is set in italic type and is separated from the meeting information by a comma. The information on meeting location is set in roman type, but it is not repeated if it is included in the book title. The entire field ends with a semicolon.

> Garrone, E.; Ugliengo, P. In *Structure and Reactivity of Surfaces,* Proceedings of the European Conference, Trieste, Italy, Sept 13–20, 1988; Zecchina, A., Cost, G., Morterra, C., Eds.; Elsevier: Amsterdam, 1988.

Abstracts are slightly different in that they usually do not have editors. The word "in" is not used before the book title.

> Prasad, A.; Jackson, P. *Abstracts of Papers, Part 2,* 212th National Meeting of the American Chemical Society, Orlando, FL, Aug 25–29, 1996; American Chemical Society: Washington, DC, 1996; PMSE 189.

When the phrase "Proceedings of" is part of the reference, include the publisher and place of publication. When a society sponsors a meeting, the society is assumed to be the publisher. If the place of the meeting and the place of publication are the same, additional publisher and place information is not required. However, many organizations such as the ACS sponsor meetings in various cities.

> Harwood, J. S. Direct Detection of Volatile Metabolites Produced by Microorganisms. *Proceedings of the 36th ASMS Conference on Mass Spectrometry and Allied Topics,* San Francisco, CA, June 5–10, 1988.

## CASSI *Citations*

Proceedings and abstracts of meetings and conferences are indexed in *CASSI.* The reference format follows that for periodicals.

> *Abstr. Pap.—Am. Chem. Soc.* **1989,** *198.*

*CASSI* gives the number of a meeting in ordinal form. Convert this number to an italic cardinal number, and use it as the volume number in the citation, unless *CASSI* has already indicated another volume number.

Journal format can be used for references to preprint papers.

> Jones, J.; Oferdahl, K. *Natl. Meet.—Am. Chem. Soc., Div. Environ. Chem.* **1989,** *29* (2), ENVR 22 (or Paper 22).

## *Material That Has No Publication Information*

RECOMMENDED FORMATS

> Author 1; Author 2; Author 3; etc. Title of Presentation (if any). Presented at Conference Title, Place, Date; Paper Number.

List the data concerning the conference (name, place, and date) separated by commas and followed by a semicolon and the paper number (if any). The entire citation is set in roman type.

Zientek, K. D.; Eyler, J. R. Presented at the 51st ASMS Conference on Mass Spectrometry and Allied Topics, Montreal, Canada, June 8–12, 2003.

Dizman, B.; Elasri, M. O.; Mathias, L. J. Presented at the 227th National Meeting of the American Chemical Society, Anaheim, CA, March 28–April 1, 2004; Paper POLY 229.

# Theses

RECOMMENDED FORMATS

Author. Title of Thesis. Level of Thesis, Degree-Granting University, Location of University, Date of Completion.

References to theses should be as specific as practical, including, at a minimum, the degree-granting institution and date.

Chandrakanth, J. S. Effects of Ozone on the Colloidal Stability of Particles Coated with Natural Organic Matter. Ph.D. Dissertation, University of Colorado, Boulder, CO, 1994.

Mäckel, H. Capturing the Spectra of Silicon Solar Cells. Ph.D. Thesis, The Australian National University, December 2004.

Kulamer, T. M.S. Thesis, Princeton University, 2004.

## Author Name Field

Cite the name in inverted form: surname first, then first initial, middle initial, and qualifiers (Jr., II). End the field with a period.

## Title Field

Thesis titles are not essential, but they are informative. They are set in roman type and end with a period.

Erickson, T. A. Development and Application of Geostatistical Methods to Modeling Spatial Variation in Snowpack Properties, Front Range, Colorado. Ph.D. Dissertation, University of Colorado, Boulder, CO, 2004.

Moore, S. Synthesis and Pharmacology of Potential Site-Directed Therapeutic Agents for Cocaine Abuse. Ph.D. Thesis, Georgia Institute of Technology, Atlanta, GA, 2004.

## Thesis Level Field

Work done at a master's level is often called a thesis. Work toward the Ph.D. (doctor of philosophy) may be called a thesis or a dissertation, depending on the policy of the degree-granting institution. The following abbreviations are standard for U.S. degrees. Many variations exist for degrees from institutions of other countries.

A.B., B.A., B.S.
A.M., M.A., M.S., M.B.A.
Ph.D., M.D.

Coghill, S. M.S. Thesis, Northwestern University, Evanston, IL, 2004.
Breton, J. C. Intégrales Multiples Stochastiques Poissonniennes. Ph.D. Dissertation, University of Lille, France, 2001.

### *University Name and Location Field*

The name of the degree-granting university is the minimum requirement for an acceptable citation. You should also include the city and state or city and country. Use the two-letter postal abbreviations for states. Spell out names of countries unless they have standard abbreviations, such as U.K. for United Kingdom.

Blättler, T. M. Covalent Immobilization of Poly(L-lysine)-*g*-poly(ethylene glycol) onto Aldehyde Plasma Polymer Coated Surfaces. Diploma Thesis, University of South Australia, 2004.
Mohamed, M. Waterjet Cutting Up to 900 MPa. Ph.D. Thesis, University of Hannover, Germany, 2004.

### *Date of Completion Field*

Indicate the date the thesis was completed by year only; month and year; or month, day, and year.

Fleissner, C. Ph.D. Thesis, New York University, March 2003.
Marshall, M. Ph.D. Thesis, University of California, San Francisco, CA, 2005.
Stover, J. Ph.D. Dissertation, Harvard University, May 24, 2001.

## Patents

RECOMMENDED FORMAT

Patent Owner 1; Patent Owner 2; etc. Title of Patent. Patent Number, Date.

The minimum data required for an acceptable citation are the name(s) of the patent owner(s), the patent number, and the date. Ensure that the patent stage (Patent, Patent Application, etc.) is indicated and that the pattern of the number (e.g., spaces, commas, dashes) follows that of the original patent document. If possible, include the title and the *Chemical Abstracts* reference (preceded by a semicolon) as well.

Sheem, S. K. Low-Cost Fiber Optic Pressure Sensor. U.S. Patent 6,738,537, May 18, 2004.
Lenssen, K. C.; Jantscheff, P.; Kiedrowski, G.; Massing, U. Cationic Lipids with Serine Backbone for Transfecting Biological Molecules. Eur. Pat. Appl. 1457483, 2004.

Petrovick, P. R.; Carlini, E. Antiulcerogenic Preparation from *Maytenus ilicifolia* and Obtaintion Process. Br. Patent PI 994502, March 6, 1999.

Langhals, H.; Wetzel, F. Perylene Pigments with Metallic Effects. Ger. Offen. DE 10357978.8, Dec 11, 2003; *Chem. Abstr.* **2005,** *143,* 134834.

Shimizu, Y.; Kajiyama, H. (Kanebo, Ltd., Japan; Kanebo Synthetic Fibers, Ltd.). Jpn. Kokai Tokkyo Koho JP 2004176197 A2 20040624, 2004.

# Government Publications

Publications of the U.S. government and those of state and local governments can be pamphlets, brochures, books, maps, journals, or almost anything else that can be printed. They may have authors or editors, who may be individuals, offices, or committees, or the author may not be identified. They are published by specific agencies, but they are usually (though not always) available through the Government Printing Office rather than the issuing agency. To enable others to find the publication, the American Library Association suggests that you include as much information as possible in the citation. The following are examples of the most commonly cited types of references.

## *Publications of Federal Government Agencies*

RECOMMENDED FORMAT

Author 1; Author 2; etc. Chapter Title. *Document Title;* Government Publication Number; Publishing Agency: Place of Publication, Year; Pagination.

The format resembles that of a serial publication in book format. Include as much information as possible.

Gebhardt, S. E.; Thomas, R. G. *Nutritive Value of Foods;* Home and Garden Bulletin No. 72; U.S. Department of Agriculture, U.S. Government Printing Office: Washington, DC, 2002.

*Agriculture Fact Book 2000;* U.S. Department of Agriculture, U.S. Government Printing Office: Washington, DC, 2000.

Dey, A. N.; Bloom, B. *Summary Health Statistics for United States Children: National Health Interview Survey, 2003;* DHHS Publication PHS 2005-1551; Department of Health and Human Services, Centers for Disease Control and Prevention, National Center for Health Statistics, U.S. Government Printing Office: Washington, DC, 2005.

*ISCORS Assessment of Radioactivity in Sewage Sludge: Modeling To Assess Radiation Doses;* NUREG-1783; EPA 832-R-03-002A; DOE/EH-0670; ISCORS Technical Report 2004-03; Interagency Steering Committee on Radiation Standards, Sewage Sludge Subcommittee, U.S. Government Printing Office: Washington, DC, 2005.

### Author Name Field

Include all author names. With multiple authors, separate the names from one another by semicolons. Always end the author field with a period. List the names in inverted form: surname first, then first initial, middle initial, and qualifiers (Jr., II). Some publications list the first 10 authors followed by a semicolon and "et al."

### Chapter Title Field

Chapter titles are set in roman type and end with a period.

### Document Title Field

Treat the formal title of the document as the title of a book. These titles are set in italic type and are separated from the next component of the reference by a semicolon, which is set in italic type.

### Government Publication Number Field

The government publication number, also called an agency report number, is important because it is unique to the publication and because some indexing services provide access by these numbers. These numbers (or number–letter combinations) are usually printed somewhere on the cover or title page of the document and are sometimes identified as a "report/accession number". Treat a report number the same as a series number; that is, it follows the book title, ends with a semicolon, and is set in roman type.

### Publishing Agency Field

The publishing agency field may take on added complexity in government publications. Often, the office or agency issuing the report as well as the Government Printing Office must be cited. The order is department or agency, administration or office, and finally U.S. Government Printing Office, all separated by commas and set in roman type. The field ends with a colon.

### Place of Publication Field

For the U.S. Government Printing Office, it is always Washington, DC. The field ends with a comma preceding the date of publication.

### Year of Publication Field

The year of publication is set in roman type and ends with a semicolon if it is followed by pagination information. It ends with a period if it is the last field.

### Pagination Field

The page numbers are set in roman type and end with a period, unless miscellaneous material is appended to the reference.

### Alternative Format

Government agency references can also be given with *CASSI* abbreviations. In that case, the format is the same as for periodicals.

Thompson, C. R.; Van Atta, G. R.; Bickoff, E. M.; Walter, E. D.; Livingston, A. L.; Guggloz, J. *Tech. Bull.—U.S. Dep. Agric.* **1957,** *No. 1161*, 63–70.

## *Other Federal Publications*

### Federal Register

The *Federal Register* is a periodical and is treated as such in citations.

Agency for Toxic Substances and Disease Registry. Update on the Status of the Superfund Substance-Specific Applied Research Program. *Fed Regist.* **2002,** *67*, 4836–4854.

U.S. Food and Drug Administration. Food Labeling: Health Claims and Label Statements for Dietary Supplements. *Fed Regist.* **1999,** *65* (195), 59855–59857.

### Code of Federal Regulations

Licensing of Government Owned Inventions. *Code of Federal Regulations*, Part 404, Title 37, 2005.

Labeling Requirements for Pesticides and Devices. *Code of Federal Regulations,* Part 156, Title 40, 1998; *Fed. Regist.* **1998,** *15*, 7.

### U.S. Code

The Public Health and Welfare. *U.S. Code,* Section 1396a, Title 42, 2000.

### U.S. Laws

Treat the name of the law as a chapter title (roman, terminated with a period). No publisher name is needed. The number and date of the law are separated by a comma. If additional publication information is given, it is preceded by a semicolon.

Domestic Quarantine Notices. *Code of Federal Regulations,* Section 301.10, Title 7, Vol. 5, 2005.

Federal Insecticide, Fungicide, and Rodenticide Act. Public Law 92-516, 1972; *Code of Federal Regulations,* Section 136, Title 7, 1990.

## *State and Local Government Publications*

RECOMMENDED FORMAT

Author 1; Author 2; etc. Chapter Title. *Document Title;* Publication Number or Type; Publishing Agency: Place of Publication, Date; Pagination.

*Annual Report 2004: Moving Forward;* Santa Barbara County Air Pollution Control District: Santa Barbara, CA, 2005.

Turner, B.; Powell, S.; Miller, N.; Melvin, J. *A Field Study of Fog and Dry Deposition as Sources of Inadvertent Pesticide Residues on Row Crops;* Report of the Environmental Hazard Assessment Program; California Department of Food and Agriculture: Sacramento, CA, November 1989.

## Technical Reports and Bulletins

Technical reports and bulletins come in many forms. Examples of some of these have already been presented. Many are in-house publications, and some are government publications. Others are reports of work in progress. The publication itself may include a phrase alluding to its status as a technical report or technical bulletin, but it may also simply be called a report or bulletin. Include whatever information is available, following the format shown for the word "Report", "Report No.", etc. Document titles are set in italic type.

RECOMMENDED FORMAT

Author 1; Author 2; etc. *Title of Report or Bulletin;* Technical Report or Bulletin Number; Publisher: Place of Publication, Date; Pagination.

Tschantz, B. A.; Moran, B. M. *Modeling of the Hydrologic Transport of Mercury in the Upper East Fork Poplar Creek (UEFPC) Watershed;* Technical Report for Lockheed Martin Energy Systems: Bethesda, MD, September 2004.

*Fourth DELOS Workshop. Evaluation of Digital Libraries;* Final Report to the National Science Foundation on Grant IIS-225626; Hungarian Academy of Sciences: Budapest, 2002.

## Data Sets

Data sets are compilations of data, such as spectra or property tables. These data sets are often published serially as loose-leaf services, but the content is not always organized in chapters as in other serial publications. The citation of a serial data set should contain the title of the data set, the publisher, the place of publication, the date of the volume, the data entry number (as opposed to the data value), and the name of the figure or other identifying information. The page number can be included in the citation if page numbers are used in the index of the data set.

References to data retrieved from a stand-alone database should cite the source as a computer program (for example, *MDL CrossFire Commander,* see p 323 f) or as an online reference book (for example, the *Kirk-Othmer Encyclopedia of Chemical Technology,* see p 305 f), with the data entry number or other identifying information included at the end of the citation. Data retrieved from an Internet-based database should cite the source as a Web site (see pp 316 ff). If the data retrieved are calculated data, also cite the software used for calculation (for example, ACD/Labs).

RECOMMENDED FORMAT FOR PRINTED DATA SETS

*Title;* Publisher: Place of Publication, Date; Data Entry Number, Figure Title or other identifying information.

*The Sadtler Standard Spectra: 300 MHz Proton NMR Standards;* Bio-Rad, Sadtler Division: Philadelphia, PA, 1994; No. 7640 (1-Chloropentane).

## Material Safety Data Sheets

Material Safety Data Sheets (MSDSs) are published by the company that manufactures the material covered on the sheet. Citations should include the title of the data sheet, which is the name of the material; the MSDS number; the manufacturing company; the location of the company; and the date on which the document was released. If the online version was used, the designation "Online" is included in brackets after the MSDS number, and the URL and date accessed are included at the end of the citation.

RECOMMENDED FORMATS

*Title;* MSDS Number; Manufacturing Company: Location of Company, Date.
*Title;* MSDS Number [Online]; Manufacturing Company: Location of Company, Date. URL (accessed Month Day, Year).

*Titanium Dioxide;* MSDS No. T3627; Mallinckrodt Baker: Phillipsburg, NJ, Nov 12, 2003.
*Acetic Anhydride;* MSDS No. A0338 [Online]; Mallinckrodt Baker: Phillipsburg, NJ, Feb 18, 2003. http://www.jtbaker.com/msds/englishhtml/a0338.htm (accessed Nov 10, 2004).

## Unpublished Materials

Material in any stage preceding actual publication falls under this general classification, as do personal communications and work not destined for publication.

RECOMMENDED FORMAT FOR MATERIAL INTENDED FOR PUBLICATION

Author 1; Author 2; etc. Title of Unpublished Work. *Journal Abbreviation,* phrase indicating stage of publication.

Various phrases indicating the stage of publication are acceptable in these references.

➤ For material accepted for publication, use the phrase "in press".

Tang, D.; Rupe, R.; Small, G. J.; Tiede, D. M. *Chem. Phys.,* in press.

➤ For material intended for publication but not yet accepted, use "unpublished work", "submitted for publication", or "to be submitted for publication".

Chatterjee, K.; Visconti, A.; Mirocha, C. J. Deepoxy T-2 Tetraol: A Metabolite of T-2 Toxin Found in Cow Urine. *J. Agric. Food Chem.*, submitted for publication, 2004.

Nokinara, K. Duke University, Durham, NC. Unpublished work, 2003.

As Gordon G. Hammes says in Chapter 1 of this book,

> Occasionally, the attribution of an idea or fact may be to a "private communication" of a colleague or fellow scientist. In such cases, permission must be obtained from the individual in question before the citation is made. Reference to unpublished material should be avoided if possible because it generally will not be available to interested readers.

RECOMMENDED FORMAT FOR MATERIAL NOT INTENDED FOR PUBLICATION

Author. Affiliation, City, State. Phrase describing the material, Year.

Henscher, L. X. University of Minnesota, Minneapolis, MN. Personal communication, 2001.

Heltman, L. R. DuPont. Private communication, 2003.

Wagner, R. L. University of Utah, Salt Lake City, UT. Unpublished work, 2004.

Messages sent by electronic mail are considered personal communications and are referred to as such.

## Electronic Sources

Electronic media continue to develop rapidly in content, organization, and presentation of information. The conventions for citing electronic resources are evolving to reflect these changes, but the basic principles of citation remain the same: present enough documentation with enough clarity to establish the identity and authority of the source and direction for locating the reference. The guidelines stress consistency both in presentation of information and in reasons for exceptions.

To date, much of the material available in electronic media corresponds to and/or is modeled after the traditional print-based sources discussed earlier in this chapter and should be cited according to those guidelines as appropriate. However, given the transient nature of electronic sources, it is important to provide additional documentation about the format or online location and the date the source was accessed.

### *Internet Sources*

Internet sources include online editions of traditional sources such as periodicals and books available through Web technology; new collective sites of information, including Internet-based databases using Web, file transfer protocol (FTP), and Telnet technologies; and electronic mailing lists and mail messages that may or may not use Web interfaces. Each source has an electronic address; for sources using the World Wide Web, this address is called the uniform resource locator (URL). As Web interfaces and supporting technologies evolve, direct addresses of items will often change to reflect new structures. Changing addresses can disrupt access to information sources that may still be available but at new locations and in modified formats. This issue can be resolved locally through use of persistent URLs. A persistent URL remains constant, but the actual location of a source is tracked through a local database that can be updated without disrupting the URL.

Information sources can also be tracked globally through coordinated efforts such as the Digital Object Identifier (DOI) system. Information providers register their sources, which are assigned unique and persistent DOIs. Each DOI is similar to a barcode that manages a complex profile of multiple pieces, formats, locations, ownership rights, and interoperability features. The identity of and access to an electronic information source is maintained through its DOI regardless of changes in location, format, or publisher. The use of DOIs is spreading among publishers as an efficient system to manage journal articles and other types of intellectual property on the Web. Further information about the DOI system can be found at http://www.doi.org (accessed April 13, 2005).

CrossRef is an application of the DOI system that links online citations across publishers. The unique identification and persistent location information in the DOI is packaged into an open URL that publishers and libraries can use to link to subscribed full-text content from reference lists. These links appear with citations in the reference lists of online articles and databases from several participating publishers.

For the purposes of citation, reference style conventions continue to use the URL as the most direct route to the location of a source. DOIs are sometimes used by publishers in place of page numbers or article numbers and should be included in citations in this context.

➤ URLs can be long and complicated, and there are conventions for splitting an address between multiple lines; see Chapter 10 (pp 156–157) for guidelines on breaking URLs and e-mail addresses.

### Online Periodicals

There are several types of periodicals online, including those based on print editions, electronic copies retrieved from databases, articles released online in advance of a full print issue, periodicals published only in electronic format, and article preprints posted in preprint servers. The reference styles for periodicals apply, with additional information concerning online location and accession date assigned as needed. As for print periodicals, article titles are desirable but not included in all ACS publications; check the publication itself.

#### RECOMMENDED FORMAT FOR ONLINE PERIODICALS BASED ON PRINT EDITIONS

Author 1; Author 2; Author 3; etc. Title of Article. *Journal Abbreviation* [Online] **Year,** *Volume,* Inclusive pagination or other identifying information. URL (accessed Month Day, Year).

Currently, the majority of the articles retrieved from online publications are based on corresponding print versions. For these articles, the basic periodical reference style is used, but if the article has been viewed only in its electronic form, the designation "Online" is included in brackets after the journal abbreviation.

Fine, L. Einstein Revisited. *J. Chem. Educ.* [Online] **2005,** *82,* 1601 ff. http://jchemed.chem.wisc.edu/Journal/Issues/2005/Nov/abs1601.html (accessed Oct 15, 2005).

#### RECOMMENDED FORMATS FOR ELECTRONIC COPIES OF ARTICLES RETRIEVED FROM A DATABASE PROVIDER

Author 1; Author 2; Author 3; etc. Title of Article. *Journal Abbreviation* [Online] **Year,** *Volume,* Article Number or other identifying information. Database Provider. URL of top page (accessed Month Day, Year).

Author 1; Author 2; Author 3; etc. Title of Article. *Title of Periodical* [Online], Complete Date, Pagination. Database Provider. URL of top page (accessed Month Day, Year).

Electronic copies of periodicals, nonscientific magazines, or newspapers retrieved from subscription database services often provide only the original text but not the original formatting or the figures. For online articles provided as content in a subscription database, use the reference style for periodicals or nonscientific magazines as appropriate, and include the name of the database provider, the URL of the top page, and the date accessed.

Hallet, V. Scanning the Globe for Organic Chemistry. *U.S. News and World Report* [Online], April 19, 2004, p 59. Business Source Premier. http://www.epnet.com/academic/bussourceprem.asp (accessed April 24, 2005).

#### RECOMMENDED FORMAT FOR ARTICLES PUBLISHED ONLINE IN ADVANCE OF PRINT ISSUES

Author 1; Author 2; Author 3; etc. Title of Article. *Journal Abbreviation* [Online early access]. DOI or other identifying information. Published Online: Month Day, Year. URL (accessed Month Day, Year).

Often, articles are ready for publication in advance of a full issue of a periodical. Several publishers offer these articles online up to weeks in advance of the print issue. They are identical to the corresponding print articles except that page numbers are often not yet available. Publishers market this service under different names; the ACS Publications Division labels them As Soon As Publishable (ASAP). For citation purposes, use the designation "Online early access" in brackets after the journal abbreviation in place of the publisher-specific term.

Also include the DOI or other identifying information, the online publication date, the URL, and the date accessed.

Pratt, D. A.; van der Donk, W. A. Theoretical Investigations into the Intermediacy of Chlorinated Vinylcobalamins in the Reductive Dehalogenation of Chlorinated Ethylenes. *J. Am. Chem. Soc.* [Online early access]. DOI: 10.1021/ja047915o. Published Online: Dec 8, 2004. http://pubs.acs.org/cgi-bin/asap.cgi/jacsat/asap/html/ja047915o.html (accessed Dec 8, 2004).

RECOMMENDED FORMAT FOR PERIODICALS PUBLISHED ONLY IN ELECTRONIC FORMAT

Author 1; Author 2; Author 3; etc. Title of Article. *Journal Abbreviation* [Online] **Year,** *Volume,* Article Number or other identifying information. URL (accessed Month Day, Year).

A periodical published only in electronic format may include additional electronic features, data, or commentaries. Use the reference style for periodicals, and include the direct URL of the article as well as the date accessed. Volume and page numbers are often not relevant. If they are not used, include the article number, DOI, or other identifying information.

Zloh, M.; Esposito, D.; Gibbons, W. A. Helical Net Plots and Lipid Favourable Surface Mapping of Transmembrane Helices of Integral Membrane Proteins: Aids to Structure Determination of Integral Membrane Proteins. *Internet J. Chem.* [Online] **2003,** *6,* Article 2. http://www.ijc.com/articles/2003v6/2/ (accessed Oct 13, 2004).

RECOMMENDED FORMAT FOR ARTICLES RETRIEVED FROM PREPRINT SERVERS

Author 1; Author 2; Author 3; etc. Title of Article. Year, Article Number. Name of Repository. URL (accessed Month Day, Year).

Ward, D. W.; Nelson, K. A. Finite Difference Time Domain (FDTD) Simulations of Electromagnetic Wave Propagation Using a Spreadsheet. 2004, arXiv:physics/0402096. arXiv.org e-Print archive. http://arxiv.org/abs/physics/0402096 (accessed Oct 13, 2004).

## Online Books

Books published online generally correspond to printed versions, and the reference styles are similar. Online location and access date should always be included when citing online books. Reference works published online are often updated with new content, and the dates on which sections were posted or updated should also be included.

RECOMMENDED FORMAT FOR ONLINE BOOKS WITHOUT EDITORS

Author 1; Author 2; Author 3; etc. *Book Title* [Online]; Series Information (if any); Publisher: Place of Publication, Year; Volume Number, Pagination. URL (accessed Month Day, Year).

Tour, J. M. *Molecular Electronics: Commercial Insights, Chemistry, Devices, Architecture and Programming* [Online]; World Scientific: River Edge, NJ, 2003; pp 177–180. http://legacy.netlibrary.com/ebook_info.asp?product_id=91422&piclist=19799,20141,20153 (accessed Nov 7, 2004).

RECOMMENDED FORMAT FOR ONLINE BOOKS WITH EDITORS

Author 1; Author 2; Author 3; etc. Chapter Title. In *Book Title* [Online]; Editor 1, Editor 2, etc., Eds.; Series Information (if any); Publisher: Place of Publication, Year; Volume Number, Pagination. URL (accessed Month Day, Year).

Oleksyn, B. J.; Stadnicka, K.; Sliwinski, J. Structural Chemistry of Enamines: A Statistical Approach. In *The Chemistry of Enamines* [Online]; Rappoport, Z., Ed.; The Chemistry of Functional Groups; Patai, S., Rappoport, Z., Series Eds.; Wiley & Sons: New York, 1994; Chapter 2, pp 87–218. http://www3.interscience.wiley.com/cgi-bin/summary/109560980/SUMMARY (accessed April 24, 2005).

RECOMMENDED FORMAT FOR ONLINE ENCYCLOPEDIAS

Article Title. *Encyclopedia Title*, edition [Online]; Publisher, Posted Online Posting Date. URL (accessed Month Day, Year).

Alkanolamines from Nitro Alcohols. *Kirk-Othmer Encyclopedia of Chemical Technology* [Online]; Wiley & Sons, Posted March 14, 2003. http://www.mrw.interscience.wiley.com/kirk/articles/alkaboll.a01/frame.html (accessed Nov 7, 2004).

## Web Sites

Aside from online periodicals and books, general Web sites containing a wide variety of information might need to be cited. Some sites are accessible by anyone, but many are accessible only by subscription. Reference styles for FTP and Telnet sites are similar to those for Web sites. Specific examples are given here for general Web sites and databases, stand-alone documents, unpublished conference proceedings, and electronic theses.

RECOMMENDED FORMAT FOR GENERAL WEB SITES

Author (if any). Title of Site. URL (accessed Month Day, Year), other identifying information (if any).

Use the title found on the Web site itself; add the words "Home Page" for clarification when needed. Data retrieved from Internet-based databases should include a data entry number. Stand-alone databases should be cited as computer programs (see p 323).

ACS Publications Division Home Page. http://pubs.acs.org (accessed Nov 7, 2004).

Chemical Abstracts Service. STN on the Web. http://stnweb.cas.org (accessed Nov 7, 2004).

International Union of Pure and Applied Chemistry Home Page. http://www.iupac.org/dhtml_home.html (accessed April 24, 2005).

Library of Congress Home Page. http://www.loc.gov (accessed April 24, 2005).

Northern Illinois University. Department of Chemistry and Biochemistry Home Page. http://www.chembio.niu.edu (accessed Nov 7, 2004).

Sheffield Chemistry Software Archive (Macintosh). ftp://ftp.shef.ac.uk/pub/uni/academic/A-C/chem (accessed April 24, 2005).

U.S. Environmental Protection Agency. http://www.epa.gov (accessed Nov 7, 2004).

#### RECOMMENDED FORMAT FOR DOCUMENTS RETRIEVED FROM INSTITUTIONAL OR AGENCY WEB SITES

Author 1; Author 2; Author 3; etc. Title of Document, Year. Title of Site. URL (accessed Month Day, Year).

If an article is contained within a large and complex Web site, such as that for a university or a government agency, the host organization and the relevant program or department should be identified before giving the direct URL of the article and accession date.

Chou, L.; McClintock, R.; Moretti, F.; Nix, D. H. Technology and education: New wine in new bottles: Choosing pasts and imagining educational futures, 1993. Columbia University Institute for Learning Technologies Web site. http://www.ilt.columbia.edu/publications/papers/newwine1.html (accessed Aug 24, 2000).

#### RECOMMENDED FORMAT FOR ONLINE UNPUBLISHED CONFERENCE PRESENTATIONS

Author 1; Author 2; etc. Title of Presentation. Presented at Conference Title [Online], Place, Date; Paper Number. Title of Site. URL (accessed Month Day, Year).

Works presented at conferences or meetings can be cited in several formats, as discussed earlier in this chapter. Generally, published abstracts or proceedings can be cited as online books or as online periodicals. Materials from oral presentations, posters, or demonstrations that do not contain publication information should be cited as follows.

Manly, S. Collective flow with PHOBOS. Presented at the 20th Winter Workshop on Nuclear Dynamics [Online], Trelawny Beach, Jamaica, March 15–20, 2004. University of Rochester, DSpace Web site. http://hdl.handle.net/1802/228 (accessed Oct 13, 2004).

#### RECOMMENDED FORMAT FOR ELECTRONIC THESES

Author. Title of Thesis. Level of Thesis [Online], Degree-Granting University, Location of University, Date of Completion. URL (accessed Month Day, Year).

Lozano, P. C. Studies on the Ion-Droplet Mixed Regime in Colloid Thrusters. Ph.D. Thesis [Online], Massachusetts Institute of Technology, Cambridge, MA, January 2003. http://theses.mit.edu/Dienst/UI/2.0/Describe/0018.mit.etheses%2f2003-1 (accessed Nov 7, 2004).

### Electronic Lists and Newsgroups

RECOMMENDED FORMAT FOR ELECTRONIC LISTS AND NEWSGROUPS

Mailing List or Newsgroup Name, other information, electronic address (accessed Month Day, Year).

Chemical Information List Server, CHMINF-L@iubvm.ucs.indiana.edu (accessed Oct 13, 2004).

Computational Chemistry List, solvent discussion in archived messages of September 2003, chemistry@ccl.net (accessed Nov 10, 2004).

Molecular Diversity for Basic Research & Drug Discovery, mol-diversity@listserv.arizona.edu (accessed Nov 10, 2004).

### Electronic Mail Messages

Whether the message was personal and sent only to you or whether it was posted in a newsgroup, it is not published. E-mail messages should be cited the same as any other personal communication. Include the year and the professional affiliation of the author.

RECOMMENDED FORMAT FOR ELECTRONIC MAIL MESSAGES

Author. Affiliation, City, State. Personal communication, Year.

Solla, L. R. Cornell University, Ithaca, NY. Personal communication, 2005.

## *CD-ROMs and DVDs*

The reference style for information published in CD-ROM or DVD format follows that for periodicals and books as appropriate, and the designation "CD-ROM" or "DVD" is included in brackets.

RECOMMENDED FORMAT FOR CD-ROM AND DVD PERIODICALS

Author 1; Author 2; Author 3; etc. Title of Article. *Journal Abbreviation* [CD-ROM or DVD] **Year,** *Volume,* pagination or other identifying information.

Fleming, S. A.; Jensen, A. W. Substituent Effects on the Photocleavage of Benzyl–Sulfur Bonds. Observation of the "*Meta*" Effect. *J. Org. Chem.* [CD-ROM] **1996,** *61,* 7044.

RECOMMENDED FORMATS FOR CD-ROM AND DVD BOOKS

Author 1; Author 2; etc. Chapter Title. In *Book Title,* Edition Number [CD-ROM or DVD]; Editor 1, Editor 2, etc., Eds.; Publisher: Place of Publication, Year; Volume Number.

Author 1; Author 2; etc. Chapter Title. *Book Title,* Edition Number [CD-ROM or DVD]; Publisher: Place of Publication, Year; Volume Number.

Vining, W. J.; Kotz, J.; Harman, P.; Vining, W.; McDonald, A.; Ward, J. *General Chemistry,* 3rd ed. [CD-ROM]; Thomson Brooks/Cole: Florence, KY, 2002.

Rowley, D.; Ramaker, D. *Standard Deviants Chemistry DVD Pack* [DVD]; Goldhil Educational: Camarillo, CA, 2000.

Many books in CD-ROM or DVD format are reference works, so they have no authors, editors, or chapter titles.

*The Merck Index*, 13.4 [CD-ROM]; Wiley: New York, 2005.
*Green Chemistry: Meeting Global Challenges* [DVD]; American Chemical Society: Washington, DC, 2003.

RECOMMENDED FORMAT FOR CONFERENCE PROCEEDINGS ON CD-ROM OR DVD

Author 1; Author 2; etc. Title of Presentation. In *Title of Conference*, Location of Meeting, Date of Meeting [CD-ROM or DVD]; Publisher: Place of Publication, Year; other identifying information.

Vasaru, G. Sources of Tritium. In *Proceedings of the 2nd International Conference on Nuclear Science and Technology in Iran*, Shiraz, Iran, April 27–30, 2004 [CD-ROM]; Conference Permanent Committee, Ed.; Shiraz University: Shiraz, Iran, 2004.

## *Computer Programs*

References to computer programs must be treated on a case-by-case basis. Five common presentations of computer programs are possible:

1. book format, with the name of the program as the title
2. technical report format
3. *CASSI* format
4. free style, as a simple listing of program title and author of program
5. thesis style

### Book Format

RECOMMENDED FORMAT

Author 1; Author 2; etc. *Program Title*, version or edition; Publisher: Place of Publication, Year.

The recommended format is the same as that for a book citation, except that there are no chapters or pages. The name of the computer program, with any descriptors, is considered the title and is set in italic type. If you wish to include additional information about a program that is important for the reader to know, you may add it at the end of the reference with or without parentheses or append it to the title in parentheses before the semicolon.

Binkley, J. S. *GAUSSIAN82;* Department of Chemistry, Carnegie Mellon University: Pittsburgh, PA, 1982.
Main, P. *MULTAN 80: A System of Computer Programs for the Automated Solution of Crystal Structures from X-ray Diffraction Data;* Universities of York and Louvain: York, England, and Louvain, Belgium, 1980.

RECOMMENDED FORMAT FOR COMMERCIAL SOFTWARE AND DATABASES

*Program Title,* version or edition; comments; Publisher: Place of Publication, Year.

References to data should include the data entry number or other identifying information at the end of the citation. The date of access can also be included if the database is updated frequently. If the data retrieved are calculated data, also cite the software used for the calculation (for example, ACD/Labs).

*Mathematica,* version 5.1; software for technical computation; Wolfram Research: Champaign, IL 2004.

*MDL CrossFire Commander,* version 7; Elsevier MDL: San Leandro, CA, 2004; BRN 635994.

*Scifinder Scholar,* version 2004.2; Chemical Abstracts Service: Columbus, OH, 2004; RN 107-21-1 (accessed Dec 20, 2005); calculated using ACD/Labs software, version 8.14; ACD/Labs 1994–2006.

## Technical Report Format

RECOMMENDED FORMAT

Author. *Title of Report;* Technical Report Number; Publisher: Place of Publication, Year; Pagination (if any).

In a citation to a computer program as a technical report, a report or technical report number is included. As with book format, the name of the computer program is considered the title of the technical report.

Beurskens, P. T.; Bossman, W. P.; Doesburg, H. M.; Gould, R. O.; van der Hark, Th. E. M.; Prick, P. A. J. *DIRDIF: Direct Methods for Difference Structures;* Technical Report 1980/1; Crystallographic Laboratory: Toernooiveld, Netherlands, 1980.

Johnson, C. K. *ORTEP-II: A Fortran Thermal Ellipsoid Plot Program for Crystal Structure Illustrations;* Report ORNL-5138; National Technical Information Service, U.S. Department of Commerce: Springfield, VA, 1976.

## *CASSI* Format

Because of the broad base from which *Chemical Abstracts* indexes work, computer programs, in the form of technical reports, may be referenced. In such cases, *CASSI* format would be appropriate.

Johnson, C. K. *Oak Ridge Natl. Lab., [Rep.] ORNL (U.S.)* **1978,** ORNL-5348.

## Free Style

When only minimal information (e.g., author and program name) is available, present the information as simply as possible.

Programs used in this study included local modifications of Jacobson's ALLS, Zalkins's FORDAP, Busing and Levy's ORFEE, and Johnson's ORTEP2.

Lozos, G.; Hoffman, B.; Franz, C. SIMI4A, Chemistry Department, Northwestern University.

### Thesis Style

Sheldrick, G. M. SHELX-76: Program for Crystal Structure Determination. Cambridge University, 1976.

## Collating References

Collate all references at the end of the manuscript in numerical order if cited by number and in alphabetical order if cited by author. Do not include items in the reference list that are not cited in the manuscript. Check the publication for which you are writing. Some publications do not allow multiple references to be listed as one numbered entry; they prefer that each numbered entry include only one unique reference.

To collate references according to the author–date style, use the following format.

1. Alphabetize in order of the first authors' surnames.

2. When the same first author is common to multiple references,
   - Group the single-author references first. List them chronologically. To distinguish among references having the same year, add a lowercase letter (a, b, c, etc.) to the year.
   - Group the two-author references next. List them chronologically. To distinguish among references having the same year, add a lowercase letter (a, b, c, etc.) to the year.
   - Group all multiple-author (three or more) references last. List them chronologically. To distinguish among references having the same year, add a lowercase letter (a, b, c, etc.) to the year.

Hamilton, F. J. *Biochemistry* **2003,** *42,* 78–86.

Hamilton, F. J. *J. Agric. Food Chem.* **2004a,** *52,* 1622–1633.

Hamilton, F. J. *J. Org. Chem.* **2004b,** *69,* 298–306.

Hamilton, F. J.; Salvo, P. A. *J. Agric. Food Chem.* **2005,** *53,* 918–924.

Hurd, R. *J. Magn. Reson.* **1999,** *87,* 422.

Mills, M. S.; Thurman, E. M. *Anal. Chem.* **2001,** *73,* 1985–1990.

O'Connor, D. J. *Environ. Eng. ASCE* **2002,** *114,* 507–522.

Rahwan, R. G., Witiak, D. T., Eds. *Calcium Regulation by Calcium Antagonists;* ACS Symposium Series 201; American Chemical Society: Washington, DC, 2004.

Scarponi, T. M.; Moreno, S. P. *Biochemistry* **2002,** *41,* 345–360.

Scarponi, T. M.; Adams, J. S. *J. Pharm. Sci.* **2003,** *92,* 703–712.
Serpone, N.; Pellizetti, E. *Photocatalysis: Fundamentals and Applications;* Wiley & Sons: New York, 2004.
Tewey, L. P.; Rodriguez, R. E.; Jennes, A. C. *J. Agric. Food Chem.* **2001,** *49,* 1879–1886.
Tewey, L. P.; Rodriguez, R. E.; Fortunato, B. D.; Jennes, A. C. *Ind. Eng. Chem. Res.* **2002a,** *41,* 465–472.
Tewey, L. P.; Hiroshi, C. Y.; Allen, P. R.; Lowe, D. L. *Biochemistry* **2002b,** *41,* 11689–11699.
Tewey, L. P.; Rolland, H. J.; Harwood, C. C. *J. Org. Chem.* **2002c,** *67,* 3548– 3556.
Tewey, L. P.; Allen, P. R.; Levy, M. S. *J. Am. Chem. Soc.* **2003,** *125,* 2520.

Do not use the Latin terms ibid. (in the same place) or idem (the same) because the actual reference source cannot be searched on electronic databases.

## Reference/Citation Managers

Software programs are available to assist with the process of collecting and collating references. With such programs, researchers can create personal electronic collections or libraries of references and tailor the formatting to any number of uses and publishing guidelines. Citations are parsed into searchable databases of component fields, and formatting templates draw on the data to produce reference lists in a variety of reference styles. The process is further enhanced by filters designed to correctly interpret the variety of incoming reference formats. Filters, fields, and templates are customizable to accommodate additional sources and styles.

Additional features have been developed to improve the convenience of these tools, including connection scripts for hundreds of public-access and subscription-based bibliographic databases and increased variety of field types to accommodate figures, cross-linking, personal annotation, etc. Plug-ins are available for word-processing packages to format citations within the text, reference lists, and lists of figures as authors write. There are also networking options for cooperative reference building and linking to full-text versions of references.

Researchers can search literature databases either directly or through a reference manager interface; import text, images, and figures from journal articles, Web sites, and other reference managers; arrange reference lists in numerous collections or libraries; search and retrieve records by any field; format footnotes, endnotes, and stand-alone bibliographies; share and co-edit these lists with colleagues; and export citations in hundreds of publication-specific styles in several languages. These software packages assist the research process from initial literature searching to writing and editing final publications.

Leading reference management programs include EndNote, Reference Manager, ProCite, RefWorks, and Biblioscape. EndNote, Reference Manager, and

ProCite are currently all owned by Thomson Scientific and are available as stand-alone software packages for both Windows and Macintosh platforms. RefWorks is a Web-based program with individual accounts that can be accessed across platforms from any point of Internet access. Biblioscape is available in a variety of stand-alone and Web-based options. For the most part, these programs cover the gamut of research disciplines and are fairly well populated with filters and templates specific to the chemistry literature. Reviews of these and other bibliographic management software tools are regularly available in the library literature.

The Thomson Scientific products were developed independently and still retain distinctive characteristics in their functionality. EndNote focuses on the reference input and output needs of the individual researcher, with hundreds of connection scripts and filters and more than 1000 citation style templates. EndNote is updated regularly and has a growing number of enhanced features available. Reference Manager has traditionally targeted collaborative reference sharing between colleagues, with networking options that allow multiple users to work on the same reference list for a project. Some of these features are now becoming available in EndNote as well. Reference Manager is only available for the Windows platform. ProCite has focused on managing reference collections with larger numbers of fields and reference types and more advanced grouping and searching techniques. ProCite has not been updated since version 5 in 2001.

RefWorks is published by RefWorks.com and emphasizes the convenience and collaborative nature of Web-based software. The program and updates are provided on the RefWorks server, and users' bibliographic data are stored there as well. Multiuser accounts are available for collaborative work. The options for filters, fields, reference types, and templates are less developed in RefWorks than in the other tools discussed here and do not include filters or templates for ACS journal styles.

Biblioscape is published by CG Information, founded by scientists specifically to manage scientific and electronic information. Biblioscape is a suite of products with different sets of features designed for a variety of users, including undergraduate and graduate students, researchers, and librarians. Options include Web access, intranet, and freeware editions. More than 1000 output styles are available, including the ACS journal styles.

➤➤➤➤➤

APPENDIX 14-1

# *CASSI* Abbreviations for the 1000+ Most Commonly Cited Journals

This appendix lists the *Chemical Abstracts Service Source Index,* or *CASSI,* abbreviations for more than 1000 of the most commonly cited journals. Note that some journals of the same name are published in more than one city. Authors should check the journal name carefully and include the city to prevent misunderstanding.

ACS Symp. Ser.
Acta Crystallogr., Sect. C: Cryst. Struct. Commun.
Acta Crystallogr., Sect. D: Biol. Crystallogr.
Acta Crystallogr., Sect. E: Struct. Rep. Online
Acta Hortic.
Acta Mater.
Acta Pharmacol. Sin.
Acta Phys. Pol., B
Adv. Exp. Med. Biol.
Adv. Mass Spectrom.
Adv. Mater. (Weinheim, Ger.)
Adv. Sci. Technol. (Faenza, Italy)
Adv. Space Res.
Adv. Synth. Catal.
AIChE J.
AIDS (London, U.K.)
AIDS Res. Hum. Retroviruses
AIP Conf. Proc.
Alcohol.: Clin. Exp. Res.
Aliment. Pharmacol. Ther.
Am. Heart J.
Am. J. Cardiol.
Am. J. Clin. Nutr.
Am. J. Hum. Genet.
Am. J. Obstet. Gynecol.
Am. J. Pathol.
Am. J. Physiol.
Am. J. Respir. Cell Mol. Biol.
Am. J. Vet. Res.
Am. Mineral.
Anal. Bioanal. Chem.
Anal. Biochem.
Anal. Chem.
Anal. Chim. Acta
Anal. Lett.
Anal. Sci.
Analyst (Cambridge, U.K.)
Anesth. Analg. (Hagerstown, MD, U.S.)
Anesthesiology
Angew. Chem., Int. Ed.
Anim. Genet.
Ann. N.Y. Acad. Sci.
Ann. Neurol.
Annu. Rep.—Conf. Electr. Insul. Dielectr. Phenom.
Annu. Tech. Conf.—Soc. Plast. Eng.
Anti-Cancer Drugs
Anticancer Res.
Antimicrob. Agents Chemother.
Antioxid. Redox Signaling

Appl. Biochem. Biotechnol.
Appl. Catal., A
Appl. Catal., B
Appl. Environ. Microbiol.
Appl. Geochem.
Appl. Microbiol. Biotechnol.
Appl. Opt.
Appl. Organomet. Chem.
Appl. Phys. A: Mater. Sci. Process.
Appl. Phys. B: Lasers Opt.
Appl. Phys. Lett.
Appl. Radiat. Isot.
Appl. Spectrosc.
Appl. Surf. Sci.
Aquaculture
Aquat. Toxicol.
Arch. Biochem. Biophys.
Arch. Environ. Contam. Toxicol.
Arch. Pharmacal Res.
Arch. Virol.
ARKIVOC (Gainesville, FL, U.S.)
Arterioscler., Thromb., Vasc. Biol.
Arthritis Rheum.
Asian–Australas. J. Anim. Sci.
Asian J. Chem.
Astron. Astrophys.
Astron. J.
Astron. Soc. Pac. Conf. Ser.
Astrophys. J.
Astrophys. J., Suppl. Ser.
Atherosclerosis (Amsterdam, Neth.)
Atmos. Chem. Phys.
Atmos. Environ.
Aust. J. Chem.
Azerb. Khim. Zh.
Bandaoti Xuebao
Behav. Brain Res.
Biochem. Biophys. Res. Commun.
Biochem. Eng. J.
Biochem. J.
Biochem. Pharmacol.
Biochem. Soc. Trans.
Biochem. Syst. Ecol.
Biochemistry
Biochemistry (Moscow, Russ. Fed.)
Biochim. Biophys. Acta
Bioconjugate Chem.
Bioinformatics
Biol. Chem.
Biol. Pharm. Bull.
Biol. Psychiatry
Biol. Reprod.
Biol. Trace Elem. Res.
Biomacromolecules
Biomaterials
Bioorg. Med. Chem.
Bioorg. Med. Chem. Lett.
Biophys. Chem.
Biophys. J.
Biopolymers
Bioresour. Technol.
Biosci., Biotechnol., Biochem.
Biosens. Bioelectron.
BioTechniques
Biotechnol. Bioeng.
Biotechnol. Lett.
Biotechnol. Prog.
Blood
BMC Bioinf.
Bone (San Diego, CA, U.S.)
Bone Marrow Transplant.
Br. J. Anaesth.
Br. J. Cancer
Br. J. Clin. Pharmacol.
Br. J. Haematol.
Br. J. Nutr.
Br. J. Pharmacol.
Brain Res.
Breast Cancer Res. Treat.
Bul. Stiint. Univ. "Politeh." Timisoara Rom., Ser. Chim. Ing. Mediului
Bull. Chem. Soc. Jpn.
Bull. Environ. Contam. Toxicol.
Bull. Exp. Biol. Med.
Bull. Korean Chem. Soc.
Bunseki Kagaku
C. R. Chim.
Cailiao Kexue Yu Gongcheng Xuebao
Can. J. Chem.
Cancer (New York, NY, U.S.)
Cancer Biol. Ther.
Cancer Cell

Cancer Chemother. Pharmacol.
Cancer Epidemiol., Biomarkers Prev.
Cancer Genet. Cytogenet.
Cancer Lett. (Amsterdam, Neth.)
Cancer Res.
Cancer Sci.
Carbon
Carbohydr. Polym.
Carbohydr. Res.
Carcinogenesis
Cardiovasc. Res.
Catal. Commun.
Catal. Lett.
Catal. Today
Cell (Cambridge, MA, U.S.)
Cell Biol. Int.
Cell Cycle
Cell Death Differ.
Cell. Mol. Life Sci.
Cell. Signalling
Cem. Concr. Compos.
Cem. Concr. Res.
Ceram. Eng. Sci. Proc.
Ceram. Int.
Ceram. Trans.
Cereal Chem.
Chem. Biol.
Chem. Commun. (Cambridge, U.K.)
Chem. Eng. J. (Amsterdam, Neth.)
Chem. Eng. News
Chem. Eng. Process.
Chem. Eng. Res. Des.
Chem. Eng. Sci.
Chem.—Eur. J.
Chem. Geol.
Chem. Heterocycl. Compd. (New York, NY, U.S.)
Chem. Ing. Tech.
Chem. Lett.
Chem. Mater.
Chem. Nat. Compd.
Chem. Pet. Eng.
Chem. Pharm. Bull.
Chem. Phys.
Chem. Phys. Lett.
Chem. Res. Chin. Univ.
Chem. Res. Toxicol.
Chem. Rev. (Washington, DC, U.S.)
Chem. Sens.
ChemBioChem
Chemosphere
ChemPhysChem
Chest
Chin. Chem. Lett.
Chin. J. Chem.
Chin. Med. J. (Beijing, China, Engl. Ed.)
Chin. Sci. Bull.
Chromatographia
Circ. Res.
Circulation
Clin. Biochem.
Clin. Cancer Res.
Clin. Chem. (Washington, DC, U.S.)
Clin. Chem. Lab. Med.
Clin. Chim. Acta
Clin. Diagn. Lab. Immunol.
Clin. Endocrinol. (Oxford, U.K.)
Clin. Exp. Allergy
Clin. Exp. Immunol.
Clin. Exp. Pharmacol. Physiol.
Clin. Immunol. (San Diego, CA, U.S.)
Clin. Infect. Dis.
Clin. Sci.
Collect. Czech. Chem. Commun.
Colloid Polym. Sci.
Colloids Surf., A
Colloids Surf., B
Combust. Flame
Commun. Soil Sci. Plant Anal.
Comp. Biochem. Physiol., Part A: Mol. Integr. Physiol.
Comp. Biochem. Physiol., Part B: Biochem. Mol. Biol.
Compos. Sci. Technol.
Comput. Chem. Eng.
Congr. Anu.—Assoc. Bras. Metal. Mater.
Corros. Sci.
Crit. Care Med.
Cryst. Growth Des.
Cuihua Xuebao
Curr. Biol.
Curr. Med. Chem.

Curr. Microbiol.
Curr. Pharm. Des.
Curr. Sci.
Cytogenet. Genome Res.
Cytokine+
Czech. J. Phys.
Dalton Trans.
Desalination
Dev. Biol. (San Diego, CA, U.S.)
Dev. Brain Res.
Dev. Cell
Dev. Dyn.
Development (Cambridge, U.K.)
Di-San Junyi Daxue Xuebao
Diabetes
Diabetologia
Diamond Relat. Mater.
Dianchi
Dianyuan Jishu
Dier Junyi Daxue Xuebao
Diffus. Defect Data, Pt. B
Dig. Dis. Sci.
Disi Junyi Daxue Xuebao
Diyi Junyi Daxue Xuebao
DNA Repair
Dokl. Biol. Sci.
Dokl. Bulg. Akad. Nauk.
Dokl. Earth Sci.
Dopov. Nats. Akad. Nauk Ukr.
Drug Metab. Dispos.
Dyes Pigm.
EAAP Publ.
Earth Planet. Sci. Lett.
Ecotoxicol. Environ. Saf.
Electroanalysis
Electrochem. Commun.
Electrochem. Solid-State Lett.
Electrochemistry (Tokyo, Jpn.)
Electrochim. Acta
Electron. Lett.
Electrophoresis
EMBO J.
EMBO Rep.
Endocrinology
Energy Fuels
Environ. Health Perspect.
Environ. Pollut. (Oxford, U.K.)
Environ. Sci. Technol.
Environ. Technol.
Environ. Toxicol. Chem.
Enzyme Microb. Technol.
Eukaryotic Cell
Eur. Food Res. Technol.
Eur. J. Biochem.
Eur. J. Cancer
Eur. J. Clin. Nutr.
Eur. J. Endocrinol.
Eur. J. Hum. Genet.
Eur. J. Immunol.
Eur. J. Inorg. Chem.
Eur. J. Org. Chem.
Eur. J. Pharm. Biopharm.
Eur. J. Pharm. Sci.
Eur. J. Pharmacol.
Eur. Phys. J. A
Eur. Phys. J. B
Eur. Phys. J. C
Eur. Phys. J. D
Eur. Polym. J.
Eur. Space Agency, [Spec. Publ.] SP
Europhys. Lett.
Exp. Biol. Med. (Maywood, NJ, U.S.)
Exp. Cell Res.
Exp. Eye Res.
Exp. Gerontol.
Exp. Hematol. (New York, NY, U.S.)
Exp. Neurol.
Expert Opin. Invest. Drugs
Expert Opin. Pharmacother.
Farmaco
FASEB J.
FEBS Lett.
Fed. Regist.
FEMS Immunol. Med. Microbiol.
FEMS Microbiol. Lett.
Fenxi Huaxue
Fenxi Kexue Xuebao
Fenxi Shiyanshi
Ferroelectrics
Fish Physiol. Biochem.
Fiz. Khim. Tverd. Tila
Fluid Phase Equilib.

Food Addit. Contam.
Food Chem.
Food Chem. Toxicol.
Food Hydrocolloids
Forensic Sci. Int.
Free Radical Biol. Med.
Free Radical Res.
Fresenius Environ. Bull.
Front. Biosci.
Front. Sci. Ser.
Fuel
Fuel Process. Technol.
Fusion Energy
Fusion Eng. Des.
Gangtie
Gaodeng Xuexiao Huaxue Xuebao
Gaofenzi Cailiao Kexue Yu Gongcheng
Gaofenzi Xuebao
Gaoneng Wuli Yu Hewuli
Gaoxiao Huaxue Gongcheng Xuebao
Gastroenterology
Gen. Comp. Endocrinol.
Gendai Iryo
Gene
Gene Expression Patterns
Gene Ther.
Genes Dev.
Genetics
Genome Res.
Genomics
Geochim. Cosmochim. Acta
Geophys. Res. Lett.
Gongcheng Suliao Yingyong
Gongneng Cailiao
Gongye Cuihua
Green Chem.
Guangpu Shiyanshi
Guangpuxue Yu Guangpu Fenxi
Guangzi Xuebao
Guisuanyan Xuebao
Gut
Gynecol. Oncol.
Haematologica
Handb. Exp. Pharmacol.
Han'guk Hwankyong Uisaeng Hakhoechi
Han'guk Sikp'um Yongyang Kwahak Hoechi
Hecheng Huaxue (1000)
Hecheng Xiangjiao Gongye
Helv. Chim. Acta
Hepatology (Philadelphia, PA, U.S.)
Heterocycles
Horm. Metab. Res.
Huagong Shikan
Huagong Xuebao (Chin. Ed.)
Huanjing Kexue Xuebao
Huanjing Wuran Zhili Jishu Yu Shebei
Huaxue Tongbao
Huaxue Xuebao
Huaxue Yanjiu Yu Yingyong
Hum. Mol. Genet.
Hum. Mutat.
Hum. Pathol.
Hum. Reprod.
Hydrobiologia
Hyomen Gijutsu
Hyperfine Interact.
Hypertension
IEEE Electron Device Lett.
IEEE J. Quantum Electron.
IEEE Trans. Electron Devices
IEEE Trans. Magn.
IEEE Trans. Nucl. Sci.
Igaku no Ayumi
Immunol. Lett.
Immunology
Ind. Eng. Chem. Res.
Indian J. Chem., Sect. A: Inorg., Bio-inorg., Phys., Theor. Anal. Chem.
Indian J. Chem., Sect. B: Org. Chem. Incl. Med. Chem.
Indian J. Environ. Prot.
Indian J. Pharm. Sci.
Infect. Immun.
Inflammation Res.
Inorg. Chem.
Inorg. Chem. Commun.
Inorg. Chim. Acta
Inorg. Mater.
Insect Biochem. Mol. Biol.
Inst. Phys. Conf. Ser.
Int. Conf. Thermoelectr.
Int. Congr. Ser.
Int. DATA Ser., Sel. Data Mixtures, Ser. A

Int. Immunol.
Int. Immunopharmacol.
Int. J. Antimicrob. Agents
Int. J. Biochem. Cell Biol.
Int. J. Cancer
Int. J. Food Microbiol.
Int. J. Heat Mass Transfer
Int. J. Hydrogen Energy
Int. J. Mass Spectrom.
Int. J. Mod. Phys. B
Int. J. Mol. Med.
Int. J. Nanosci.
Int. J. Oncol.
Int. J. Parasitol.
Int. J. Pharm.
Int. J. Quantum Chem.
Int. J. Syst. Evol. Microbiol.
Integr. Ferroelectr.
Intermetallics
IP.com J.
ISIJ Int.
Izv. Akad. Nauk, Ser. Fiz.
Izv. Vyssh. Uchebn. Zaved., Khim. Khim. Tekhnol.
J. Agric. Food Chem.
J. Allergy Clin. Immunol.
J. Alloys Compd.
J. Am. Ceram. Soc.
J. Am. Chem. Soc.
J. Am. Coll. Cardiol.
J. Am. Oil Chem. Soc.
J. Am. Soc. Mass Spectrom.
J. Am. Soc. Nephrol.
J. Anal. Appl. Pyrolysis
J. Anal. At. Spectrom.
J. Anal. Chem.
J. Anim. Sci. (Savoy, IL, U.S.)
J. Antimicrob. Chemother.
J. AOAC Int.
J. Appl. Crystallogr.
J. Appl. Electrochem.
J. Appl. Microbiol.
J. Appl. Phys.
J. Appl. Physiol.
J. Appl. Polym. Sci.
J. Appl. Spectrosc.
J. Bacteriol.
J. Biochem. (Tokyo, Jpn.)
J. Biol. Chem.
J. Biomed. Mater. Res., Part A
J. Biomed. Mater. Res., Part B
J. Biomol. NMR
J. Biosci. Bioeng.
J. Biotechnol.
J. Bone Miner. Res.
J. Cardiovasc. Pharmacol.
J. Catal.
J. Cell. Biochem.
J. Cell Biol.
J. Cell. Physiol.
J. Cell Sci.
J. Ceram. Soc. Jpn.
J. Chem. Ecol.
J. Chem. Educ.
J. Chem. Eng. Data
J. Chem. Eng. Jpn.
J. Chem. Inf. Comput. Sci.
J. Chem. Phys.
J. Chem. Res., Synop.
J. Chem. Technol. Biotechnol.
J. Chin. Chem. Soc. (Taipei, Taiwan)
J. Chromatogr., A
J. Chromatogr., B: Anal. Technol. Biomed. Life Sci.
J. Clin. Endocrinol. Metab.
J. Clin. Invest.
J. Clin. Microbiol.
J. Colloid Interface Sci.
J. Comp. Neurol.
J. Comput. Chem.
J. Comput. Electron.
J. Controlled Release
J. Coord. Chem.
J. Cryst. Growth
J. Dairy Sci.
J. Electroanal. Chem.
J. Electrochem. Soc.
J. Electron. Mater.
J. Endocrinol.
J. Environ. Eng. (Reston, VA, U.S.)
J. Environ. Monit.
J. Environ. Qual.

J. Environ. Radioact.
J. Environ. Sci. Health, Part A: Toxic/ Hazard. Subst. Environ. Eng.
J. Essent. Oil Res.
J. Eur. Ceram. Soc.
J. Exp. Biol.
J. Exp. Bot.
J. Exp. Med.
J. Exp. Theor. Phys.
J. Fluorine Chem.
J. Food Prot.
J. Food Sci.
J. Gen. Virol.
J. Geophys. Res., [Atmos.]
J. Hazard. Mater.
J. Hepatol.
J. Heterocycl. Chem.
J. Histochem. Cytochem.
J. Hypertens.
J. Immunol.
J. Immunol. Methods
J. Inclusion Phenom. Macrocyclic Chem.
J. Indian Chem. Soc.
J. Infect. Dis.
J. Inorg. Biochem.
J. Invest. Dermatol.
J. Korean Ceram. Soc.
J. Korean Phys. Soc.
J. Leukocyte Biol.
J. Lipid Res.
J. Liq. Chromatogr. Relat. Technol.
J. Low Temp. Phys.
J. Lumin.
J. Magn. Magn. Mater.
J. Magn. Reson.
J. Mass Spectrom.
J. Mater. Chem.
J. Mater. Process. Technol.
J. Mater. Res.
J. Mater. Sci.
J. Mater. Sci. Lett.
J. Mater. Sci.: Mater. Electron.
J. Mater. Sci.: Mater. Med.
J. Mater. Sci. Technol. (Shenyang, China)
J. Med. Chem.
J. Med. Genet.
J. Med. Virol.
J. Membr. Sci.
J. Metastable Nanocryst. Mater.
J. Microbiol. Biotechnol.
J. Microbiol. Methods
J. Mol. Biol.
J. Mol. Catal. A: Chem.
J. Mol. Catal. B: Enzym.
J. Mol. Cell. Cardiol.
J. Mol. Evol.
J. Mol. Liq.
J. Mol. Spectrosc.
J. Mol. Struct.
J. Nat. Prod.
J. Natl. Cancer Inst.
J. Neurochem.
J. Neuroimmunol.
J. Neurophysiol.
J. Neurosci.
J. Neurosci. Res.
J. Non-Cryst. Solids
J. Nucl. Mater.
J. Nutr.
J. Opt. Soc. Am. B
J. Optoelectron. Adv. Mater.
J. Org. Chem.
J. Organomet. Chem.
J. Pathol.
J. Pharm. Biomed. Anal.
J. Pharm. Pharmacol.
J. Pharm. Sci.
J. Pharmacol. Exp. Ther.
J. Pharmacol. Sci. (Tokyo, Jpn.)
J. Photochem. Photobiol., A
J. Phys. A: Math. Gen.
J. Phys. B: At., Mol. Opt. Phys.
J. Phys. Chem. A
J. Phys. Chem. B
J. Phys. Chem. Solids
J. Phys.: Condens. Matter
J. Phys. D: Appl. Phys.
J. Phys. G: Nucl. Part. Phys.
J. Phys. IV
J. Phys. Org. Chem.
J. Phys. Soc. Jpn.
J. Physiol. (Oxford, U.K.)

J. Plant Physiol.
J. Polym. Sci., Part A: Polym. Chem.
J. Polym. Sci., Part B: Polym. Phys.
J. Power Sources
J. Quant. Spectrosc. Radiat. Transfer
J. Radioanal. Nucl. Chem.
J. Raman Spectrosc.
J. Rheumatol.
J. Sci. Food Agric.
J. Sep. Sci.
J. Solid State Chem.
J. Steroid Biochem. Mol. Biol.
J. Surg. Res.
J. Therm. Anal. Calorim.
J. Thromb. Haemostasis
J. Univ. Chem. Technol. Metall.
J. Urol. (Hagerstown, MD, U.S.)
J. Vac. Sci. Technol., A
J. Vac. Sci. Technol., B: Microelectron. Nanometer Struct.—Process., Meas., Phenom.
J. Vet. Pharmacol. Ther.
J. Virol.
J. Virol. Methods
JAERI—Conf
JETP Lett.
Jiegou Huaxue
Jikken Igaku
Jingxi Huagong
Jingxi Huagong Zhongjianti
Jinshu Xuebao
Jisuanji Yu Yingyong Huaxue
Jixie Gongcheng Cailiao
Jpn. J. Appl. Phys., Part 1
Jpn. J. Appl. Phys., Part 2
Kagaku to Kogyo (Tokyo, Jpn.)
Kagaku to Kyoiku
Kagaku to Seibutsu
KEK Proc.
Key Eng. Mater.
Kidney Int.
Kogyo Zairyo
Kongop Hwahak
Korean J. Chem. Eng.
Lab. Invest.
Lancet
Langmuir
Lect. Notes Phys.
Leuk. Lymphoma
Leuk. Res.
Leukemia
Life Sci.
Liq. Cryst.
Low Temp. Phys.
Lung Biol. Health Dis.
Macromol. Chem. Phys.
Macromol. Rapid Commun.
Macromol. Symp.
Macromolecules
Magn. Reson. Chem.
Mar. Pollut. Bull.
Mater. Chem. Phys.
Mater. Lett.
Mater. Res. Bull.
Mater. Res. Soc. Symp. Proc.
Mater. Sci. Eng., A
Mater. Sci. Eng., B
Mater. Sci. Eng., C
Mater. Sci. Forum
Mater. Sci. Technol.
Mater. Trans.
Meas. Sci. Technol.
Meat Sci.
Mech. Dev.
Med. Hypotheses
Meded.—Fac. Landbouwkd. Toegepaste Biol. Wet. (Univ. Gent)
Metab., Clin. Exp.
Metall. Mater. Trans. A
Meteorit. Planet. Sci.
Methods Enzymol.
Methods Mol. Biol. (Totowa, NJ, U.S.)
Methods Mol. Med.
Microbes Infect.
Microbiology (Reading, U.K.)
Microchim. Acta
Microelectron. Eng.
Microelectron. Reliab.
Microporous Mesoporous Mater.
Miner. Eng.
Mod. Phys. Lett. A
Mol. Biochem. Parasitol.

Mol. Biol. Cell
Mol. Biol. Evol.
Mol. Brain Res.
Mol. Cancer Ther.
Mol. Cell
Mol. Cell. Biochem.
Mol. Cell. Biol.
Mol. Cell. Endocrinol.
Mol. Cell. Neurosci.
Mol. Cryst. Liq. Cryst.
Mol. Ecol.
Mol. Ecol. Notes
Mol. Endocrinol.
Mol. Genet. Genomics
Mol. Genet. Metab.
Mol. Immunol.
Mol. Med. (Tokyo, Jpn.)
Mol. Microbiol.
Mol. Pharmacol.
Mol. Phylogenet. Evol.
Mol. Phys.
Mol. Plant–Microbe Interact.
Mol. Psychiatry
Mol. Reprod. Dev.
Mol. Ther.
Mon. Not. R. Astron. Soc.
Monatsh. Chem.
Mutat. Res.
N. Engl. J. Med.
Nano Lett.
Nanotechnology
NASA Conf. Publ.
Nat. Biotechnol.
Nat. Cell Biol.
Nat. Genet.
Nat. Immunol.
Nat. Mater.
Nat. Med. (New York, NY, U.S.)
NATO Sci. Ser., II
NATO Sci. Ser., IV
NATO Sci. Ser., Ser. I
Nature (London, U.K.)
Naunyn-Schmiedeberg's Arch. Pharmacol.
Nephrol., Dial., Transplant.
Neuron
Neurobiol. Aging
Neurobiol. Dis.
Neurochem. Int.
Neurochem. Res.
Neurology
Neuropharmacology
Neuropsychopharmacology
NeuroReport
Neurosci. Lett.
Neuroscience (Oxford, U.K.)
New J. Chem.
New Phytol.
Nippon Kessho Seicho Gakkaishi
Nippon Kikai Gakkai Ronbunshu, B-hen
Nongyao
Nongye Huanjing Kexue Xuebao
Nucl. Eng. Des.
Nucl. Fusion
Nucl. Instrum. Methods Phys. Res., Sect. A
Nucl. Instrum. Methods Phys. Res., Sect. B
Nucl. Phys. A
Nucl. Phys. B
Nucl. Phys. B, Proc. Suppl.
Nucleic Acids Res.
Nucleosides, Nucleotides Nucleic Acids
Oncogene
Oncol. Rep.
Opt. Commun.
Opt. Lett.
Opt. Mater. (Amsterdam, Neth.)
Opt. Spectrosc.
Org. Biomol. Chem.
Org. Lett.
Org. Process Res. Dev.
Organohalogen Compd.
Organometallics
Orient. J. Chem.
Oxford Monogr. Med. Genet.
Pain
Pediatr. Res.
Peptides (New York, NY, U.S.)
Pfluegers Arch.
Pharm. Chem. J.
Pharm. Res.
Pharmacol., Biochem. Behav.
Pharmacol. Res.
Pharmazie

Philos. Mag.
Phosphorus, Sulfur Silicon Relat. Elem.
Photochem. Photobiol.
Photochem. Photobiol. Sci.
Phys. At. Nucl.
Phys. Chem. Chem. Phys.
Phys. Fluids
Phys. Lett. A
Phys. Lett. B
Phys. Plasmas
Phys. Rev. A: At., Mol., Opt. Phys.
Phys. Rev. B: Condens. Matter Mater. Phys.
Phys. Rev. C: Nucl. Phys.
Phys. Rev. D: Part. Fields
Phys. Rev. E: Stat., Nonlinear, Soft Matter Phys.
Phys. Rev. Lett.
Phys. Solid State
Phys. Status Solidi A
Phys. Status Solidi B
Phys. Status Solidi C
Physica B (Amsterdam, Neth.)
Physica C (Amsterdam, Neth.)
Physica E (Amsterdam, Neth.)
Physiol. Behav.
Physiol. Genomics
Physiol. Plant.
Phytochemistry (Elsevier)
Planta
Plant Cell
Plant Cell Physiol.
Plant J.
Plant Mol. Biol.
Plant Physiol.
Plant Sci. (Amsterdam, Neth.)
Plant Soil
Planta Med.
Plasma Phys. Controlled Fusion
Plast. Massy
Pol. J. Chem.
Polyhedron
Polym. Degrad. Stab.
Polym. Eng. Sci.
Polym. Int.
Polym. J. (Tokyo, Jpn.)
Polym. Mater. Sci. Eng.
Polym. Prepr. (Am. Chem. Soc., Div. Polym. Chem.)
Polymer
Poult. Sci.
Poverkhnost
Powder Technol.
Pramana
Prepr.—Am. Chem. Soc., Div. Pet. Chem.
Prepr. Ext. Abstr. ACS Natl. Meet., Am. Chem. Soc., Div. Environ. Chem.
Prepr. Symp.—Am. Chem. Soc., Div. Fuel Chem.
Proc.—Annu. Conf., Am. Water Works Assoc.
Proc.—Electrochem. Soc.
Proc. Natl. Acad. Sci. U.S.A.
Proc. SPIE—Int. Soc. Opt. Eng.
Proc.—Water Qual. Technol. Conf.
Process Biochem. (Oxford, U.K.)
Prog. Org. Coat.
Prostaglandins, Leukotrienes Essent. Fatty Acids
Prostate (New York, NY, U.S.)
Protein Expression Purif.
Protein Sci.
Proteins: Struct., Funct., Bioinf.
Proteomics
Psychopharmacology (Berlin, Ger.)
Publ. Australas. Inst. Min. Metall.
Pure Appl. Chem.
Quim. Nova
Radiat. Phys. Chem.
Radiat. Prot. Dosim.
Ranliao Huaxue Xuebao
Rapid Commun. Mass Spectrom.
React. Kinet. Catal. Lett.
Recents Prog. Genie Procedes
Regul. Pept.
Rengong Jingti Xuebao
Reproduction (Bristol, U.K.)
Res. Discl.
Rev. Chim. (Bucharest, Rom.)
Rev. Mex. Astron. Astrofis., Ser. Conf.
Rev. Roum. Chim.
Rev. Sci. Instrum.
RILEM Proc.

Rinsho Men'eki
RNA
Russ. Chem. Bull.
Russ. J. Appl. Chem.
Russ. J. Coord. Chem.
Russ. J. Electrochem.
Russ. J. Gen. Chem.
Russ. J. Genet.
Russ. J. Org. Chem.
Saibo Kogaku
Sci. Total Environ.
Science (Washington, DC, U.S.)
Scr. Mater.
Sekitan Kagaku Kaigi Happyo Ronbunshu
Semicond. Sci. Technol.
Semiconductors
Sens. Actuators, A
Sens. Actuators, B
Sep. Purif. Technol.
Sep. Sci. Technol.
Sepu
Shandong Daxue Xuebao, Yixueban
Shengwu Yixue Gongchengxue Zazhi
Shijie Huaren Xiaohua Zazhi
Shipin Kexue (Beijing, China)
Shiyou Huagong
Shiyou Lianzhi Yu Huagong
Shock
Soc. Automot. Eng., [Spec. Publ.] SP
Soil Biol. Biochem.
Soil Sci. Soc. Am. J.
Sol. Energy Mater. Sol. Cells
Solid State Commun.
Solid-State Electron.
Solid State Ionics
Solid State Sci.
Spec. Publ.—R. Soc. Chem.
Spectrochim. Acta, Part A
Spectrochim. Acta, Part B
Steroids
Stroke
Structure (Cambridge, MA, U.S.)
Stud. Surf. Sci. Catal.
Supercond. Sci. Technol.
Surf. Coat. Technol.
Surf. Interface Anal.
Surf. Sci.
Symp.—Int. Astron. Union
Synlett
Synth. Commun.
Synth. Met.
Synthesis
Talanta
Tanpakushitsu Kakusan Koso
Tech. Phys.
Tech. Phys. Lett.
Tetrahedron
Tetrahedron: Asymmetry
Tetrahedron Lett.
Tetsu to Hagane
Text. Res. J.
Tezhong Zhuzao Ji Youse Hejin
THEOCHEM
Theor. Appl. Genet.
Theriogenology
Thermochim. Acta
Thin Solid Films
Thromb. Haemostasis
Thromb. Res.
Tissue Antigens
Tissue Eng.
Tokyo Daigaku Genshiryoku Kenkyu Sogo Senta Shinpojumu
Top. Catal.
Toxicol. Appl. Pharmacol.
Toxicol. Lett.
Toxicol. Sci.
Toxicology
Toxicon
Trans. Am. Foundry Soc.
Trans. Nonferrous Met. Soc. China
Transition Met. Chem. (Dordrecht, Neth.)
Transplant. Proc.
Transplantation
Trends Opt. Photonics
Trends Pharmacol. Sci.
Tsvetn. Met. (Moscow, Russ. Fed.)
Ukr. Khim. Zh. (Russ. Ed.)
Vaccine
Vacuum
VDI—Ber.
Vet. Microbiol.

Virology
Virus Res.
Vysokomol. Soedin., Ser. A Ser. B
Water, Air, Soil Pollut.
Water Res.
Water Sci. Technol.
Wear
World J. Gastroenterol.
Wuji Cailiao Xuebao
Wuji Huaxue Xuebao
Wuli Huaxue Xuebao
Wuli Xuebao
Xibao Yu Fenzi Mianyixue Zazhi
Xiyou Jinshu
Xiyou Jinshu Cailiao Yu Gongcheng
Yaoxue Xuebao
Yingyong Huaxue
Yingyong Shengtai Xuebao
Youji Huaxue
Z. Anorg. Allg. Chem.
Z. Kristallogr.—New Cryst. Struct.
Z. Metallkd.
Z. Naturforsch., B: Chem. Sci.
Z. Naturforsch., C: J. Biosci.
Zairyo
Zavod. Lab., Diagn. Mater.
Zh. Fiz. Khim.
Zh. Neorg. Khim.
Zhengzhou Daxue Xuebao, Yixueban
Zhongcaoyao
Zhongguo Bingli Shengli Zazhi
Zhongguo Dongmai Yinghua Zazhi
Zhongguo Gonggong Weisheng
Zhongguo Jiguang
Zhongguo Jishui Paishui
Zhongguo Shenghua Yaowu Zazhi
Zhongguo Shengwu Gongcheng Zazhi
Zhongguo Shengwu Huaxue Yu Fenzi Shengwu Xuebao
Zhongguo Shouyi Xuebao
Zhongguo Suliao
Zhongguo Xinyao Zazhi
Zhongguo Xitu Xuebao
Zhongguo Yaolixue Tongbao
Zhongguo Yaoxue Zazhi (Beijing, China)
Zhongguo Yiyao Gongye Zazhi
Zhongguo Yiyuan Yaoxue Zazhi
Zhongguo Youse Jinshu Xuebao
Zhonghua Yixue Yichuanxue Zazhi
Zhongliu Fangzhi Zazhi

➤➤➤➤➤

APPENDIX 14-2

# A Sample *CASSI* Entry

**J***ournal of the* **Am***erican* **Chem***ical* **Soc***iety.* JACSAT. ISSN 0002–7863 (Absorbed Am. Chem. J.). In English; English sum. History: v1 1879+. *w* **126 2004**. *ACS Journals* or *Maruzen.*

AMERICAN CHEMICAL SOCIETY. JOURNAL. WASHINGTON, D. C.

Doc. Supplier: CAS.

AAP; AB 1905+; ABSR; ARaS; ATVA; AU–M 1893–1918,1920–1926,1928+; AkU 1879–1906,1919+; ArU; ArU–M 1923+; AzTeS; AzU 1889+; C; CL; CLSU; CLSU–M 1895–1897,1905,1908+; CLU–M; CLU–P; CMenSR 1916+; CPT; CSf; CSt; CSt–L; CU; CU–A; CU–I 1920+; CU–M; CU–Riv 1907+; CU–RivA; CU–RivP; CU–S; CU–SB; [etc.]

In this example,

- **J***ournal of the* **Am***erican* **Chem***ical* **Soc***iety* is the complete publication title with its abbreviated form indicated by boldface type (*J. Am. Chem. Soc.*).
- **JACSAT** is the CODEN, a six-character, unique title abbreviation used to represent titles in manual or machine-based information systems. The CODEN source is the *International CODEN Directory,* administered by Chemical Abstracts Service. The sixth character of each CODEN is a computer-calculated check character that ensures the reliability of the CODEN in computer-based systems.
- **ISSN 0002–7863** is the International Standard Serial Number (ISSN), assigned by the Library of Congress.
- **Absorbed Am. Chem. J.** is a reference to former titles and to any variant forms of the selected title.
- **In English; English sum.** is the language of the publication, summaries, and tables of contents.
- **History: v1 1879+** is the history of the publication. Volume 1 began in 1879. The + following the year indicates that the publication is still in existence under that title.
- ***w*** means weekly. The frequency of publication could also be *a* for annually, *ba* for biennially (every two years), *bm* for bimonthly (every two months), *bw* for biweekly (every two weeks), *d* for daily, *m* for monthly, *q* for quarterly, *sa* for semiannually (two times per year), *sm* for semimonthly (two times per month), or *sw* for semiweekly (two times per week).

- **126 2004** is the volume–year correlation (i.e., the first volume number of that year, which is the most recent covered by that edition of *CASSI;* volume 126 was the first volume number of 2004).
- ***ACS Journals* or *Maruzen*** is the publisher or source address or abbreviation.
- **AMERICAN CHEMICAL SOCIETY. JOURNAL. WASHINGTON, D. C.** is the AACR entry. This is the abbreviated entry as catalogued according to the *Anglo-American Cataloguing Rules* (2nd ed.). It is included here because of its predominance in library collection records.
- **Doc. Supplier: CAS** means that articles are available through the CAS Document Delivery Service.
- **AAP; AB 1905+; ABSR;** etc., is the library holdings information. Libraries are identified by their *National Union Catalog* symbols, and holdings are shown by inclusive years.

➤➤➤➤➤ CHAPTER 15

# Figures

Betsy Kulamer

This chapter discusses methods of preparing and submitting the figures and other illustrations that accompany a scientific paper for publication. The past 10 years have seen a radical technological shift in the way figures are created by authors and handled by publishers. Whereas authors previously submitted figures that were *camera-ready*—that is, ready for the printer's camera to photograph before making printing plates—today many authors create figures using computer software, and most publishers encourage electronic versions of figures. This chapter presents guidelines for working with figures using the computer tools currently in widespread use; it also includes recommendations regarding the use of color, fonts, and scanners, as well as good style for citing and captioning figures.

Technical requirements and certain style points differ from publisher to publisher and from journal to journal, so bear in mind that this chapter presents general guidelines. ACS journals have some specific requirements that may differ from the general guidelines presented here, and these are noted throughout the chapter. As always, before you finalize figures for your manuscript, you should consult your publisher's guidelines for submitting artwork. For ACS journals, consult recent issues as well as the Guide, Notes, or Instructions for Authors that appear in each journal's first issue of the year or at https://paragon.acs.org/paragon/index.jsp (see "Author Information"). For ACS books, consult the brochure "How To Prepare Your Manuscript for the ACS Symposium Series" or see "Info for Authors" at https://pubs.acs.org/books.

## Box 15-1. Do I Need a Figure or a Table?

Do I want the basic point to be communicated at a glance?
*Use a figure.*

Do I want the reader to see trends and relationships?
*Use a figure.*

Do I want the reader to see exact numbers?
*Use a table.*

Do I want to communicate a lot of information with words?
*Use a table.*

## When To Use Figures

A *figure* is an illustration used in scientific or scholarly publishing. Figures can be graphs of data, photographs, sketches, flow charts, and so on. Figures can play a major role in highlighting, clarifying, and summarizing data and results and can substantially increase the readers' comprehension of the text by communicating visually. For example, line graphs show trends. Bar graphs compare magnitudes. Pie charts show relative portions of a whole. Photographs can provide absolute proof of findings. In general, figures should be used when the picture really is worth a thousand (or so) words.

However, figures can decrease a reader's comprehension, and they can cause outright confusion if they are poorly rendered or cluttered, if they do little more than repeat data already presented in text, or, worse, if they present information at odds with the text. An excessive number of figures can dilute the value of any individual figure: when presented with too many figures, a reader may look carefully at none of them. Figures should not be used to present data that would be better presented in a table. See Box 15-1 and Chapter 16, "Tables."

The cost of publishing figures has decreased substantially over the past two decades, partly because authors are able to prepare better quality figures (so the publisher does not have to) and partly because of the shift to electronic methods of Web and print publishing that are less cumbersome than earlier technologies, which required a high-quality, pristine paper original of all art. The cost of publishing color figures has also decreased substantially because most color photographs can now be provided in electronic form and no longer require expensive color separations before being put on a printing press. The cost of using color in Web publications is inconsequential (although color should still be used carefully to enhance your meaning, not just to grab attention). See Box 15-2.

➤➤➤➤➤

### Box 15-2. Do I Need To Use Color?

Do I need color to make the picture comprehensible to the reader?

*Color should be used only when it is essential to understand the chemical nature of the material in the picture. Otherwise the picture should be prepared in black and white.*

Do I need color in my line graph, bar graph, or pie chart?

*Color is rarely required in these figures. Lines with varying dash styles in line graphs and distinct grayscale shades in bar graphs or pie charts work as well.*

Do I need color to catch the reader's eye?

*No.*

Do I need color to organize related information for the reader?

*No. Judicious use of fonts and careful placement of words and objects usually does just as well.*

## How To Cite Figures

All figures must be "called out", that is, mentioned or discussed by name and number in the text. (If you cannot find a graceful, logical place to put the callout, then you might consider deleting the figure.) Follow these guidelines in citing figures.

➤ Capitalize the word "Figure" when it is followed by the figure number.

➤ Number figures sequentially with arabic numerals in order of discussion in the text (Figure 1, Figure 2, etc.).

➤ Designate parts of a figure by using a combination of the arabic numeral and a sequence of consistent labels, usually (but not always) letters: Figure 1a, Figure 1b; Figure 1A, Figure 1B; Figure 1-I, Figure 1-II. Do not cite, for example, Figure 4 and Figure 4A.

EXAMPLES OF FIGURE CALLOUTS IN TEXT

> The block copolymers may contain a small but detectable fraction of impurities, as shown in Figures 1 and 2.
>
> Figures 3–5 show the production of acid-reactive substances in three different oils.
>
> The deuterium-labeled substrate gave rise to the partial $^{1}$H NMR spectra shown in Figure 2a,b.

As seen in Figure 3b–d, the catalytic wave shows a small but distinct decrease upon addition of the nucleophile.

Parts a and b of Figure 4 illustrate that the voltammetric plateau current depends on the number of enzyme monolayers.

Curves c–e of Figure 5 were obtained for various methyl groups in the protein.

POOR EXAMPLES OF FIGURE CALLOUTS IN TEXT

Figure 2c and Figure 2D show... *should read* Figure 2c,d shows...

Figure 3 and Figure 4 show... *should read* Figures 3 and 4 show...

Chemical structures and schemes should not be numbered as figures; they should be labeled according to separate sequences. See Chapter 17, Chemical Structures.

## How To Prepare Figures

At one time, almost all publishers required that figures be submitted as camera-ready artwork. Today, most publishers accept, many encourage, and a few require that figures be submitted in electronic format. Most scientists store their data on computers and create their figures using computer software, and good-quality document scanners are in common use. Thus, the submission of electronic files is simpler than ever before and has the advantage of keeping original data and photographs in the author's own hands. Satisfactory results in the final publication are not guaranteed, however, and the burden rests on authors to prepare figures to suit the requirements of the medium in which they will be published.

### *Publication Medium*

The traditional method for publishing books, journals, and other documents is by *print publishing*, that is, the application of ink on paper. Newer avenues include publishing on the Internet, usually called *Web publishing*, as well as other formats that are viewed on a computer monitor, such as CD-ROMs. The requirements of ink-on-paper and pixels-on-screen publishing are quite different and may affect your choices regarding figure creation and final electronic format. For printed materials to be satisfactory, the figure files must have much detail and therefore must be prepared at a higher resolution, that is, a higher number of pixels per inch (ppi). Paper printers also use a scale called dots per inch or dpi, which you may need to use sometimes. Materials viewed on a computer monitor do not require the same level of detail but—on the Web especially—they must download quickly, so they are prepared at a low resolution. Only for printed materials is color significantly more expensive than black-and-white, so in print, color is used conservatively. Conversely, on a computer monitor, color costs nothing additional and is used widely.

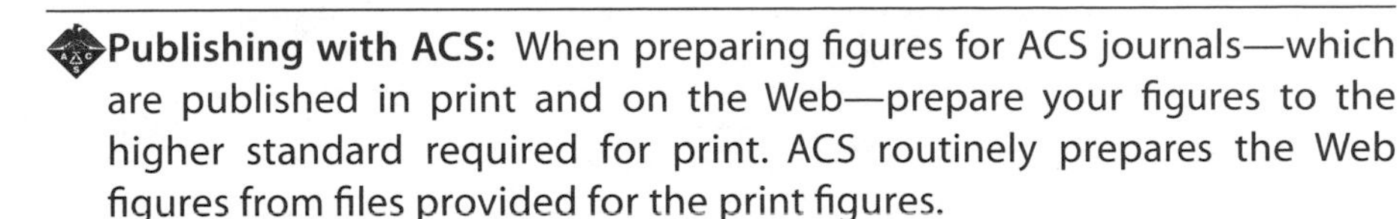

**Publishing with ACS:** When preparing figures for ACS journals—which are published in print and on the Web—prepare your figures to the higher standard required for print. ACS routinely prepares the Web figures from files provided for the print figures.

## *Types of Figure Artwork*

The two broad categories of figure artwork are line art and continuous-tone art. *Line art* consists of black markings on a white background, with no additional colors and no shades of gray (see Figure 15-1A). Line art contains lines, solid shapes, and type (letters or words). Examples of line art include line graphs, flow charts, and scatter plots. (Structures and schemes are usually line art as well; see Chapter 17.) *Continuous-tone art* (also called *halftone art*) contains shades of gray (or color) (see Figure 15-1B). Examples of continuous-tone art include photographs (whether black-and-white or color) and drawings or sketches with midtones.

In print publishing, line art is straightforwardly rendered by ink on paper. To print continuous-tone art, the shading must be converted into dots through a process called *screening*. The size and spacing of the dots produces the illusion of shading to the reader's eye. Halftones do not reproduce type or fine lines well because screening distorts the edges of the letters and lines. On a computer monitor, all text and images are converted to uniformly sized pixels.

Until recently, publishers strongly encouraged authors to submit figures as line art, with halftones reserved for photographs. Currently available software and the prevalence of electronic file submission is bringing a third category of art into more common use. *Grayscale line art* is black-and-white art that contains gray shading alongside fine lines and type (see Figure 15-1C). Grayscale line art may be used for pie charts, bar graphs, and area graphs, where different sections are distinguished by different shades of gray.

A fourth category of art is a combination of a photograph with some type, or a *combo* (see Figure 15-1D). An example might be a photomicrograph with a unit-of-measure legend in one corner; another might be a photograph with callouts identifying parts of an image, such as a piece of equipment. If there is only a small amount of type (as in the first example), then a combo can be handled like other halftones. When there is more type (as in the second example), then a combo requires special handling, because screened type loses legibility very quickly.

**Publishing with ACS:** For ACS journals, line art should be created and saved at 1200 ppi. Grayscale art, including grayscale line art and combos, should be created and saved at 600 ppi. Halftones, whether black-and-white or color, should be created and saved at 300 ppi.

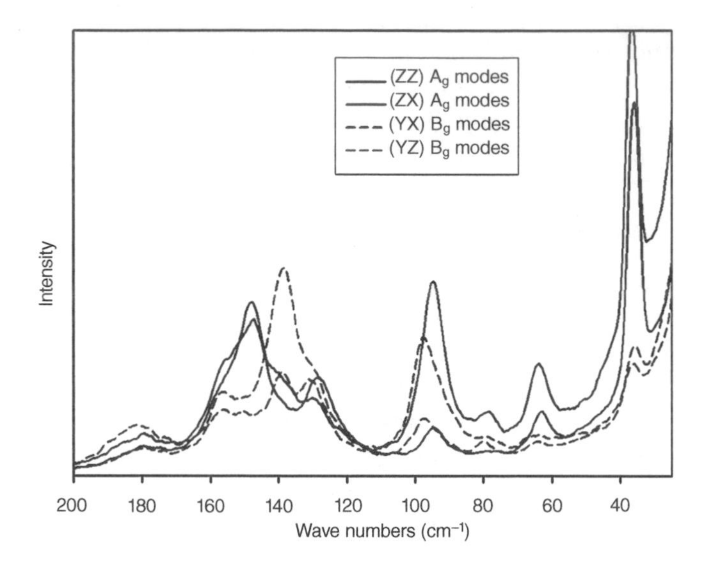


B

**Figure 15-1.** Examples of categories of figure art: (A) line art, (B) halftone or continuous-tone art, (C) grayscale line art, and (D) a combo.

*Sources:* (A) Reprinted from Stevens, L. L.; Haycraft, J. J.; Eckhardt, C. J. *Cryst. Growth Des.* **2005,** *5,* 2060–2065. Copyright 2005 American Chemical Society. (B) Photograph courtesy of Raj Mehta. Reprinted from *Reagent Chemicals,* 10th ed. Copyright 2006 American Chemical Society. (C) Reprinted from Watanabe, K.; Niwa, S.; Mori, Y. H. *J. Chem. Eng. Data* **2005,** *50,* 1672–1676. Copyright 2005 American Chemical Society. (D) Adapted from Odelius, K.; Plikk, P.; Albertsson, A.-C. *Biomacromolecules* **2005,** *6,* 2718–2725. Copyright 2005 American Chemical Society.

C

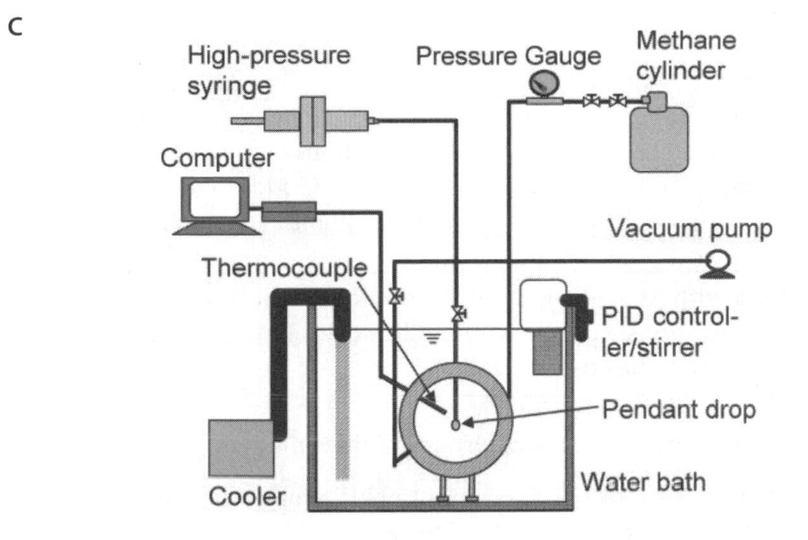


D

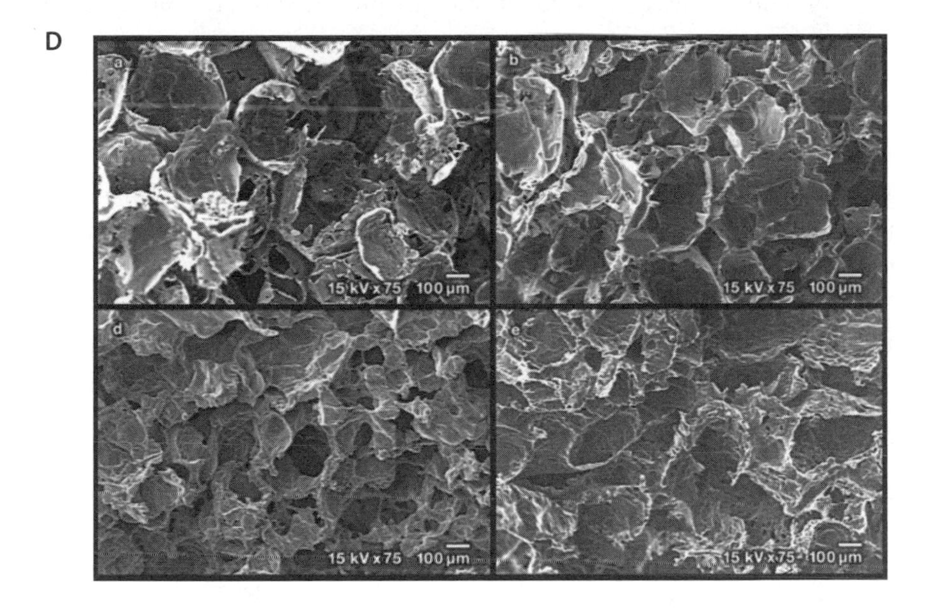


**Figure 15-1.** *Continued.*

## *Use of Color*

The use of color in figures depends partly on the medium and partly on a particular publisher's preferences. As mentioned earlier, color can be used in Web publishing at virtually no cost, so the only restrictions have to do with scientific effectiveness and good taste. Color in print publishing adds to the cost, so before preparing figures in color, you should ascertain a publisher's policies on color.

---

**Publishing with ACS:** ACS journals do not require authors to pay the cost of color illustrations. Authors who publish color figures in ACS books are asked to pay half the cost.

---

In publications, color is produced by one of three modes:

- Black-and-white, or B&W: For print publications, where only one color of ink (usually black) is used.
- Cyan–magenta–yellow–black, or CMYK (also called four-color): For print publications, where all shades of color are produced by layering screened dots of four standardized inks.
- Red–green–blue, or RGB: For Web publications and others viewed on computer monitors, where all shades of color are produced by overlapping red, green, and blue light.

Most publishers can switch electronically from CMYK to RGB, although it is always better to use the publisher's preferred format. Switching from one of the color formats to B&W may produce unsatisfactory results, however. If the gray tones that result are hard to distinguish from each other, a figure that was intelligible in color can become meaningless in B&W. For this reason, it is always better to prepare art that will be reproduced in black and white as B&W.

## *Electronic File Formats*

Many software packages are available today for drawing, scanning, and manipulating figures. When a figure is saved in a format that can be read or reopened only by the software that produced it, the file is called an *application file.* Examples include files with names that end with the suffix .cdx, which can be opened only in Cambridgesoft ChemDraw, or files with names that end in the suffix .psd, which can be opened only in Adobe Photoshop. Application files often pose problems in both Web and print publishing because the publisher (or reader) might not have the necessary software to open or view the file. Many publishers will not accept application files for figures.

There are a handful of file formats, however, that can be opened in many different software packages. Drawing and scanning programs generally provide the option of saving to one or more of these formats. For example, ChemDraw and

Photoshop will both export figures as TIFFs, a format frequently used for print publications. The most universally accepted of these formats are the following:

### For Print

- Encapsulated PostScript, or EPS. PostScript is a programming language developed by Adobe to describe pages, graphics, and fonts; an EPS file is coded to be embedded in a large PostScript file.
- Tagged Image File Format, or TIFF. TIFFs describe an image by dividing it into a grid of pixels and assigning a value to each square. The quality of a TIFF image depends on its resolution (in ppi), that is, the size of the original grid.

### For Web

- Graphics Interchange Format, or GIF. GIF is a pixel-based format like TIFF, except the resolution is preset at 72 ppi, which is appropriate for the Web but seldom acceptable for print.
- Joint Photographic Experts Group, or JPEG. JPEG is another pixel-based format like TIFF but capable of higher resolutions than GIFs. JPEGs, however, are actually compressed files, and they lose small amounts of digital information every time they are opened, recompressed, and saved, so JPEG images tend to degrade with repeated processing. Only "fresh", high-quality JPEGs are suitable for print publication.

### For Print and Web

Many publishers now accept properly created Portable Document Files, or PDFs. A PDF document is created when an application file produced by any of a number of software programs is "distilled", using software such as Adobe's Acrobat Distiller. The PDF document can be viewed on any platform using viewing software such as Adobe's Acrobat Reader, regardless of whether the viewer has the original application software.

---

**Publishing with ACS:** For ACS journals, figure art that will be submitted in PDF format should be distilled using Adobe Acrobat Distiller's "Press Quality" setting.

---

With so many different software packages and file formats, there is no single "right" way to prepare figures for publication. A few common paths are listed here:

- A line-art figure is drawn in an application such as Adobe Illustrator, Macromedia Freehand, or Corel Draw and exported as an EPS or a TIFF file.
- A B&W photograph is scanned with the scanner's software and saved as a grayscale TIFF.

**Table 15-1.** Image File Formats and Their Suitability by Publication Medium and Type of Figure[a]

| | | *Publication Medium* | | | *Best Used To Represent* | | | |
|---|---|---|---|---|---|---|---|---|
| *File Format* | *Optimal ppi*[b] | *Web* | *B&W Printing* | *Color Printing* | *Line Art* | *Photo-graphs* | *Grayscale Line Art* | *Combos* |
| ***For Web*** | | | | | | | | |
| GIF | 72 | H | X | X | R | R | R | R |
| RGB JPEG | 72 | R | X | X | R | R | R | R |
| ***For Print*** | | | | | | | | |
| B&W JPEG | 300 | X | R | X | X | R | R | R |
| CMYK JPEG | 300 | X | X | R | X | R | R | R |
| | 600–1200 | X | X | R | R | X | X | X |
| Bitmap TIFF | 1200 | X | H | X | H | X | X | X |
| Grayscale TIFF | 600 | X | H | X | X | H | R | R |
| CMYK TIFF | 300 | X | X | H | X | H | R | R |
| EPS | N/A | X | R | H | H[c] | X | H[c] | H[c] |

[a]H indicates highly recommended; R indicates recommended; X indicates not recommended.

[b]The optimal number of pixels per inch (ppi) for art prepared to the size at which it will be published. It is always permissible to supply art with a higher ppi than what is listed here.

[c]EPS files are the best way to submit grayscale line art and combos, but the fonts used must be embedded or converted to outlines.

- A color digital photograph cropped in Photoshop is saved as a 72 ppi RGB GIF file for the Web or a 300 ppi CMYK TIFF file for print.
- Grayscale line art is drawn in an application, such as Illustrator or Freehand, and saved as an EPS file.

Table 15-1 lists image file formats commonly accepted by publishers and their suitability for different publishing media and types of artwork.

Several software packages that are in common use can produce satisfactory GIFs for Web publishing but do not export images in a form that is acceptable for the more rigorous requirements of print. Corel WordPerfect and Microsoft Word, Excel, and PowerPoint cannot save figures as valid PostScript files or export directly to TIFFs. Box 15-3 offers suggestions for preparing publishable figures in that program.

## *Sizing Artwork*

Always determine the size at which your figures will be printed and prepare the final electronic version to those dimensions. Consult guidelines from the publisher first. If there are no guidelines, use the following:

- For a Web publication, size your artwork so that it is visually appealing and all parts of the graphic are easily readable on your monitor at 100%.

## Box 15-3. Tips for Submitting Figures Created in Excel

Microsoft Excel and PowerPoint are widely used to create graphs and other figures, but they are unable to export to TIFF or any PostScript file format. Therefore, if figures prepared with these software programs are destined for print publication, they must be printed on paper first and then scanned. (It may also work to submit the figure as a PDF.) Moreover, the default settings of these programs will not produce satisfactory figures. The following tips help you get the best results from Excel.

- Do not use color to differentiate areas of the figure. Use black only for a color, and use gray tones to differentiate areas of the figure.
- Be sure that the weight of all lines is at least 0.5 point. Some lines in Excel default to 0.12 point, which is too thin for scanning. If the figure will be printed large and then reduced, calculate the width of lines accordingly (see the chart below).
- Be sure that the type and symbols also will scale down accordingly.
- It is better for tick marks to extend into the figure, rather than extend outside.
- Print the figure on clean, opaque, white paper, using a fresh ink or toner cartridge. Do not use a dot matrix printer. Set the laser or inkjet printer to a resolution of at least 600 ppi.
- Scan the figure to the correct size, and save it as a 600 ppi bitmap TIFF.

If your publisher requests the Excel application file, be sure that the source data are embedded (not linked) in the file you send.

Line width for a typical figure created in Excel, printed landscape on 8.5 × 11 in. paper, and destined to be printed at a width of 20 or 27 picas:

| | *20 picas (single-column journal)* | *27 picas (6 x 9 in. book)* |
|---|---|---|
| Approximate reduction | 38% | 50% |
| Type size in Excel | 14 pt | 14 pt |
| Type size after reduction | 5 pt | 7 pt |
| Line width in Excel | 1.5 pt | 1 pt |
| Line width after reduction | 0.5 pt | 0.5 pt |

**Table 15-2.** Column Dimensions Most Common in ACS Publications

| | *Column Width* | | | *Page Length* | | |
|---|---|---|---|---|---|---|
| *Publication* | *picas* | *inches* | *centi-meters* | *picas* | *inches* | *centi-meters* |
| Books, trim size | | | | | | |
| 6 × 9 in. | 27 | 4½ | 11 | 42 | 7 | 17½ |
| 7 × 10 in. | 33 | 5½ | 13½ | 51 | 8½ | 21½ |
| 8 × 11 in. | | | | 56 | 9 | 23 |
| single column | 20 | 3¼ | 8¼ | | | |
| double column | 42 | 7 | 17½ | | | |
| Journals and magazines, two-column format | | | | 60 | 10 | 25½ |
| single column | 20 | 3¼ | 8¼ | | | |
| double column | 42 | 7 | 17½ | | | |
| Magazines, three-column format | | | | 60 | 10 | 25½ |
| single column | 13 | 2 | 5 | | | |
| double column | 27½ | 4½ | 11 | | | |
| triple column | 42 | 7 | 17½ | | | |

- For a printed publication, you can determine the column width and page length by measuring a sample copy of the journal or a similar book from the same publisher.
- For ACS journals, try to design figures to fit the width of one column. Table 15-2 gives the column widths and page lengths of many ACS publications.
- Do not put unnecessary frames or boxes around your figures.
- Remember that the length that you measure or take from Table 15-2 is the *maximum* space available for the figure *plus* captions or notes. Therefore, the art must be small enough to leave space for the caption.
- Finally, the width and depth of a figure should not exceed the needs of that figure. Be economical; do not waste space.

## Working with Line Art

If you are creating line-art figures with a computer program, save them in an appropriate format (see Table 15-1) or print them on white, high-quality, smooth, laser-printer paper and scan them. If line-art figures are being drawn by hand, use black ink on white, high-quality, smooth, opaque paper, and then scan them. Box 15-4 provides tips for successful scanning.

Here are some ways to improve the quality of the line-art figures you submit:

➤ Select software that will yield the best results for the line art you wish to create.

## Box 15-4. Tips for Successful Scanning

- If possible, use a flatbed scanner, where the art rests flat on a glass surface and is scanned by a moving scanner's eye, rather than a sheet-fed scanner, where the art is rolled past the scanner's eye. Flatbed scanners are less likely to distort your figure.
- Use your scanner's optical (that is, actual) resolution capability, not the interpolated (that is, calculated) one.
- Scan photographs at a minimum resolution of 300 ppi.
- Scan line art at a resolution of minimum 600 ppi, but 800, 1000, or 1200 ppi is even better. (Line art for ACS publications must be 1200 ppi.)
- If possible, determine the final size of your figure in print and calculate the percentage of expansion or reduction. Then scan the photograph at that percentage.
- If you are not sure about the final size, scan your photograph at a larger size (say, 150% with 300 ppi) or higher resolution (say, 100% with 450 ppi). Scan your line art at 150% with 600 ppi or at 100% with 1000 ppi. It is better to submit image files with too much information than too little.
- Do not adjust contrast or color according to what you see on your monitor. Instead, use the automatic adjustments that come with the scanning software or use the calibration curves that come with image-editing software, such as Photoshop.
- Do not apply any screening in the scanning or editing software, even if you know the figure will be screened before it is published. Let the publisher and the publisher's printer determine the appropriate screening.

➤ Create the line art to the exact size at which it will be published.

➤ Keep line-art figures clear and simple. Keep words to a minimum. Lengthy explanations should go in the caption or a note accompanying the caption.

➤ Scale the length, width, type, symbols, and lines of the art proportionally; keep the symbols, lines, and type at uniform density, or darkness.

- Make the lines at least 0.5 point wide and usually not more than 1.5 points wide.

➤➤➤➤➤

## Box 15-5. Type Size and Font

In publishing, type is measured in points; space is measured in picas. There are 72 points to an inch; there are 6 picas to an inch. The size of the type you are reading on this page is 10 points. The column is 28 picas wide, and the text page is 44 picas long.

You can use a pica ruler to measure space, but it is hard to find a type gauge for measuring type; you must compare it to known type sizes. For example, in 10 point type, no character actually is 10 points high.

The font (or typeface) is the style or design of the letters. There are literally hundreds of fonts, but plain, simple fonts such as Helvetica or Times Roman are best for scientific art.

Type also comes in different weights. Most of the type you are reading is lightface; **this is boldface type.** Type may be *italic* or roman. Generally, you should use lightface, roman type for figures.

This is 12 point Times Roman.
$^{14}C_6H_6$ shows subscripts and superscripts.

This is 12 point Helvetica.
$^{14}C_6H_6$ shows subscripts and superscripts.

This is 14 point Times Roman.
$^{14}C_6H_6$ shows subscripts and superscripts.

This is 14 point Helvetica.
$^{14}C_6H_6$ shows subscripts and superscripts.

- Select a type size of 7–10 points; for ACS journals, select a type size of 5–6 points.
- Make the symbols at least the size of a lowercase letter "oh" (about 2 mm) (see Box 15-5).
- Use a clear type font, preferably Helvetica or Times Roman (see Box 15-6).
- Disparities between the size of the type and the symbols make for illegible or unattractive art; if the type is the right size, the symbols will appear too small or too large, and vice versa. Likewise, the proportion between the overall size of the figure and the type and symbols should be appropriate (see Figure 15-2).

➤ Use simple, common symbols that would not be confused with each other and would be readily available in any publishing house, for example, ○, ●, □, ■, △, ▲, ▽, ▼, ◇, ◆, +, ×, ☆, and ★. (Even if you provide a legend within the figure, you may wish to refer to the symbols in your text.)

## Box 15-6. Tips Regarding Fonts

The fonts used to create some types of electronic figure files occasionally cause problems for publishers, in both print and Web publications. The problems can include character substitution, distorted spacing between letters, and copyright infringement.

The software behind a font is covered by copyright law. Whereas some copyright holders allow open access to their fonts, others exercise some sort of limit. Publishers respect the copyright on fonts and will not use a font without proper permission.

Only some electronic files embed font software in the files.

- GIFs, TIFFs, JPEGs, and other pixel-based files embed only the images of letters, not the font software itself, in their files. Type in these files can be erased, but it is not truly editable.
- EPS and PDF files must have font software fully embedded to be usable. Type in these files is fully editable.

TrueType is a font format designed by Microsoft for maximum legibility on a computer monitor, and it prints well on dot-matrix, inkjet, and laser printers. Microsoft permits free use of the fonts it provides (such as Times New Roman and Arial), but other companies that provide TrueType fonts may have use restrictions. TrueType fonts perform well in Web publishing, but they often pose technical problems in print publishing. TrueType fonts are fine for producing GIFs, TIFFs, and JPEGs, as well as PDFs intended for the Web.

PostScript is a font format designed for high-end graphic arts. These fonts are designed for maximum legibility on paper and for maximum compatibility with high-quality, high-output printing equipment. PostScript fonts are generally covered by copyright and have limitations on their use. They seldom pose technical problems in print publishing, but the copyright issues may limit their usefulness in Web publishing. PostScript fonts are best for EPS files and PDFs intended for print.

OpenType fonts are a relatively new format intended to combine the virtues of both TrueType and PostScript fonts.

STIX (scientific and technical information exchange) fonts were developed by a group of six scientific publishers to establish a comprehensive set of fonts that contains essentially every character that might be needed to publish a technical article in any scientific discipline. STIX fonts are available as a free download, under license, at www.stixfonts.org. A white paper by Tim Ingolsby on the STIX fonts project is available at www.aipservices.org/newsroom/white_papers/pdf/STIX-fonts.pdf.

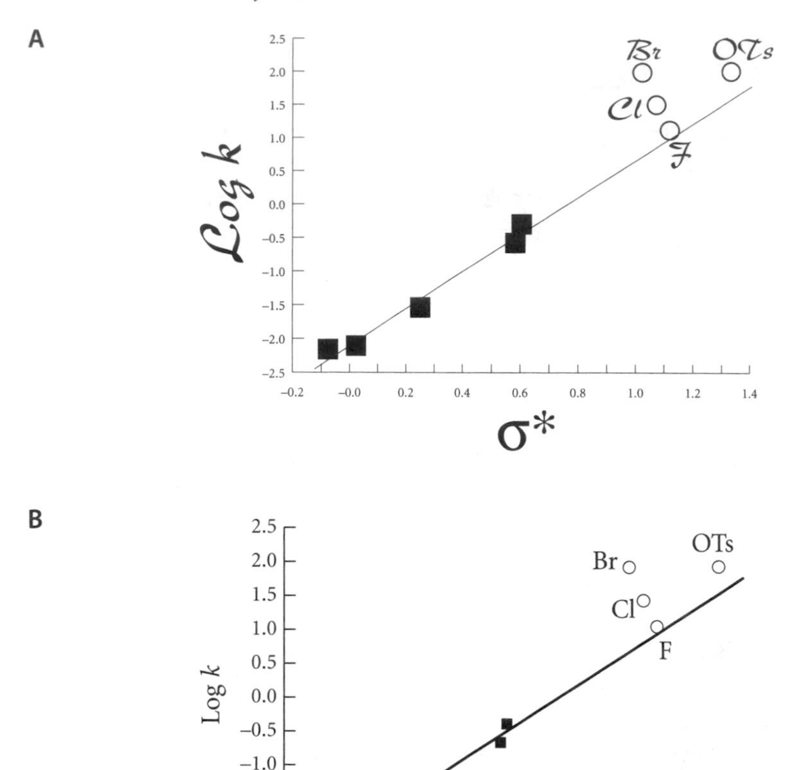


**Figure 15-2.** (A) Example of a poorly rendered line graph, with lines that are too thin, a type font that is too ornate, and type and symbols that are sized disproportionally to the figure. (B) Example of the same line graph properly rendered.

Source: Adapted from Alunni, S.; De Angelis, F.; Ottavi, L.; Papavasileiou, M.; Tarantelli, F. *J. Am. Chem. Soc.* **2005**, *127*, 15151–15160. Copyright 2005 American Chemical Society.

➤ Lines may be differentiated by the use of varying line styles, such as solid, dashed, dotted, and various weights (but never less than 0.5 point wide).

➤ You may combine curves plotted on the same set of axes, but do not put more than five curves in one figure. (If you have more than five curves, consider splitting the figure into two parts or two separate figures.) Label all curves clearly. Leave sufficient space between curves; they should not overlap so much that they become indistinguishable.

A

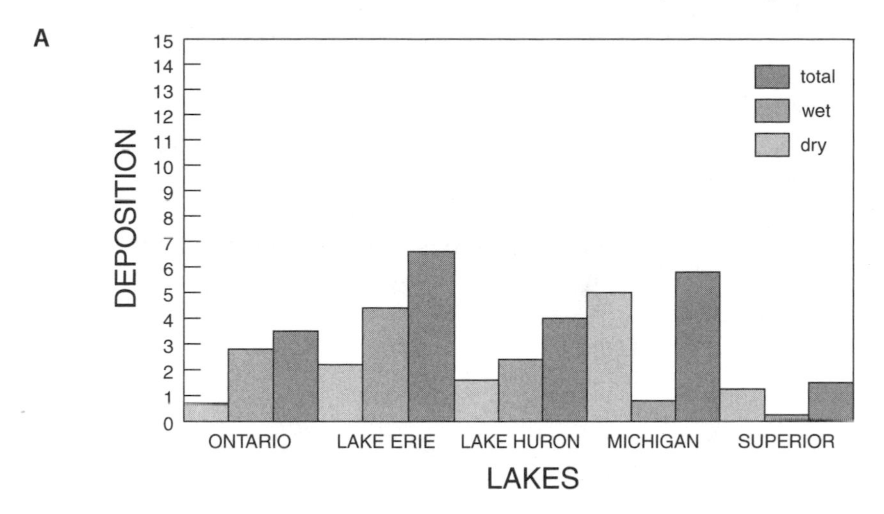


B

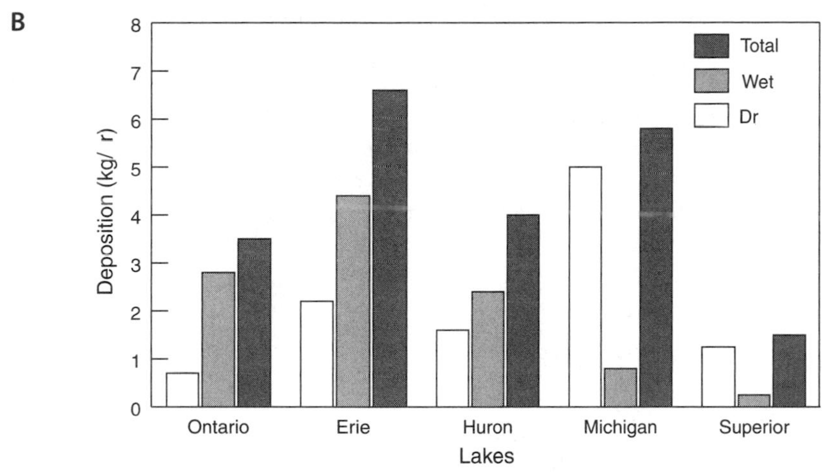


**Figure 15-3.** (A) Example of a poorly rendered bar graph, with excess white space at the top, shading that is difficult to distinguish, and poor and incomplete labels. (B) Example of the same bar graph properly rendered.

Source: Adapted from Ma, J.; Venkatesh, S.; Li, Y.-F.; Daggupaty, S. *Environ. Sci. Technol.* **2005,** *39,* 8123–8131. Copyright 2005 American Chemical Society.

➤ Keep line-art figures compact; draw axes only long enough to define the contents. For example, if the highest data point on the curve is 14, then the scale should extend no longer than 15. Furthermore, the origin or lowest point on the axes does not have to be zero. For example, if the lowest data point on the curve is 4, then the axis can start at 3. Put grid marks on the axes to indicate the scale divisions.

➤ Use complete and consistent axis labels (see Figure 15-3).

- Label each axis with the parameter or variable being measured and the units of measure in parentheses.

- Use initial capital letters only, not all capitals: Time (min), Reaction Temperature (°C), Thickness (μm).
- Place all labels outside and parallel to the axes. Numbers and letters on the abscissa and ordinate should read from left to right and from bottom to top, respectively.
- Do not place arrowheads on the ends of the axis lines.
- Label the tick marks on an axis in type that is one or two points smaller than the axis labels (but not smaller than 7 points in general, 5 points for ACS journals).

➤ Do not draw a box around line-art figures.

➤ For insets, labels, or legends within the figure, use initial capital letters and use the same size type as for the axis labels.

➤ Save electronic figure files with a minimum of white space around the edges.

➤ If you are printing line-art figures for subsequent scanning, always check that the toner or ink cartridge is fresh, and never use a dot-matrix printer.

➤ If you print a hard copy to submit with the electronic file, print it with one figure per page and be sure that the figure number appears on the page.

## Working with Photographs and Other Halftone Figures

Good reproduction of photographs begins with good photographs. When you are selecting photographs for figures to accompany your text, consider the following points:

➤ As much as possible, the photograph should show what is important to your text, and only what is important. Clutter and unrelated objects should be cropped out. (This is easy to do with electronic images.)

➤ Photographs always lose some detail in reproduction, so be sure to take the photograph in good focus and, for digital photographs, with sufficient resolution. See Figure 15-4.

➤ Photographs in print will tend to lose contrast and become muddy or gray, so take a few steps to ensure that your photograph has good contrast and tonal definition:

- Adequately light objects that you are photographing.
- Place objects on light or dark backgrounds that are plain or solid.

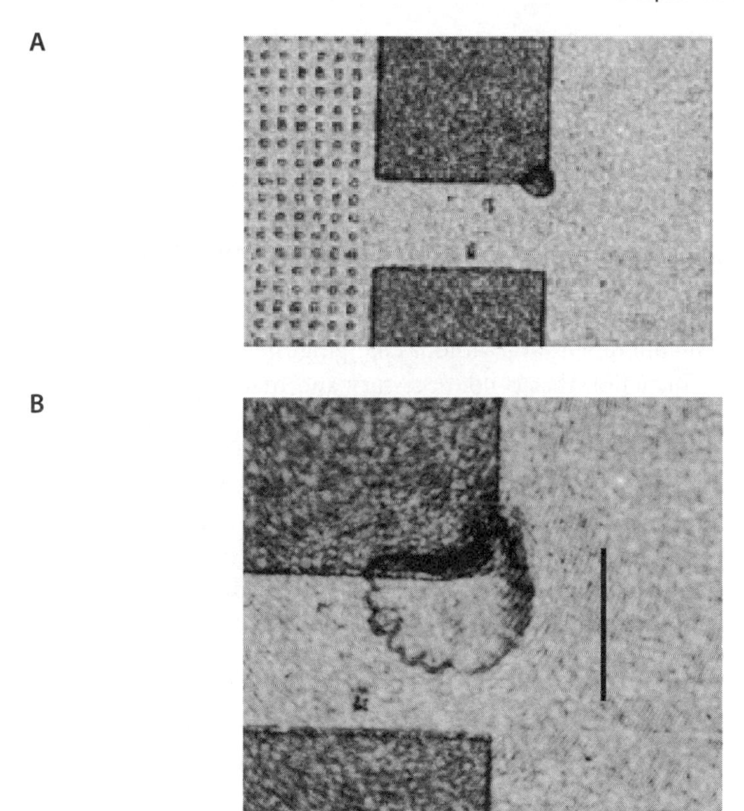


**Figure 15-4.** (A) Example of a photograph with a resolution that is too low for print, so the pixels show as squares, instead of blending to show tones. (B) A similar photograph with a true resolution of 300 ppi.

Source: Adapted from Balss, K. M.; Avedisian, C. T.; Cavicchi, R. E.; Tarlov, M. J. *Langmuir* **2005,** *21,* 10459–10467. Copyright 2005 American Chemical Society.

- Take several photographs at different settings, so that you can select the one with the best contrast and tonal definition.

➤ Photographs can have tricky copyright and permissions issues. Be sure you understand whether permission is needed to publish the photograph. (See the section called "Common Misconceptions about Copyright" in Chapter 7.)

➤ If you are scanning photographic prints, see Box 15-4.

➤ If you are using a digital camera, select a setting that collects enough digital information. The lowest setting might be sufficient for GIFs, but the higher settings will be necessary for print publication.

When you are working with the electronic files of photographs and other halftones, keep the following in mind:

➤ Do not adjust the ppi upward without adjusting the overall dimensions downward. Simply changing the ppi setting from 72 to 300, even if your software allows it, does not put digital information in that was not there before. The file might indicate that the resolution is 300 ppi, but the art will still print as fuzzily as a 72 ppi picture.

➤ You can adjust the ppi downward without changing the overall dimensions. This throws away information that is not necessary and makes a file more portable if, for instance, you scanned at a higher resolution because you were not sure how much you would crop it.

➤ You can switch from RGB to CMYK color modes and vice versa.

➤ If you switch from either RGB or CMYK to B&W, you will probably need to adjust the contrast of your photograph to get the black areas black and the white areas white.

If you are providing photographic prints instead of electronic files, the following points are important to remember:

➤ Submit a good-quality photographic print, preferably not the only original you have. Some publishers try to return original art, but many do not.

➤ Label photographs by writing on an adhesive label (such as a mailing label) first and then applying the label to the back of the photograph, or by writing on a separate sheet of paper and taping it to the back. Writing directly on the back of the photographic print might damage the front: ballpoint pens can break the smooth surface of the photograph, and felt tip pens can bleed through.

➤ Handle photographs with care. Most flaws on a photographic print will be reproduced in the published photograph. You can damage photographs by paper clipping, folding, stapling, or using tape on them. Depending on the amount and location of the damage, you might have to reprint the photograph.

➤ Do not submit prescreened illustrations, that is, illustrations that were scanned from a print source (which is already screened) or scanned using a screen setting.

➤ Do not submit color prints to be reproduced in black and white. Use a professional photo lab to prepare B&W prints from color originals because photo labs use special paper to make B&W prints from color originals.

## Working with Grayscale Line Art

Grayscale line art for print publication of such figures as pie charts, bar graphs, and area graphs should be submitted only in an electronic format, preferably EPS. EPS files use mathematical formulas to produce lines and shading, so lines and type are never screened, and gray areas are screened only when the printing plates are made. (Rescreening screened art creates a distortion called moiré.) As a result, the lines in EPS files remain crisp even when the gray tones have been screened.

➤ Always fully embed the fonts or the font outlines in an EPS file. Otherwise, a different font may be substituted at the publisher's, with unsatisfactory results. Be sure that you have rights to embed the font—you will be safest if you use Times Roman or Helvetica.

➤ Use screen shadings that are different by at least 20 percentage points. For example, you could fill in the sections of a bar graph with 10%, 30%, 50%, 70%, and 100% black. (0%, 20%, 40%, 60%, 80%, and 100% also work well.)

➤ If you choose to screen lines within a line graph (never the axes), assign the lines a 1.5 or 2 point weight, and do not use symbols to mark points on the lines.

## Working with Combinations of Type and Photographs

As mentioned in the earlier section, "Types of Figure Artwork", a photograph that has a simple measuring scale or a few single-letter labels can be adequately handled like other halftones. However, if the photograph (or other halftone artwork) must be accompanied by text to convey meaning, then you should prepare a combo.

One way to do this is to prepare the photograph as a grayscale or color TIFF and then import the photographic image file into a drawing program, such as Illustrator or Freehand. Use the drawing program to add type, lines, axes, tick marks, and so on, and save the file as an EPS with an embedded TIFF.

An alternative is to create the combo in a word-processing program, such as Word or WordPerfect, and then export the page as a PDF file. Simply assembling the type and photograph in the word-processing program and submitting the application (.doc or .wpd) file is seldom acceptable to publishers other than ACS journals.

## How To Submit Illustrations

If you follow the recommendations presented in this chapter, you should be well on your way to trouble-free submission of your figure files. Keep the following points in mind for submitting to publishers other than ACS journals.

**Publishing with ACS:** For manuscripts submitted to ACS journals through Paragon, illustrations should always be embedded in the text document before submission—that is, illustrations should *not* be submitted as separate files from the text. Illustrations submitted through the Paragon Plus environment can be embedded in the text or supplied separately. Before being embedded, however, the illustrations should be prepared outside the word-processing program to the correct size and resolution as recommended in this chapter. Each illustration should be "inserted" (not "pasted") into the text document so that it maintains its original resolution, and it should not be adjusted for size after insertion. Graphics taken from Web sites should not be used because, at 72 ppi, their resolution is not adequate for print publication.

### *For All Figures*

➤ Submit a clean photocopy or printout of all figures to accompany your manuscript.

➤ Unless a publisher instructs otherwise, do not put the figures in the midst of text; put them on separate pages at the end.

➤ Put one figure on each page, and label the page with the figure number only. Submit captions as part of your text.

### *For Electronic Files*

➤ Be sure that the art is oriented correctly in the file.

➤ The file name should be brief and consistent, preferably something like Smith_fig02.tiff or Jones_chap5_fig10.eps.

➤ Unless the publisher requests that you submit via a Web site, write the files to a CD, DVD, Zip disk, or 3.5 in. floppy disk. Label the disk with the author name, manuscript title, and general contents, such as manuscript number, figures, tables, and so on.

### *For Camera-Ready Art*

➤ Note the figure number and the first author's surname either on the back or on the front of each piece of artwork, about 1 in. (2.5 cm) clear of the image area. Do not write on the front or back of the image area; write only in the margins.

➤ Indicate the top of the illustration with the word "top" if the correct orientation is not obvious.

➤ Do not fold or roll artwork or photographs; protect art with cardboard or heavy paper for transport. Do not staple, clip, or punch holes in photographs or any artwork.

## How To Prepare Figure Captions

Every figure must have a caption that includes the figure number and a brief, informative description, preferably in nonsentence format.

GOOD EXAMPLES OF FIGURE CAPTIONS

> Figure 2. Mass spectrum obtained when laboratory ambient air containing 2.5 ppm of **1** was introduced into the MS system.
>
> Figure 4. Change in carotenoid contents during maturation of three varieties of grapes: (A) Concord grapes; (B) Thompson seedless; and (C) Chilean red.
>
> Figure 6. Variable-temperature NMR spectra of **3d** in $CD_2Cl_2$ solution at 500 MHz.
>
> Figure 7. Reaction rate constants as a function of proton affinity for the reactions shown in eqs 5–7: $k_{exp}$, experimental; $k_c$, calculated.
>
> Figure 1. Specificity of bovine muscle LDH antibodies in a sandwich ELISA. Data represent the averages of three replicates.

If more information is necessary, use complete sentences and standard punctuation. The caption should be understandable without reference to the text (this is essential in Web publishing because the figure may open in a separate frame) and should not include material that is at odds with the text. Use similar wording for captions of related figures.

If a figure contains symbols that require long explanations or has more than four or five symbols, then the key to the symbols will be large and give the artwork a cluttered appearance. In this case, put the key to the symbols in the caption. Make sure that the symbols and abbreviations in the caption agree with those in the figure itself and in the text and that the symbols are typographically available.

Submit the figure captions separately from the artwork, typed double-spaced on a page at the end of the text. Verify that the numbers in the caption agree with the numbers on the figures and in the electronic file names. Bear in mind that figures and captions are handled differently in both Web and print production; captions are usually typeset in a font and size that follow the style of the publication, whereas figures are used as is. If you place the caption on the art, the editors usually will delete it and have the caption typeset according to the publication's style. At best, this creates additional work and an opportunity for errors to occur; at worst, the presence of captions with the text and on the art may cause confusion if they are inconsistent.

## Reproducing Previously Published Materials

To reproduce a figure, photograph, or table that has been published elsewhere, you (the author) must obtain permission in writing from the copyright owner (usually the publisher), and you must submit the written permissions along with your final manuscript. Even if you were the author of the previously published figure or table, you still need written permission from the copyright owner. The only exception is for a work of the U.S. government. See Box 15-7 for examples of credit lines. See also Chapter 7, "Copyright Basics", for further information.

➤ If you construct a figure or a table from data that were previously published as text or use data from a table to create an original figure, you do not need permission, but you should reference the source of the data (e.g., "Data are from ref 7."). However, if you are using a portion of a table that has been previously published, even very small portions such as a few data points, permission is needed from the copyright holder.

➤ If you adapt or use only part of a figure or table, permission is still needed. The credit line for adapted material is similar to the credit lines in Box 15-7 except "reprinted" is replaced with "adapted".

> Adapted with permission from ref XX. Copyright Year Copyright Owner's Name.
>
> Adapted with permission from Author Names (Year of Publication). Copyright Year Copyright Owner's Name.

➤ If you are thinking about using a previously published figure or table, consider carefully whether citing it as a reference would be adequate.

## Box 15-7. Credit Lines in ACS Publications

In ACS publications, credit lines for art reproduced from previously published work appear at the end of the caption in parentheses in one of two formats and follow three possible wordings, depending on the original source:

### *Format 1*

**Most publishers:** Reprinted with permission from ref XX. Copyright Year Copyright Owner's Name.

> Reprinted with permission from ref 10. Copyright 2003 American Pharmaceutical Association.

**Published by ACS:** Reprinted from ref XX. Copyright Year American Chemical Society.

> Reprinted from ref 12. Copyright 2005 American Chemical Society.

**Published by the U.S. government:** Reprinted from ref XX.

> Reprinted from ref 23.

### *Format 2*

**Most publishers:** Reprinted with permission from Author Names (Year of Publication). Copyright Year Copyright Owner's Name.

> Reprinted with permission from Camiola and Altieri (2006). Copyright 2006 American Institute of Physics.

**Published by ACS:** Reprinted from Author Names (Year of Publication). Copyright Year American Chemical Society.

> Reprinted from Fitzgerald and Cheng (2004). Copyright 2004 American Chemical Society.

**Published by the U.S. government:** Reprinted from Author Names (Year of Publication).

> Reprinted from Takanishi and Schmidt (2006).

Be sure to check the author guidelines for your publication.

➤➤➤➤➤ CHAPTER 16

# Tables

Betsy Kulamer

This chapter presents guidelines for preparing the tables that accompany a scientific paper for publication. Tables are handled in many ways like figures, so this chapter focuses on the ways in which tables are different and briefly discusses the preparation of tables using word-processing programs.

## When To Use Tables

Use tables when the data cannot be presented clearly as narrative, when many precise numbers must be presented, or when meaningful interrelationships can be better conveyed by the tabular format. Tables should supplement, not duplicate, text and figures. (If you are not sure whether you need a table or figure, see Box 15-1.) Examples of material that is best handled as narrative in text are results of IR absorption and NMR chemical shift studies, unless they are major topics of discussion. In many instances, one table with representative data, rather than several tables, is all that is needed to convey an idea.

## How To Cite Tables

Like figures, all tables must be called out, that is, mentioned or discussed by name and number in the text.

➤ Capitalize the word "Table" when it is followed by the table number.

➤ Number tables sequentially with arabic or roman numerals, depending on the publication's style, in order of discussion in the text: Table 2 or Table IV.

➤ Discuss tables sequentially, so that Table 1 is discussed before Table 2, Table 2 before Table 3, and so on.

For good examples of a callout in text, see Chapter 15, Figures, pp 345–346.

## How To Prepare Tables

There are two kinds of tables: informal (or in-text) and formal. An *informal table* consists of three to five lines and is no more than four columns wide; it cannot exceed the width of a text column. Informal tables may be placed in text following an introductory sentence, and each column should have a heading. They are not given titles or numbers, nor do they contain footnotes.

A *formal table* should consist of at least three interrelated columns and three rows. If you have only two columns, try writing the material as narrative. If you have three columns, but they do not relate to each other, perhaps the material is really a list of items and not a table at all (see the discussion of lists at the end of this chapter). If your table has unusual alignment and positioning requirements, perhaps it should really be a figure. It is important to understand these differences because tables are more expensive to produce than text; the larger the table, the higher the cost. A well-constructed, meaningful table is worth the expense, but anything else is wasteful and does not enhance your paper.

Tables should be simple and concise; arrange all data for optimal use of space. If you have many small tables, consider combining some. Combining is usually possible when the same column is repeated in separate tables or when the same type of material is presented in several small tables. Use consistent wording for all elements of similar or related tables. Be consistent with symbols and abbreviations among tables and between tables and text.

➤ The table width will depend on the widths of its individual columns.
- Generally, tables having up to 6 columns will fit in a single journal column; tables having up to 13 columns will fit in the double-column spread. Tables that exceed the double-column spread will be rotated 90° and set lengthwise on the page.
- In books, tables having up to 8 columns can fit in the page width; tables having 9–12 columns will be set lengthwise on the page. Larger tables can span two pages.
- In all publications, extremely wide tables can cause composition difficulties. In such cases, consider presenting the material as two or more smaller tables.

➤ The style for the individual parts of tables (i.e., the use of capital and lowercase letters and whether the entries are centered or flush left) varies among publications. Consult a recent edition of the journal or the journal's instructions for authors.

➤ Keep sections of multipart tables at similar widths. Widely divergent section widths within a table waste space and detract from general appearance.

Effective tables are well-designed, so think carefully, first, about the data you need to present and, second, about the best way to present it visually on a page. Sometimes, what looks fine on a letter-size sheet of paper is not practicable for a journal or book page. Sometimes, what you originally conceived as the column headings works better as the row headings. (In general, you should have more row headings than column headings.) Understanding the parts of a table will help you design your tables effectively; they are identified in Figure 16-1.

### Title

➤ Give every formal table a brief, informative title that describes its contents in nonsentence format. The title should be complete enough to be understood without referring to the text. Place details in table footnotes, not in the title.

➤ Begin the table title with the word "Table" and its number, and then continue with the title.

### Column Headings

Every column must have a heading that describes the material below it. A column heading should not apply to the entire table; information that describes all of the columns belongs in a general table footnote. If a column heading applies to more than one column, use a rule below it that spans the columns to which it applies; this is called a *straddle rule*. Below the rule, give the specific headings for each column. A unit of measure alone is not an acceptable column heading, unless the column heading appears under a straddle rule.

➤ Be as succinct as possible, keep column headings to two lines if possible, and use abbreviations and symbols whenever practical.

➤ Be consistent with the text and with other column headings.

➤ Define nonstandard abbreviations in table footnotes. Name the variable being measured, and indicate the unit of measure after a comma or slash or within enclosing marks. Use the same style within and among all tables.

### Column Entries

In many tables, the leftmost column is the *stub* or *reading column*. Usually, all other columns refer back to it. Stub entries should be consistent with the text as well as logical and grammatically parallel. Main stub entries may also have subentries, which should be indented.

Stub column

Straddle head

Table title

**Table 2. Conditioned WRA and Mechanical Strength of Plain-Weave Cotton Fabric Treated with Different Cross-Linking Agents[a]**

Column headings

Table text

| *Cross-Linker Concentration* | *Catalyst Concentration* | *Curing Condition* | *WRA (deg, w + f) No. of Laundering Cycles* 1 | 5 | 10 | 20 | *Flex Abrasion Retention (no. of cycles, warp)* |
|---|---|---|---|---|---|---|---|
| *Unmodified* | | | | | | | |
| 8% PMA | | | | | | | |
| Dried | 2% $NaH_2PO_2$ | 180 °C, 1.5 min | 273 | 261 | 261 | 260 | 309 |
| Soaked | 3% $NaH_2PO_2$ | 182 °C, 1.3 min | 274 | 260 | 260 | 262 | 310 |
| 6% BTCA | 4% $NaH_2PO_2$ | 180 °C, 1.5 min | 287 | 276 | 273 | 270 | 148 |
| *Modified* | | | | | | | |
| 10.5% modified DMDHEU | precatalyzed | 165 °C, 1.5 min | 278 | 273 | 269 | 264 | 68 |
| control | | 180 °C, 1.5 min | 190 | | | | 868 |

Table footnote

[a]The concentrations of PMA, BTCA, and $NaH_2PO_2$ are calculated on the basis of 100% active ingredient; the concentration of DMDHEU is based on the weight of the commercial product, which contains 55% solid. The wet pickup of the treated fabric is approximately 105–110%.

**Figure 16-1.** Parts of a table.

Source: Adapted from Chen, D.; Yang, C. Q.; Qiu, X. *Ind. Eng. Chem. Res.* **2005,** *44,* 7921–7927. Copyright 2005 American Chemical Society.

➤ Material in columns can be aligned in various ways; use only one type of alignment per column. Words are usually aligned on the left, and numbers are usually aligned on the decimals, unless they do not have the same units, in which case they are aligned on the left. Use numbers on both sides of a decimal point; numbers less than 1 should have a zero to the left of the decimal point. Columns that are made up of numbers and words together or columns that contain a variety of sizes or types of information might call for alignment on the left, right, or center, depending on the publication's style.

➤ Do not use ditto marks or the word "ditto" to indicate the same entry in a column; repeat the entry.

➤ Define nonstandard abbreviations in table footnotes.

➤ Try to keep all entries at similar lengths by placing any explanatory material in table footnotes. If you use a dash as a column entry, explain it in a footnote the first time it is used (e.g., "—, too low to be measured.").

➤ Make sure that all of the columns are really necessary. If there are no entries in most of a column, it probably should be deleted and replaced with a general table footnote. Alternatively, if the entries in the entire column are the same, the column should be replaced with an appropriate table footnote, such as "In all cases, the value was *x*."

## *Footnotes*

Table footnotes include explanatory material referring to the whole table and to specific entries. Examples of information that should be placed in general footnotes referring to the whole table are the following: units of measure that apply to all entries in the table, explanations of abbreviations and symbols used frequently throughout the table, details of experimental conditions if not already described in the text or if different from the text, general sources of data, and other literature citations.

Information that should be placed in specific footnotes includes units of measure that are too long to fit in the column headings, explanations of abbreviations and symbols used with only one or two entries, statistical significance of entries, experimental details that apply to specific entries, and different sources of data.

In some publications, such as books, general footnotes and sources are not cited with superscripts; they are labeled "Note" and "Source", respectively. Specific footnotes are cited with superscripts. In other publications, all footnotes are cited with superscripts. Check the directions for the publication to which you are submitting your paper.

➤ Where superscripts are needed, use superscript lowercase italic letters in alphabetical order, starting from the top of the table and proceeding from left to right.

➤ Write footnotes as narrative and use standard punctuation. Short phrases such as "ND, not determined." and "$x = 23$." are acceptable.

➤ Label each footnote with its superscript letter and group the footnotes together at the end of the table. All footnotes must have a callout in the table title or text.

## *Using Word-Processing Software*

When you prepare your tables using word-processing software, a few techniques ensure a smoother transition to either Web or print publication.

➤ In Microsoft Word or WordPerfect, use the software's table feature, rather than aligning columns using the tab key. Entries arranged with the table feature are more likely to be properly aligned in publication than entries that have been tabbed.

➤ Set up the table in 10 or 12 point type, although 8 point type can be used if necessary. If you need to use type smaller than 8 points to fit your table on a letter-size page, it probably will not fit comfortably on a book or journal page.

➤ Double-space the text in the table.

➤ When you use the table feature, put only one row of entries in each row of the table. Do not put multiple entries in a single cell by using the hard return.

➤ Avoid using hard returns to add space between rows of the table. If you wish to show more space than is apparent with double-spacing, use the line formatting feature of the word-processing program instead.

## How To Submit Tables

If you follow the recommendations presented in this chapter, you should experience trouble-free submission of your tables. Keep the following points in mind:

➤ Place formal tables after the references at the end of the text file, each on its own page.

➤ Place informal tables in place within the text.

➤ Submit a printout of tables along with the printout of text if the publisher requests one.

➤ If a table must contain structures or other art or special symbols, or if a table has special alignment and positioning requirements, be sure that these are evident on the printout.

---

**Publishing with ACS:** In manuscripts submitted to ACS journals through Paragon, tables should always be embedded in the text document before submission, that is, tables should *not* be submitted as separate files from the text. Tables in manuscripts submitted through the Paragon Plus environment can be embedded in the text or supplied separately.

---

## When To Use Lists

Sometimes you may need to give numerous examples of items, such as chemical names. In such cases, if there are too many to run into the text, they can be set as a *list* in some publications. Put the entries in alphabetical order, unless there is a reason to do otherwise. A list of names is not truly a figure and not really a table. Give the list an unnumbered title. In ACS journals, lists may be handled as informal tables or even as charts.

POTENTIALLY CARCINOGENIC MEDICINES

| | | |
|---|---|---|
| azacitidine | cyclophosphamide | methotrexate |
| azathioprine | cytarabine | nitrofurazone |
| chloramphenicol | dacarbazine | phenacetin |
| chlornaphazine | fluorouracil | phenoxybenzamine |

➤➤➤➤➤ CHAPTER 17

# Chemical Structures

Antony Williams

The communication of chemical structures is inherently visual, so the language of both chemistry and biochemistry would be sparse without the ability to accurately represent chemistry in a graphical format. Scientific manuscripts today benefit from the visually appealing output of software used to generate structure representations conveying the details of molecular connectivity as well as the details associated with chemical structures, reactions, and schemes. This chapter discusses the use of chemical structures; the related reactions, charts, and schemes, in a scientific manuscript; and methods of preparing and submitting structures for publication. The chapter presents general guidelines; authors should consult with their publishers for specific instructions.

## When To Use Structures

A chemical *structure* is a pictorial representation of the bonding of atoms in a molecule. Chemical structures appear within text at the point at which they are discussed. Structures are numbered sequentially with either arabic or roman numerals; consult the publisher's guidelines to determine if one is preferred over the other. Generally, there is no need to provide graphical representations of structures for materials that can be accurately represented on one line or in the form of text. For scientific publications, structures should be included for clarity of communication only.

For example, papers describing previously unreported syntheses or reaction sequences, structure–activity relationships, or newly discovered chemical compounds make good use of chemical structure images. Simple chemical struc-

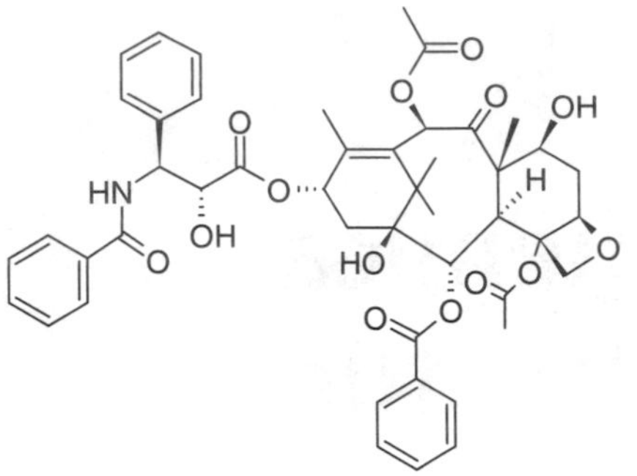

**Figure 17-1.** Example of a chemical structure.

tures can often be represented by a line formula, such as $C_6H_5COOH$ or 1,2,4-$Br_3C_6H_3$, or by the systematic name of the compound, such as benzoic acid or 1,2,4-tribromobenzene, but complex chemical structures depend on structural representations exhibiting the atom–atom connectivity, including the order and stereochemistry of the bonds. Even though a systematic name can accurately describe the chemical structure of paclitaxel, specifically, (2a*R*,4*S*,4a*S*,6*R*,9*S*, 11*S*,12*S*,12a*R*,12b*S*)-6,12b-diacetoxy-9-{[(2*R*,3*S*)-3-(benzoylamino)-2-hydroxy-3-phenylpropanoyl]oxy}-4,11-dihydroxy-4a,8,13,13-tetramethyl-5-oxo-2a,3,4,4a, 5,6,9,10,11,12,12a,12b-dodecahydro-1*H*-7,11-methanocyclodeca[3,4]benzo[1,2-*b*]oxet-12-yl benzoate), most chemists would prefer to visualize the structure (shown in Figure 17-1).

---

**Reminder:** Simple chemical structures can often be represented by a line formula or by the systematic name of the compound, but complex chemical structures depend on structural representations exhibiting the atom–atom connectivity, including the order and stereochemistry of the bonds.

---

A chemical *reaction* is a pictorial representation of a change or process. Reactions can be represented with either line formulas or structures; they are placed in text and numbered much like mathematical equations. See pp 272–274 of Chapter 13 for information on how to format chemical reactions.

Groups of structures are called *charts*. Charts contain representations of multiple structures to facilitate discussion of them. Groups of reactions are called *schemes*; as shown in Figure 17-2, several reactions can be presented together to show, for example, a step-by-step process. Schemes show action; charts do not.

## How To Cite Structures

➤ Structures are identified in text with boldface numerals (either arabic or roman), boldface alphabet letters (capital or lowercase), or a combination of

**Sc eme 1**

$S_n$ (Closed form) → $S_0^*$ (EZZ, ZEZ, EEZ)
<2 ps
≈ 20 ps
*h*ν
Ph
Ph O
Closed form
Δ
Ph
Ph HO
EZZ (sub ns)
Δ
Δ
h Ph
H
O
ZEZ (730 ns)
*h*ν
Δ
EZE, ZEE, EEE
(>1 ms)
H Ph
Ph
O
EEZ (48 μs)
Major isomer

**Figure 17-2. Example of a scheme representing a group of reactions.**
Representing work reported by Kodama, Y.; Nakabayashi, T.; Segawa, K.; Hattori, E.; Sakuragi, M.; Nishi, N.; Sakuragi, H. *J. Phys. Chem. A* **2000,** *104,* 11478.

these. (These numbers should also appear as labels on the structures themselves.) For several structures or for a series of structures, use a consistent sequence of labels. Typical series may include the following:

| | | |
|---|---|---|
| **1, 2, 3** | **a, b, c** | **1A, 1B, 1C** |
| **I, II, III** | **1a, 1b, 1c** | **IA, IB, IC** |
| **A, B, C** | **Ia, Ib, Ic** | |

If a compound is mentioned several times within the text, refer to it only by its label (e.g., "as shown by the reaction with **9**"). Do not include or label structures that are not discussed in the text.

➤ Chemical reactions set as equations should be labeled with lightface roman letters, numerals, or combinations in parentheses in the right margin (see Chapter 13). Depending on the text, it may be appropriate to use one set of labels for both chemical reactions and mathematical equations, or two separate and distinct numbering sequences (for instance, eqs 1, 2, 3, … and reactions I, II, III, …). The numbering sequences for structures and for reactions should always remain separate.

---

**Reminder:** If a compound is mentioned several times within the text, refer to it only by its label.

---

➤ Charts are numbered consecutively as Chart 1, Chart 2, and so on, with either arabic or roman numerals. Charts should be labeled with "Chart" and a number; they also may have brief titles and footnotes.

➤ Schemes are numbered consecutively as Scheme 1, Scheme 2, and so on, with either arabic or roman numerals. Schemes should be labeled with "Scheme" and a number; they also may have brief titles and footnotes.

## How To Prepare Structures

The quality and size requirements for structures (and, by extension, reactions, schemes, and charts) are similar to those for other illustrations (see Chapter 15). The requirements associated with the representation of structures are supported by most of the software programs for drawing chemical structures (described in Box 17-1). The following guidelines should be used when creating the graphics for chemical structures.

➤ Arrange structures in horizontal rows within the width of a column.

➤ Make the size of the rings and the size of the type proportional. The published size of six-membered rings should be approximately ¼ in. (6.35 mm) in diameter; the published size of five-membered rings should be slightly smaller. The type size should be 5–8 points.

➤ Keep oddly shaped rings and the shapes of bicyclic structures consistent throughout a manuscript. In multiring structures such as steroids, use partial structures that show only the pertinent points.

➤ In three-dimensional drawings, use dashed lines for lines in the background that are crossed by lines in the foreground to give a greater three-dimensional effect. Make lines in the foreground heavier, as shown below.

➤ Center the compound labels (either numbers or letters) just below the structures. If a series of related compounds are being discussed, draw only one parent structure, use a general designation (e.g., R or Ar) at the position(s) where the substituents differ, and specify modifications below the structure.

---

✎ **Reminder:** Take care that the structure graphic inserted into your text file is of sufficient resolution, that is, 300 ppi minimum, with 600 or 1200 ppi even better.

---

➤ For structures mentioned in tables, provide labeled structures as separate artwork (in text or charts), and use only their labels in the table.

➤ Center the reaction arrows vertically on the midline of the structure height and align the centers of all structures. The midline for all structures is the center of the "tallest" structure on the same line.

An example of a correctly drawn reaction is represented below. As shown, the arrow and the one-line structure are centered, top to bottom, in the height of the full structure. The reaction arrow and the plus sign have an equal amount of space on both sides.

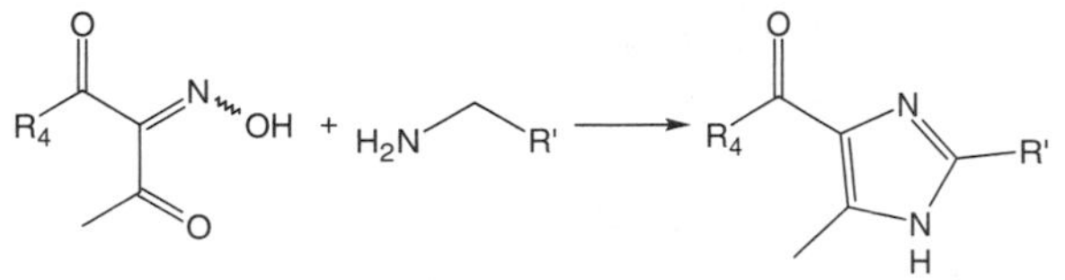

➤ Do not waste space, either vertical or horizontal. Use the full column width before starting a new line. A compact presentation is most effective. Avoid using vertical arrows unless it is necessary to portray a cyclic or "square" reaction scheme. Schemes generally read from left to right. As long as the proper sequence is maintained, it does not matter on which line any given structure appears. If a reaction continues to the next line, keep the arrow or other operator on the top line. Refrain from the use of double-column-width charts or schemes unless absolutely necessary because these charts and schemes consume a significant amount of space.

➤ Do not place circles around plus or minus signs.

➤ For most chemical structure drawing programs, it works best to use the copy-and-paste feature to insert the structure drawing into a Microsoft Word document. However, you should take care that the structure graphic inserted into your text file is of sufficient resolution, that is, 300 ppi minimum, with 600 or 1200 ppi even better. The author bears the responsibility to ensure the appropriate insertion.

## *Software for Creating Chemical Structures*

A number of chemical structure drawing and rendering programs are available to create structure images of excellent quality and accuracy for inclusion in scientific publications. The most popular commercial structure-drawing packages today include ISIS/Draw, ChemDraw, and ChemSketch (see Box 17-1). Other commercial packages include ChemWindow (now known as DrawIt) and Chemistry 4-D Draw. At this time, Web-based applets do not support the production of graphical output of sufficient quality for most books and journals; other software, such as JMol, should be considered as structure-rendering engines only. Box 17-2 contains parameter settings for ACS publications.

## Box 17-1. Software for Drawing Chemical Structures

The following is a selection of commercially available software for drawing chemical structures suitable for publication.

ChemDraw: Cambridgesoft, 100 CambridgePark Drive, Cambridge, MA 02140. Commercial: http://www.chemdraw.com.

Chemistry 4-D Draw: Cheminnovation, 7966 Arjons Dr, #A-201, San Diego, CA 92126. Commercial: http://www.cheminnovation.com/products/chem4d.asp.

ChemSketch: Advanced Chemistry Development, 110 Yonge Street, 14th Floor, Toronto, Ontario M5C 1T4, Canada. Commercial: http://www.acdlabs.com/products/chem_dsn_lab/chemsketch/. Freeware: http://www.freechemsketch.com.

DrawIt (formerly ChemWindow): Bio-Rad, Informatics, Sadtler Group, 3316 Spring Garden Street, Philadelphia, PA 19104-2596. Commercial: http://www.knowitall.com/.

ISIS/Draw: Elsevier MDL, 14600 Catalina Street, San Leandro, CA 94577. Commercial: http://www.mdli.com/products/framework/isis_draw/index.jsp. Freeware: http://www.mdli.com/downloads/downloadable/index.jsp. MDL will ultimately discontinue ISIS/Draw and replace it with Mol Draw.

The cost of these software packages can vary considerably based on academic or industrial usage. Fortunately, the cost should not be a barrier to the inclusion of appropriate renderings of chemical structures because several freeware structure-drawing packages are now available for download from the World Wide Web. These include freeware versions of both ISIS/Draw and ChemSketch. A detailed comparison of capabilities for both the freeware and commercial versions of the packages has been made by Tamas Gunda (http://dragon.klte.hu/~gundat/rajzprogramok/dprog.html). This review is an objective comparison, updated on an almost annual basis, of the capabilities of five popular drawing packages [ISIS/Draw, ChemDraw, DrawIt (ChemWindow), ChemSketch, and Chemistry 4-D Draw].

**Reminder:** The cost should not be a barrier to the inclusion of appropriate renderings of chemical structures because several freeware structure-drawing packages are now available for download.

It is preferable to use one of the drawing packages that offer journal-based templates containing the appropriate bond widths, bond lengths, fonts, and

## Box 17-2. Parameter Settings for ACS Publications

At present, ACS journals request that chemical structures be prepared according to the guidelines below. (The guidelines are also available at https://paragon.acs.org/paragon/index.jsp. Click on "Read Author Information", and then select the appropriate journal for specific information.) The parameters below are specifically for ChemDraw, using the ACS 1996 document settings; authors using other drawing packages should adapt these parameters to their systems. Most commercial and freeware packages allow these settings to be reproduced either by manual settings or by some sort of journal template feature.

| | | |
|---|---|---|
| **Drawing Settings** | chain angle | 120 degrees |
| | bond spacing | 18% of width |
| | fixed length | 14.4 pt (0.2 in.) |
| | bold width | 2.0 pt (0.0278 in.) |
| | line width | 0.6 pt (0.0083 in.) |
| | margin width | 1.6 pt (0.0222 in.) |
| | hash spacing | 2.5 pt (0.0345 in.) |
| **Text Settings** | page setup | US/Letter/Paper |
| | scale | 100% |
| | font | Helvetica (Mac), Arial (PC) |
| | size | 10 pt |
| **Preferences** | units | points |
| | tolerances | 3 pixels |

other settings recommended by a particular journal. Using these templates aids in the production of a chemical structure drawing acceptable to the publishers.

## How To Submit Chemical Structures

Many common graphical formats can be generated using structure-drawing packages, including TIFF, GIF, and BMP; follow the guidelines for submitting figures presented in Chapter 15. Also, because file formats continue to develop and the preferences of a given publisher can change, authors should check with the publisher's or journal's author guidelines for acceptable software and file formats. Box 17-3 shows guidelines for submitting structures to ACS journals.

## Box 17-3. Guidelines for Submitting Chemical Structures to ACS Journals

- In manuscripts submitted to ACS journals through Paragon, structures should be embedded in text documents. In manuscripts submitted through the Paragon Plus environment, structures can be embedded in the text or supplied separately.
- Submit graphical images of chemical structures to ACS journals as embedded TIFF files, using a resolution of 300 pixels per inch (ppi) for color and 1200 ppi for black-and-white images.
- ACS journals accept application files (native formats) for several common software programs, including .skc files from ISIS/Draw and .cdx files from ChemDraw.
- Information on submitting chemical structure connection tables is available at https://paragon.acs.org/paragon/application?pageid=content&mid=preferredsoftware.html&parentid=authorchecklist&headername=Preferred%20Software.
- CML is accepted by the ACS Paragon System for submission of supporting information.

The visual appeal of a structure or reaction can differ from one software package to another, so authors may wish to interchange data between programs to obtain the best graphical display required for publication. Such an interchange is normally performed using a specific format of the atom–atom connection table known as a *molfile*. (Information on the MDL molfile format is available at http://www.mdl.com/downloads/public/ctfile/ctfile.pdf.) Such collections may, for example, describe molecules, molecular fragments, substructures, substituent groups, polymers, alloys, formulations and mixtures, and unconnected atoms. Most chemical structure drawing software packages available today allow both import and export of a molfile. An alternative generic file format is CML, the chemical markup language, which takes advantage of the strength of Web-based technologies and is discussed in Chapter 8.

## The Future of Representing Chemical Structures

Despite the vast array of tools available today for structure drawing and representation, each tool is lacking in some way for the representation of complete structure space. Small organic molecules have been well supported, but the needs of the inorganic, organometallic, and polymer chemist have commonly lagged behind. Chemical structure drawing programs will continue to develop version by version as software vendors deliver an ongoing array of functionality for their customers.

The need to introduce additional flexibility for structure representation will be ongoing, specifically as standards are set for preferred drawing styles by organizations such as IUPAC. A scoping exercise initiated by the IUPAC commission has defined the preferred drawing styles for structure representation to ensure accurate communication of chemical structures. (See http://www.iupac.org/projects/2003/2003-045-3-800.html. The stereo part of this project is now available as provisional recommendations at http://www.iupac.org/reports/provisional/abstract05/brecher_310705.html.) It is likely that publishers will embrace the opportunity to standardize the structure representations within their publications to try to achieve greater homogeneity.

Whereas chemical structures will always be the primary vehicle for visual communication, it is possible that journal articles will be reporting a unique label representing a chemical structure in the near future, especially because this label will already allow the searching of chemical structures contained in publications indexed using Web-based search engines. For more on this, see Appendix 8-1.

➤➤➤➤➤ CHAPTER 18

# Selected Bibliography

## References on Scientific Communication

### *Technical Writing*


Barrass, R. *Scientists Must Write: A Guide to Better Writing for Scientists, Engineers, and Students,* 2nd ed.; Routledge: London, 2002.

Cook, C. K. *Line by Line: How To Edit Your Own Writing;* Houghton Mifflin: Boston, MA, 1986.

Day, R. A. The Development of Research Writing. *Scholarly Publishing* January 1989, pp 107–115.

Day, R. A. *Scientific English: A Guide for Scientists and Other Professionals,* 2nd ed.; Oryx Press: Phoenix, AZ, 1995.

Eisenberg, A. *Writing Well for the Technical Professions;* Harper & Row: New York, 1989.

King, L. S. *Why Not Say It Clearly: A Guide to Scientific Writing,* 2nd ed.; Little, Brown: Boston, MA, 1991.

O'Connor, M. *Writing Successfully in Science;* Chapman & Hall: New York, 1992.

Rathbone, R. R. *Communicating Technical Information: A New Guide to Current Uses and Abuses in Scientific and Engineering Writing,* 2nd ed.; Addison-Wesley: Reading, MA, 1985.

Schoenfeld, R. *The Chemist's English,* 3rd ed.; VCH: Weinheim, Germany, 1990.

Shaw, H. *Errors in English and Ways To Correct Them,* 4th ed.; HarperPerennial: New York, 1993.

Strunk, W., Jr.; White, E. B. *The Elements of Style,* 3rd ed.; Allyn & Bacon: New York, 1995.

Zinsser, W. *On Writing Well: The Classic Guide to Writing Nonfiction,* 6th ed.; Harper-Perennial: New York, 1998.

Zinsser, W. *Writing To Learn;* Collins: New York, 1993.

## Style and Usage

Bernstein, T. M. *The Careful Writer: A Modern Guide to English Usage;* The Free Press: New York, 1998.

Berry, T. E. *The Most Common Mistakes in English Usage;* McGraw-Hill: New York, 1971.

Copperud, R. H. *American Usage and Style: The Consensus;* Van Nostrand-Reinhold: New York, 1980.

Flesch, R. F. *The ABC of Style: A Guide to Plain English*; HarperCollins: New York, 1980.

Hodges, J. C.; Horner, W. B.; Webb, S. S.; Miller, R. K. *Harbrace College Handbook*, 13th ed.; Harcourt, Brace, Jovanovich: New York, 1998.

*Merriam-Webster's Collegiate Dictionary,* 11th ed.; Merriam-Webster Inc.: Springfield, MA, 2003. (Available as a package with a CD and online; see http://www.m-w./com for information on the online product.)

*The New York Public Library Writer's Guide to Style and Usage;* Sutcliffe, A., Ed.; HarperCollins: New York, 1994.

van Leunen, M.-C. *A Handbook for Scholars,* 2nd ed.; Oxford University Press: New York, 1992.

*Webster's Third New International Dictionary, Unabridged*; Merriam-Webster, Inc.: Springfield, MA, 2002. (Available as a package with a CD and online; see http://www.m-w./com for information on the online product.)

*Webster's New World Dictionary of the American Language,* 2nd ed.; Simon & Schuster: New York, 1980.

### Style Manuals

*AIP Style Manual,* 4th ed.; American Institute of Physics: New York, 1990.

*American Medical Association Manual of Style,* 9th ed.; Flanagin, A., et al., Eds.; Lippincott Williams & Wilkins: Baltimore, MD, 1997.

*ASM Style Manual for Journals and Books;* American Society for Microbiology: Washington, DC, 1992.

*The Chicago Manual of Style,* 15th ed.; University of Chicago Press: Chicago, IL, 2003.

*Merriam-Webster's Standard American Style Manual;* Merriam-Webster: Springfield, MA, 1994.

*The Microsoft Manual of Style for Technical Publications,* 3rd ed.; Microsoft Press: Redmond, WA, 2003.

*Publication Manual of the American Psychological Association,* 5th ed.; American Psychological Association: Washington, DC, 2001.

*Scientific Style and Format: The CBE Manual for Authors, Editors, and Publishers,* 6th ed.; Cambridge University Press: New York, 1994.

Skillin, M. E.; Gay, R. M. *Words into Type,* 3rd ed.; Prentice-Hall: Englewood Cliffs, NJ, 1974.

*U.S. Government Printing Office Style Manual 2000,* 29th ed.; Government Printing Office: Washington, DC, 2000.

*Wired Style: Principles of English Usage in the Digital Age;* Hale, C., Ed.; HardWired: San Francisco, CA, 1997.

### Mathematics and Numbers

Swanson, E.; O'Sean, A.; Schleyer, A. *Mathematics into Type: Copy Editing and Proofreading of Mathematics for Editorial Assistants and Authors;* American Mathematical Society: Providence, RI, 1999.

Taylor, B. N. *Guide for the Use of the International System of Units (SI);* National Institute of Standards and Technology: Gaithersburg, MD, 1995.

### Preparing Illustrations

Briscoe, M. H. *Preparing Scientific Illustrations: A Guide to Better Posters, Presentations, and Publications,* 2nd ed.; Springer Verlag: New York, 1996.

*Pocket Pal: A Graphic Arts Production Handbook,* 18th ed.; Graphic Arts Technical Foundation: Alexandria, VA, 2000.

# References on Chemistry

## *Conventions in Chemistry*

*National Institute of Standards (NIST) Special Publication 330, 2001 Edition,* Engl. version of *The International System of Units (SI),* 7th ed.; Bureau International des Poids et Mesures: Sèvres, France, 1998. http://physics.nist.gov/Pubs/SP330/sp330.pdf.

*Quantities and Units;* International Organization for Standardization: Geneva, Switzerland, 1993. http://www.iso.org/iso/en/prods-service/ISOstore/store.html.

*Quantities, Units and Symbols in Physical Chemistry,* 2nd ed.; Blackwell Science, Oxford, U.K., 1993 (commonly known as the "green book"). http://www.iupac.org.

## *Biochemistry*

*Biochemical Nomenclature and Related Documents,* 2nd ed.; Liébecq, C., Ed.; Portland Press: London, 1992 (commonly referred to as the "white book"). http://www.chem.qmw.ac.uk/iupac/bibliog/white.html.

*Enzyme Nomenclature;* Academic: New York, 1992. http://www.chem.qmw.ac.uk/iubmb/enzyme/.

## *Chemical Abstracts*

Advice on *Chemical Abstracts* nomenclature is available from the Manager of Nomenclature Services, Department 64, Chemical Abstracts Service, P.O. Box 3012, Columbus, OH 43210. A name-generation service is available through CAS Client Services, Chemical Abstracts Service, P.O. Box 3343, Columbus, OH 43210; e-mail address, answers@cas.org; and URL, http://www.cas.org/Support/client.html.

## *Combinatorial Chemistry*

Maclean, D.; Martin, E. J. On the Representation of Combinatorial Libraries (A Perspective) *J. Comb. Chem.* **2004,** *6,* 1–11.

Maclean, D.; Baldwin, J. J.; Ivanov, V. T.; Kato, Y.; Shaw, A.; Schneider, P.; Gordon, E. M. Glossary of Terms Used in Combinatorial Chemistry (IUPAC Technical Report) *Pure Appl. Chem.* **1999,** *71,* 2349–2365; Reprinted *J. Comb. Chem.* **2000,** *2,* 562–578; Translated to German *Angew. Chem.* **2002,** *114,* 893–906.

## *Drug Names*

*The USP Dictionary of USAN and International Drug Names;* U.S. Pharmacopeial Convention: Rockville, MD, 2005 (updated annually).

## General Chemistry

*Chemical Abstracts Index Guide;* American Chemical Society: Columbus, OH; Appendix IV (updated periodically).

*Compendium of Chemical Terminology,* 2nd ed.; McNaught, A. D., Wilkinson, A., Eds.; Blackwell Science: Oxford, U.K., 1997.

*CRC Handbook of Chemistry and Physics,* 86th ed.; Lide, D. R., Ed.; CRC Press: Boca Raton, FL, 2005.

*Kirk-Othmer Encyclopedia of Chemical Technology,* concise, 4th ed.; Wiley & Sons: New York, 2003.

*The Merck Index: An Encyclopedia of Chemicals, Drugs, and Biologicals,* 13th ed.; O'Neil, M. J., Smith, A., Heckelman, P. E., Budavari, S., Merck, Eds.; John Wiley & Sons: New York, 2001.

## Inorganic Chemistry

*Nomenclature of Inorganic Chemistry, Recommendations 2005;* Connelly, N. G., Hartshorn, R. M., Damhus, T., Hutton, A. T., Eds.; Royal Society of Chemistry: Cambridge, U.K., 2005.

Block, B. P.; Powell, W. H.; Fernelius, W. C. *Inorganic Chemical Nomenclature: Principles and Practice;* American Chemical Society: Washington, DC, 1990.

## Nomenclature

The International Union of Pure and Applied Chemistry (IUPAC) Nomenclature Documents Home Page on the World Wide Web can be found at http://www.chem.qmw.ac.uk/iupac/. This site contains information on IUPAC itself, the names and publishers of references on chemical nomenclature, and many of the recommendations.

## Organic Chemistry

IUPAC. *Nomenclature of Organic Chemistry: Sections A, B, C, D, E, F, and H;* Rigaudy, J., Klesney, S. P., Eds.; Pergamon: Elmsford, NY, 1979 (commonly referred to as the "blue book"). http://www.acdlabs.com/iupac/nomenclature.

*A Guide to IUPAC Nomenclature of Organic Compounds, Recommendations 1993;* Panico, R., Powell, W. H., Richer, C., Eds.; Blackwell Scientific Publications: Oxford, U.K., 1994. http://www.acdlabs.com/iupac/nomenclature.

*Ring Systems Handbook;* American Chemical Society: Columbus, OH, 2003 (and supplements).

Fox, R. B.; Powell, W. H. *Nomenclature of Organic Compounds: Principles and Practice,* 2nd ed.; American Chemical Society: New York, 2001.

## Polymer Chemistry

IUPAC. *Compendium of Macromolecular Nomenclature;* Blackwell Scientific Publications: Oxford, U.K., 1991. http://www.iupac.org/publications/books/author/metanomski.html.

# Index

In page references, *b* indicates boxes, *f* indicates figures, *s* indicates schemes, and *t* indicates tables.